In loving memory of my dear, departed parents.

Conceptual operational B-70B Valkyries with conceptual operational AGM-48A Skybolt air-launched ballistic missiles. *Photo/Illustration by Erik Simonsen*

Quarto is the authority on a wide range of topics.

Quarto educates, entertains and enriches the lives of our readers—enthusiasts and lovers of hands-on living.

www.quartoknows.com

© 2016 Quarto Publishing Group USA Inc.
Text © 2016 Steve Pace

First published in 2016 by Zenith Press,
an imprint of Quarto Publishing Group USA Inc.
400 First Avenue North, Suite 400, Minneapolis, MN 55401 USA.
Telephone: (612) 344-8100 Fax: (612) 344-8692

quartoknows.com
Visit our blogs at quartoknows.com

Zenith Press titles are also available at discounts in bulk quantity for industrial or sales-promotional use. For details contact the Special Sales Manager at Quarto Publishing Group USA Inc., 400 First Avenue North, Suite 400, Minneapolis, MN 55401 USA.

ISBN: 978-0-7603-4950-2

Library of Congress Cataloging-in-Publication Data

Pace, Steve.
  The big book of X-bombers and X-fighters : USAF jet-powered experimental aircraft and their propulsive systems / by Steve Pace.
     pages cm
  Includes bibliographical references and index.
  ISBN 978-0-7603-4950-2 (hbk.)
  1. Fighter planes--United States--History. 2. Research aircraft--United States--History. 3. Bombers--United States--History. 4. Jet planes, Military--United States--History. 5. Aeronautics, Military--Research--United States--History. I. Title. II. Title: USAF jet-powered experimental aircraft and their propulsive systems.
  UG1242.F5P322 2016
  623.74'630973--dc23

                    2015036871

Acquiring Editors: Dennis Pernu and Elizabeth Demers
Project Manager: Jordan Wiklund
Art Director: James Kegley
Layout: Simon Larkin

10 9 8 7 6 5 4 3 2

Printed in China

THE BIG BOOK OF
X-BOMBERS AND X-FIGHTERS

USAF JET-POWERED EXPERIMENTAL AIRCRAFT
AND THEIR PROPULSIVE SYSTEMS

STEVE PACE
FOREWORD BY WALTER J. BOYNE USAF COL (RET)

ZENITH
PRESS

CONTENTS

FOREWORD

It takes a brave author to write a comprehensive review of every turbojet-powered Air Force fighter and bomber, produced or not. Fortunately, Steve Pace is just such an author, with extensive experience and a command of the sources for data necessary to complete such a monumental task.

When we review the remarkable array of aircraft that have been produced over the years, we are struck not only by the progress in aviation that has been achieved, but also by the significant changes experienced within our country's aviation industry.

Once there were many famous aircraft manufacturers, bearing the names of Chance Vought, Bill Boeing, Glenn Curtiss, Donald Douglas, James McDonnell, Jack Northrop, Glenn Martin, among others. Yet the very progress that each of these great firms injected into the development of aviation forced more and more either out of business or into mergers.

Some suffered the fate of the great North American Aviation firm, producer of legendary fighters such as the P-51 and F-86. First, a merger with Rockwell-Standard changed its name to North American Rockwell, then to Rockwell International, which maintained an aircraft division called North American Aircraft Operations. There followed another series of mergers in which the name disappeared completely, absorbed into Boeing's Defense division.

McDonnell disappeared in the same mysterious way, its great name, tradition, reputation—even sales backorders—disappearing without a trace. It happened in Great Britain, too, as Hawker, Avro, and others all ceased to exist. It is a shame when money triumphs over tradition.

Others among the great firms followed different paths. With fewer Department of Defense contracts available, they turned their interests to the field of space or other areas, leaving the aircraft production arena altogether.

The joy to the reader, whether student or buff, will come with the steady, reassuring history that Pace's words and hundreds of illustrations provide. Together, they clearly show the aerodynamic design changes that allowed, first fighters, then bombers to reach out past the speed of sound in level flight.

The author covers all of the well-known types in detail and provides the same kind of service to little-known programs (such as Have Blue, Tacit Blue, and Bird of Prey) that played such significant roles in the creation of the very latest bombers and fighters.

For the general reader, this book gives a detailed explanation of how turbojet-powered aircraft evolved from the fledgling and underpowered Bell P-59 of 1942 to the magnificent Northrop Grumman B-2 and the Lockheed Martin F-22 of today.

Their development, of course, depended entirely on the improvement of jet engines. The power of these engines increased tremendously, from an initial 1,250-lbf (pounds force) to the 43,000-lbf available today. At the same time, their reliability improved at an even greater rate, so that engine failure is a rare anomaly today. Specific fuel consumption has been reduced, along with weight. Sadly, since more engines tend to look alike, their histories are not followed as closely as are the histories of the airframes. Yet the author has made room here for a broad treatise on the propulsive systems that propelled the aircraft discussed in his book, telling a part of the story of the dynamos that are buried within the airframes of these fabulous bombers and fighters.

The research that led to laminar flow, swept wings, variable engine inlets, and drag reduction are all here, along with the more exotic studies that allowed stealth to become a reality.

This is a book that can be savored by picking it up and opening to any page, where the reader will get the real story of a historic or virtually unknown aircraft. But for the buff or the student, I recommend going through it page by page from the start, letting the cascade of historical facts build as you read. This method provides a unique foundation, one that clearly shows the interrelationships of the designs that are not apparent at a glance. And while it shows how the industry has changed over the years, it also shows how the industry has clung to engineering truths and integrity.

There is steady progress through the book, although fate and/or finances meant that some avenues (such as the North American XB-70) were abandoned before their time. It also shows how threats and the answers to those threats have changed over the years—the engineer had to contemplate the challenge of dropping nuclear warheads in one era and the launching of precision-guided conventional munitions in another.

One of my own personal standards in reading and rating a book is how the author's interest is reflected in the research and writing. I'm not revealing any secrets when I say that there are too many books on aircraft and aviation published by authors who are not grounded in the subject and who simply perform a quick review of current literature, slap some photos together, and move on to another subject.

This book is a perfect example of the opposite case. Steve Pace is bound up in aviation as a subject—it's in his blood, and he would have written this book even if he were the only one who would ever read it. The author takes us through a magnificent period in aviation history, one that could never have existed before and will never exist again. He has given us much to learn—and to love.

Walter J. Boyne
Silver Spring, Maryland, USA
USAF Col. (Retired)
Inducted into National Aviation Hall of Fame in 2011
Former director of the National Air and Space Museum
Best-selling author and historian

INTRODUCTION

On October 1, 1942, a peculiar airplane moved across a dry lakebed in southeast California's Mojave Desert. In a few moments, it lifted off, rose to an altitude of about twenty-five feet, then settled onto the lakebed once more, rolling slowly to a stop.

Only a few officials, mechanics, and technicians were present to witness the event, but afterward they could say they'd seen history in the making. For this was the very first time an airplane had flown in America without an attached, spinning propeller or a piston engine turning that propeller. Nestled inside this airplane's fuselage were two gas turbine (turbojet) engines, the propulsion source for a new kind of airplane.

This turbojet-powered craft was one of three super-secret XP-59A airplanes built for the US Army Air Forces by the Bell Aircraft Corporation in Buffalo, New York. Its test pilot, Robert Morris "Bob" Stanley, "test-hopped" it three more times on that momentous day—first to about fifty feet, then to seventy-five feet, and finally to one hundred feet—each time with no significant difficulties.

The next day, with dignitaries and officials in attendance to witness what is considered the first *official* flight of the airplane, Bob Stanley took the controls of the premier XP-59A—nicknamed Squirt—and made the first formal flight

of the aircraft, to an altitude of six thousand feet. He landed twenty minutes later, reporting no problems. Stanley's second flight that day brought him to ten thousand feet at a speed of three hundred miles per hour, again with no reported problems.

Since running time on these early turbojet engines was limited, and there were only thirty-five minutes left on them, Stanley suggested that one of the officials present perform the day's final test flight. Colonel Laurence C. "Bill" Craigie (assistant chief of experimental Aircraft Projects, USAAF Air Materiel Command, at Wright Field in Dayton, Ohio) jumped at the opportunity and flew Squirt for about twenty minutes before landing. Unbeknownst to the American public—or to other nations, enemy or ally—the Jet Age had dawned in America.

The road to jet aviation (at least in the military) in the United States was to be long and winding, with many bumps and thumps . . . but this was the beginning. From these two significant days in October 1942, it would be another ten years before the fleet of US combat aircraft were transformed from piston-powered, propeller-driven aircraft to turbojet-powered aircraft. By October 1952, the US Air Force had no fewer than seven turbojet-powered bomber and fighter types in frontline service, with many more under development.

Established as a separate branch of the US Armed Forces on September 18, 1947, the US Air Force (USAF) built on the work of its predecessors, the US Army Air Corps and US Army Air Forces, as the sole military branch responsible for the introduction of turbojet-powered combat aircraft to protect the military interests of the United States. The US Navy would follow suit, but not for several years.

In the more than seventy years spent developing jet-powered combat aircraft (specifically bombers and fighters), the USAF and its predecessors have generated numerous aerial warfighters, some of which spawned variations and offshoots of their own.

All of these jet-powered combat types, their variants, and their spinoffs originated in the planning, experiments, development, and demonstrations discussed in this book. While many of the test aircraft proved successful, went into production, and then entered service, a number of others didn't make the grade and were cancelled. All of these test aircraft, whether successful or not, are part of the story told here.

Following behind the growth of turbojet-powered fighter aircraft, bombardment aircraft featuring turbojet engines appeared nearly five years after the first turbojet-powered fighter. For the most part this was by design, since high-speed fighters were needed to defend America from attack, as well as to escort and protect the nation's current fleets of piston-powered, propeller-driven bombers. This took priority over the development of high-speed turbojet-powered bombers. The first turbojet-powered bomber finally did take to the air, three-and-a-half years after the first turbojet-powered fighter had become operational (that is, combat ready).

More important to the success of turbojet- and turbofanjet-powered bombers was the development of appropriate engines to propel them. This was critical and ran parallel with the development of fighter aircraft, since both employed engine types of this kind. There were serious attempts to incorporate turbopropjet engines, which proved incompatible on combat aircraft such as the bombers and fighters covered in this book. The turbojet and turbofanjet engines prevailed. This book tells the story of how these fighters and bombers actually were created, which ones proved to be winners, and which ones wound up on the scrapheap of military aircraft history.

Douglas XB-43. *AFTC/HO*

ONE

"FIRE THEM UP!"

You don't fly this airplane — you just hint to it where you want it to go.

— Milo Garrett Burcham, chief test pilot, Lockheed XP-80

By September 30, 1942, the US Army Air Forces (USAAF) had budding fleets of bombers and fighters in various stages of design, development, and construction—all of which were piston-powered and propeller-driven aircraft.

All except for one.

During its earliest history, this new kind of airplane was known to only a handful of key personnel. An experimental fighter with two engines, it looked like most other fighters of the period, with one significant difference: its engines didn't have any propellers.

11

Instead, its fuselage hid two gas turbine (turbojet) engines. Not driving propellers at all, these engines could produce forward thrust with their powerful exhaust gases, measured in pounds-force (lbf).

The design wasn't new. In fact, the engines were close copies of the UK W.2B, a jet engine designed by Whittle/Power Jets, developed and manufactured in a joint effort by Rover and Rolls-Royce. For the highly classified US program, the propeller-less engines were manufactured under license in the United States by the General Electric (GE) Corporation.

Nothing was left to chance when it came to keeping this program a secret. In fact, when the first airplane was brought out in the open, it was disguised with a bogus wooden four-bladed propeller attached to its nose.

The test aircraft was brought to North Base, a location on the edge of a dry lakebed that was part of California's Muroc Army Air Field. Over several weeks, the plane went through preflight evaluations of its systems and its unique propulsion units.

Finally, it was declared ready for flight. On October 1, 1942, its pretend propeller was removed and the command was given: "Fire them up!"

BELL XP-59A, YP-59A AIRACOMET, AND XP-59B

I didn't go very fast; I didn't get very high . . .

— Col. Laurence C. "Bill" Craigie (Lt. Gen., USAF, Retired)

The Bell Aircraft Corporation P-59 Airacomet holds the unique distinction of being the first turbojet-powered airplane built and flown in the United States. Designed, developed, manufactured, and tested during World War II, it was a single-place, twin-engine pursuit aircraft produced for war—even though it was never used in combat.

A lengthy flight test program coupled with intense gunnery evaluation was disheartening: the three experimental XP-59As and thirteen service-test YP-59As proved to be woefully underpowered, short ranging, and unstable. In fact, contemporary piston-powered and propeller-driven pursuit aircraft outperformed them. However, the P-59 *did* contribute significantly to jet airframe and jet powerplant development in America.

When the first XP-59A lifted off Muroc Lake in 1942, America entered the Jet Age. That historic event was not immediately reported to the American press, however, because only a select group of personnel knew about the airplane—and they were sworn to secrecy. It was not until January 7, 1944, that the formerly classified information was made public, when Bell Aircraft Corporation founder and president Lawrence Dale "Larry" Bell issued the following statement:

> We believe that the hundreds of successful flights made by Bell's jet-propelled ships open a new chapter in American aviation history. Bell Aircraft has built the first American fighter planes powered by jet propulsion engines constructed by the General Electric Corporation from British designs. They prove a new scientific principle—that is, that airplanes can fly without propellers. Once a principle is proved, count on the engineering genius of the Allied powers to develop it into greater performance records, not only to help speed the day of victory but to pave the way towards new achievements in the postwar aviation world.

Photographed from the roof of a North Base hangar, this is a rare color image of the first XP-59A. On the wing, Alvin M. "Tex" Johnston talks with pilot Jack Woolams. *NMUSAF*

Larry Bell was indeed correct. Photographs of the Bell XP-59A "Rocket Ship," as it was dubbed by the media, were not released until September 1944, eight months after Bell's statement and nearly two years after the Airacomet's first flight. Some twenty-five years later, the United States had a real rocket ship, but it didn't land on some dry lake in a desert—it landed on the moon.

On June 20, 1941, Maj. Gen. Henry Harley "Hap" Arnold was named commanding general of the newly established US Army Air Forces. About a month earlier, on May 15, the Gloster Aircraft Company G.40, piloted by Gloster's chief test pilot, Flight Lt. Phillip E. G. "Gerry" Sayer, made its first flight in England. It was another historic event, to be sure, because it was the United Kingdom's first jet-propelled airplane, and in essence the progenitor of the Bell XP-59A Airacomet. General Arnold was there to witness the event, and he was most impressed.

After the flight of Gloster's G.40 Pioneer, Arnold quickly collected a group of technical experts to evaluate the potential for a US jet-propelled aircraft. He included USAAF Col. Alfred J. Lyon, head of the USAAF Technical Staff in London, and D. Roy Shoults, a General Electric turbo-supercharger specialist on loan to the Technical Staff, in a briefing on the status of jet propulsion in Great Britain. Later, the group visited the Power Jets Limited and Gloster Aircraft Company facilities. They were joined by USAAF Maj. Donald J. Keirn, a propulsive system engineer, who came from the Wright Field Powerplant Laboratory, along with USAAF Maj. Carl Brandt, another assignee to the Technical Staff.

In September 1941, some four months after the G.40's first flight, the London-based USAAF Technical Staff presented its recommendations in Washington, D.C., to a high-level group of military commanders and staff. Present at this meeting were US Secretary of Defense Robert A. Lovett; USAAF Gen. Carl Andrew "Tooey" Spaatz, chief of the Air Staff; USAAF Gen. Oliver P. Echols, Chief of the Material Division; USAAF Col. Benjamin W. "Ben" Chidlaw, Gen. Echols's assistant; and, of course, Gen. Arnold. A representative of the General Electric Corporation was also in attendance.

The group decided to build three airplanes and fifteen engines in order to evaluate the operational potential for jet-propelled aircraft flight. Because of the low lbf generated by the Power Jets Limited engine, the group favored a twin-engine pursuit design. They selected the Bell Aircraft Corporation as the logical choice to build the airframe, while General Electric was chosen to build the powerplant. The choice of the latter was dictated by that firm's extensive background in the turbo-supercharger field, while several factors worked in favor of Bell:

- The company's Buffalo, New York, facility was fairly close to General Electric's Schenectady, New York, and Lynn, Massachusetts, facilities.
- Its strong emphasis on research and development (R&D).
- Bell's capacity at the time, when all other fighter-producing firms had limited factory space.

As Col. (later Gen.) Chidlaw said at the time, "Larry Bell was there to ride herd on the program!"

On September 30, 1941, the USAAF Material Command issued Bell Aircraft USAAF contract number W535-AC-21931 for three Model 27 XP-59A aircraft.

XP-59A

With contract and a rudimentary set of specs for Power Jet's Model W.2B turbojet engine in hand, Larry Bell and his chief engineer, Harland M. Poyer, immediately initiated design work. The team had eight months to complete the first example, by no later than May 30, 1942. With Bell's Edgar Peter "E. P." Rhodes named project engineer, the USAAF Material Command assigned Secret Project MX-397 to the XP-59A program.

At the time, Bell's contract called for it to build a pusher-type, piston-powered, propeller-driven pursuit—its Model 20, designated XP-59. On the books, this was to be an improved version of two pusher-types—its Model 16, the XP-52, and Model 19, the improved XP-52—both of which had been cancelled. Although the XP-59 designation had been retained, the "A" suffix identified the top-secret program's real purpose: to build a jet airplane, the XP-59A.

General Electric perpetrated a similar diversion. For those outside the project, the "Model I-A" (pronounced "eye-A") turbojet engine in development appeared to be nothing more than a new turbo supercharger for piston engines.

On October 2, 1941, Maj. Keirn returned to the United States with a team of Power Jets gas turbine engine designers and technicians, including none other than RAF Air Cdre. Frank Whittle—the creator of turbojet propulsion in Great Britain. As promised, they brought with them a complete set of W.2B engine blueprints and the prototype Model W.1X turbojet engine, tucked under false floorboards inside their B-24 bomber.

Bell XP-59A number one in three-view.
G. De Chiara © 2015

Some six months earlier, the W.1X, non-flight-rated turbojet, had powered the Gloster G.40 during its preliminary taxi trials. General Electric received the Power Jets material immediately, a signal example of the remarkable cooperation between the United Kingdom and the United States.

In November, General Electric ran its first test of the Power Jets W.1X engine on American soil. Then, following intense months of preliminary design and engineering work at Bell, construction of XP-59A number one finally began on the second floor of a building in downtown Buffalo, New York. The start date, January 7, 1942, was telling: exactly one month after Japan's sneak attack on Pearl Harbor. As it took shape, no one suspected what was going on behind the building's windows, which had been welded shut and painted over.'

Everything came together over that winter, and the crew completed its first W.2B-derived GE Type I for testing by spring. Installed within a special test cell—affectionately nicknamed "Fort Knox," in recognition of its eighteen-inch-thick walls—this first engine was test fired on April 18, 1942. Even though it stalled well below full speed, this was exactly what the British had experienced. In response, the crew called upon their vast turbo supercharger knowledge, with GE engineers returning to the drawing board and modified key elements. Exactly one month later, on May 18, the test log recorded these historic words:

> Everybody working to finish Type I so that it could go into Fort Knox. We did a great deal of checking before attempting to start; a great deal of trifling troubles were found and modified—but after many attempts, Type I ran.

Then a triumphant line, printed at the top of the log sheet in bold letters with heavy underscoring, reads:

> Type I runs at 11:05 p.m. Operator – Donald F. "Truly" Warner

As an improved version of the Power Jets W.2B, the uninstalled, stationary GE Model I-A (Type I) turbojet engine produced 1,250-lbf. It was 70.5 inches long, 44 inches in diameter, with one compression stage, one turbine stage, and a pressure ratio of 3:1. General Electric's initial run of its Type I (Model I-A) occurred just twenty-eight weeks after its stateside work had begun.

At the same time, Bell's workers were progressing with the airframe into which the GE Model I-A engines would be fitted. They operated under a basic, prudent concept: the XP-59A would be a single-seat, midwing monoplane sitting on tricycle landing gear, with broad, underslung engine air inlets for the turbojet engines that were mounted below and encompassing the wing roots. The engines themselves would be installed below and behind the pilot, with exhaust exiting just aft of the wing's trailing edge on either side of the fuselage. The airplane would have rounded wingtips and tail planes. Except for propulsion units and associated air inlets, ducting, and exhaust pipes, the craft appeared to have been configured to mount and use a standard inline or radial piston engine. In fact, it resembled a conventional piston-powered, propeller-driven pursuit—without a propeller.

In reality, the XP-59A airframe was not as revolutionary as its propulsive system. Instead, it offered familiar and fully predictable power, perfectly suited for the role employed by turbojet engine testbed air vehicles. The XP-59A initial design didn't include armament, though this was eventually installed.

The US military deemed development of this airplane critical to the Allied war effort, though it was a gamble: no one could say how the airplane would perform in combat conditions. Fortunately, it was never put to the test as a frontline fighter airplane.

Two GE Model I-A turbojet engines propelled the Bell Model 27 in its stationary configuration. Because of this—and a limited, 2,500-lbf total output—Bell's engineers could only achieve optimal performance if they kept the airplane's weight down while also streamlining its fuselage and flying surfaces. This challenge conflicted with the primary goal of getting the plane airborne as soon as possible, effectively putting an end to any aerodynamic breakthroughs. Nevertheless, in February 1942, just one month after construction on the number one XP-59A had begun, Bell offered to build thirteen service test YP-59As with the improved GE Model I-14 (1,400-lbf) turbojet engine. Approved a month later, the contract for three experiment XP-59As was amended to include the thirteen additional airplanes.

With the May 30, 1942, deadline for completion of the first XP-59A airplane looming, it was clear that neither the airframe nor its two powerplants would be ready in time. When confronted with this reality, the USAAF extended the deadline, both out of sympathy for the overtaxed crew and as an acknowledgement of the complexity of work they were performing.

As expected, the first XP-59A was about two months late; it was delivered on August 1, 1942, minus its engines, and the first pair of GE Model I-A turbojet engines arrived twenty days later. In the meantime, Robert Morris "Bob" Stanley, chief of flight test operations at Bell, appointed himself XP-59A flight test program director, having arrived at the Muroc AAF's top-secret North Base facility in the Mojave Desert on August 10. There he busied himself making arrangements for the classified flight test program.

While its official name would be *Airacomet*, the development crew dubbed the XP-59A Squirt. With the airframe and powerplants mated in Buffalo, the XP-59A was loaded into a railroad boxcar, minus its wings and horizontal stabilizers, for transportation to Muroc AAF. In an effort to prevent possible damage to its engine bearings, the crew slowly rotated them with compressed air during the rail journey west. The airplane left Buffalo on September 12 and arrived at North Base seven days later, where it was offloaded and taken into a secret hanger at North Base for assembly and ground tests.

This left-hand side view of an XP-59A illustrates one aspect of this secret project: there's no USAAF serial number stenciled on its vertical tail. Serial numbers were applied later, though they were out of sequence—the next thirteen YP-59As were numbered 42-108771 to 42-108783, while the preceding trio of XP-59As were numbered higher (rather than lower) with the serial numbers 42-108784, 42-108785, and 42-108786. *NMUSAF*

After Squirt was reassembled, a series of tests evaluated its electrical, hydraulic, and fuel systems. Then, on September 26, the two GE Model I-A engines were engaged for the first time, and both were paced successfully through a trio of five-minute run-ups. With this most important accomplishment judged satisfactory, the crew began taxi testing under power. Test pilot Stanley completed several low- medium- and high-speed taxi runs on October 1, 1942, and actually left the ground several times during the process, ascending to fifty, seventy-five, and one hundred feet. Squirt wanted to fly!

Since several unofficial flights had already been made, Stanley declared Squirt more than ready for a real flight. He suggested that flight testing should begin at once, but his boss, Larry Bell, and the other dignitaries hadn't made it to the facility. For this reason, the first *official* flight was put off until the following day, scheduled for 8:00 a.m.

Though last-minute engine maintenance delayed the event, at 12:56 p.m. on October 2, Stanley took off. He retracted the landing gear, climbed to an altitude of six thousand feet, leveled off, and evaluated the Squirt's basic flying characteristics. Satisfied with how it handled, he landed twenty minutes later with no major squawks.

Stanley's second flight of the day also lasted twenty minutes, and he attained an altitude of ten thousand feet with a speed of three hundred miles per hour. Since GE had limited their running time to three hours—including run-ups, taxiing, and flight—only thirty-five minutes remained for engine running time after this second flight. Stanley suggested that USAAF Col. Laurence C. "Bill" Craigie go up on the third test.

Colonel Craigie, assistant chief of Experimental Aircraft Projects at Wright Field at the time, described the day's events:

> Bob Stanley attempted nothing spectacular on his two flights. He merely felt out this new type of air vehicle to see if it had any particularly unusual traits. After his second flight he said to me, "Bill, we've only got about twenty-five minutes left before we run into the three-hour limitation. Why don't you fly it?" Naturally, I jumped at the chance.
>
> And that is how I happened to be the first military pilot to fly a jet airplane. I learned many years later that Col. Benjamin W. "Ben" Chidlaw (my boss at the time), Col. Ralph P. Swofford Jr., and Col. Donald J. "Don" Keirn had at one point, probably when they picked Muroc AAF to be the test site, agreed to match coins or draw straws to see who would be the first USAAF officer to fly America's first jet airplane. But they neglected to tell me about their little plan, which, as things worked out, wasn't feasible anyway because they were far away from Muroc AAF at the time.
>
> My flight was quite uneventful. I didn't go very fast, I didn't get very high; my most vivid memory of that flight was the extreme quiet at the moment the plane lifted off. Takeoff is, of course, the time of maximum noise and vibration in the case of piston-powered and propeller-driven aircraft. The XP-59A, with its rotary action gas turbine engines, did not have "up and down" piston devices.

All in all, the airplane had performed well: Squirt was airborne some sixty minutes on October 2, 1942. No one outside the group who witnessed the test runs that day knew that the American Jet Age had begun.

When the Bell XP-59A Airacomet came on the scene, the USAAF already had a trio of top-notch pursuit aircraft—the Lockheed P-38 Lightning, Republic P-47 Thunderbolt, and North American P-51 Mustang. Each fighter could reach a top speed of 400-plus mph. Squirt was expected to do at least 500 mph, but this mark eluded Bell's and General Electric's efforts for the duration of the Airacomet program. It boiled down to the crafts' weight: the total of 2,500-lbf from its two engines was inadequate for a fully fueled airplane weighing about thirteen thousand pounds. Early flights demonstrated disappointing speeds—370 mph, 380 mph, 390 mph, and then just 396 mph during Squirt's ninth test hop on October 30, 1942.

Bell test pilot "Tex" Johnston is all smiles after completing his first flight in an XP-59A. He was later assigned to the Bell X-1 test program, then promoted to chief test pilot at Boeing on the XB/YB-52 test program. In 1955 he barrel-rolled the 707 prototype known as the "Dash 80" not once but *twice* over Seattle's Lake Washington during the Seafair Gold Cup hydroplane race. *AFTC/HO*

XP-59A SPECIFICATIONS

Propulsive system:
Two non-afterburning centrifugal-flow 1,250-lbf General Electric Model I-A (Type I) turbojet engines

Length	38 feet, 2 inches
Height	12 feet, 4 inches
Wingspan	45 feet, 6 inches
Empty weight	7,500 pounds
Gross takeoff weight	13,000 pounds
Maximum speed	404 miles per hour
Ceiling	35,000 feet
Maximum range	600 miles
Armament	Not installed

If adrenaline were thrust, though, Bob Stanley would have gone 600mph.

On the same day as Stanley's initial test hops, October 1, 1942, USAAF contract number W535-AC-21931 was approved by the US War Department for the three experimental XP-59A airplanes. They were assigned USAAF serial numbers 42-108784 through 42-108786.

XP-59A number two (42-108985) arrived at Muroc AAF in early January 1943 and made its first flight on February 15, 1943, piloted by Bell test pilot Jack Valentine Woolams. XP-59A number three (42-108786) arrived in mid-March and made its first flight on April 18, 1943, also piloted by Woolams.

These three experimental pursuit aircraft remained at Muroc AAF for several years, undergoing numerous test flights while simultaneously serving as familiarization aircraft for pilots training to fly jet airplanes. Apart from a March 7, 1945, taxiing accident involving XP-59A number two (42-108785) on base at Muroc AAF, all three airplanes were relatively safe to operate, albeit somewhat unstable. To help correct this flaw, Bell designers polished the plane's aerodynamics with refinements that would be applied to all subsequent Airacomets, beginning with the YP-59A.

Only the first of the three XP-59A airplanes (42-108784) has survived to our time. That aircraft is now part of the Smithsonian Institution National Air and Space Museum collection, on display in the National Mall building in Washington, D.C.

YP-59A

Recall that Bell was awarded a contract to produce thirteen service test YP-59A airplanes. The plan was to power the YP-59A airplanes with the interim GE Model I-14 (1,400-lbf) turbojet engine until the uprated 1,650-lbf GE Model I-16 (now designated J31) turbojet engine was produced.

These airplanes (Bell Model 27, the same as XP-59A) were to be delivered under USAAF contract number W535-AC-21931, which was approved by the U.S. War Department on March 26, 1942. Assigned USAAF serial numbers 42-108771 to 42-108786, the last three serial numbers—42-108784 to 42-108786—were actually never used by any of the YP-59A airplanes, since they would have added up to sixteen rather than thirteen. Instead, the latter three serial numbers appeared on the three XP-59A airplanes, in what may have been a further attempt to keep the program as secret as possible. The serial numbers weren't even stenciled onto the aircraft until later.

ABOVE: The notable external differences between the XP-59A (top) and YP-59A: the YP-59A's wingtips are squared off, whereas the wingtips of the XP-59A are rounded. *AFTC/HO*

LEFT: One of the thirteen prototype YP-59A airplanes on a test flight. The YP-59As were used for everything from armament to performance evaluations. *NMUSAF*

ABOVE: Service test YP-59A—number eight (42-108778) of the thirteen built—here prepares to undergo US Navy flight test evaluations at the Naval Air Test Center (NATC) at NAS PAX River in Maryland. This particular YP-59A—the Navy's first jet—was issued USN Bureau Number 63960. A second prototype, YP-59A (42-108779), was also evaluated by the USN (BuNo 63961). Both examples were redesignated YF2L-1. *USN*

RIGHT: Bell test pilot Bob Stanley was the first to fly the Airacomet on October 1 and 2, 1942. Here he poses for the photographer after a subsequent test flight. *NMUSAF*

The few aerodynamic alterations that the YP-59A airplanes received were an attempt to fix their tendency to "snake" during flight. These changes included the addition of a ventral fin, squared-off wingtips, and vertical stabilizers, though the rounded tips on the horizontal stabilizers were retained. While they helped to a degree, this wasn't enough to satisfy the USAAF: some of the power-robbing glitches were chased out of the General Electric J31 turbojet engine program, though several remained to cause exhaust pipe overheating and numerous compressor stalls.

The number two YP-59A (42-108772) was ready for testing before number one, and it was the first to fly at Muroc AAF on August 15, 1943. Piloted by Bell's Frank H. "Bud" Kelley Jr. in an all-out test, the plane reached a speed of 404 mph at 31,500 feet. YP-59A number one (42-108771) followed number two into the air in September.

The YP-59A's marginally better performance was still far from acceptable. Kelley took over from Stanley on November 11, 1942, as the latter had been called back to Buffalo by Larry Bell to resume his duties as chief of flight test activities. Kelley made his first check flight five days later, on November 16, 1942.

On December 15, 1943, Bell test pilot Jack Woolams established an unofficial US record altitude flight to 46,700 feet in an YP-59A, this time fully armed. This aircraft differed from the XP-59A in that it came with nose-mounted armament comprised of a single 37mm M4 cannon (forty-four rounds of ammunition) and three .50-caliber M2 heavy machine guns (two hundred rounds of ammunition each—six hundred rounds total).

The thirteenth and last YP-59A arrived for delivery in late June 1944.

YP-59A number seven (USAAF serial number 42-108777) survives today at the Planes of Fame Air Museum in Chino, California. This extremely rare and historic airplane is currently undergoing full restoration to flyable condition, complete with refurbished J31 turbojet engines.

XP-59B

Little is known about the proposed Bell XP-59B beyond its intended use as an experimental pursuit aircraft powered by a single turbojet engine under USAAF contract number W535-AC-26614. E. P. Rhodes was the project engineer on this plan. It was assigned Bell model number 29 and the USAAF AMC assigned Secret Project MX-398 to the program.

Its single engine—most likely the de Havilland Halford H.1B Goblin later employed by the Lockheed XP-80 or a General Electric J31 (GE Model I-16)—was to be fed its required air supply via wing root air inlets on either side of its fuselage. Not much else is known about this "paper" airplane, which never got off the drawing board: Bell wasn't allowed to proceed on the program due to the company's inability to deliver in time for USAAF testing.

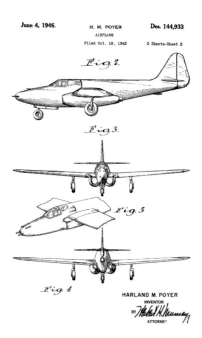

ABOVE: Some of Harland Poyer's sketches of the three-engine P-59. *United States Patent Office*

RIGHT: The business end of a YP-59A shows its landing light, gun camera, 37mm cannon, and trio of .50-caliber machine guns. *AFTC/HO*

XP-59 AND YP-59A
PRODUCTION BREAKDOWN

XP-59A:	3–42-108784 to 42-108786
YP-59A:	13–42-108771 to 42-108783

SUMMARY

Although production P-59As and P-59Bs were actually delivered in time to participate in World War II, the aircraft weren't sent into combat. The main reason for this is that the P-59 in its final production guise was neither fast enough nor maneuverable enough to challenge the best of the enemy's piston-powered, propeller-driven fighter aircraft. For this reason, it was relegated to the role of pilot training and transition aircraft for US fighter pilots learning to operate combat-capable, turbojet-powered fighters.

The Bell P-59 Airacomet program supplied US Army Air Force, US Navy, and US Marine Corps aviators a workable training aid for America's upcoming fleets of turbojet-powered combat aircraft. In this regard, these planes were unique.

The lackluster performance of the production P-59As and P-59Bs remains a disappointing aspect of the planes' history. Although available for combat during the latter part of World War II, their top speed was only 409 mph, which meant they could never have seriously challenged the era's faster piston-powered, propeller-driven fighters. Moreover, according to armament evaluations, they were not stable enough to make good gun-firing platforms.

Regardless of its missing out on combat missions, the Bell P-59 Airacomet was America's first turbojet-powered airplane, and for this reason alone it holds a special place in aviation history.

TOP: For just four months, from June to September 1943, the National "star and bar" marking were outlined in red as shown here on XP-59A number one during a flight test. *AFTC/HO*

ABOVE: Color view of right side of XP-59A number one in its original livery. *AFTC/HO*

Left-hand (port) view of the XP-79B, clearly showing its left wingtip bellow, used in conjunction with the right wingtip bellow to provide air to the bellows-boosted ailerons while in flight. *NMUSAF*

NORTHROP XP-79 AND XP-79B

In 1939, John Knudsen "Jack" Northrop founded Northrop Aircraft, Inc. (now Northrop Grumman Corporation), a company that would grow into one of the world's most steadfast supporters of flying wing aircraft. Long before phrases like "state of the art" and "quantum leap" came into common parlance, Jack Northrop was hard at work designing and creating the shape of wings to come. Early on, in his continuing effort to bolster his flying wing aircraft business, he met with USAAF Air Materiel Command in September 1942 to propose a small, light flying wing pursuit interceptor propelled by a rocket motor.

As he described it, Northrop's P-79 "Rocket Wing" would be powered by a single, liquid-fueled 2,000-lbf Aerojet Engineering Corporation (later Aerojet Rocketdyne) XCALR-2000A-1 rocket motor, an engine that was in fact under development for the USAAF at the time. Capable of being throttled, it featured four chambers—two providing 750-lbf and two 250-lbf—while burning liquid fuel at a ratio of 3.5:1 by weight of oxidizer (red, fuming nitric acid) and fuel (aniline) (that is, 8,400 pounds of fuel).

For additional takeoff power, two 1,000-lbf droppable, Aerojet rocket-assisted takeoff (RATO) units would be employed, with solid fuel that could be jettisoned automatically when thrust became zero. USAAF Air Materiel Command received Northrop's proposal favorably.

For aerodynamic evaluation of Northrop's proposed Model N-14 rocket-powered pursuit interceptor, the USAAF funded three glider aircraft in December 1942 under USAAF AMC Secret Project MX-334. Northrop would provide the three gliders (Model N-12), made mostly from wood, for flight tests within three months, under USAAF contract number W535-AC-36137.

This activity was quickly followed with an order for three rocket-powered experimental XP-79 airplanes (USAAF AMC Secret Project MX-365) in December 1942 under USAAF contract number W535-AC-36997. Suddenly, Northrop was committed to the development and construction of one of the most interesting aircraft to emerge during World War II.

In addition to its unique flying wing configuration and rocket propulsion system, the proposed XP-79 was optimized for positive 12g and negative 12g maneuvering. To accomplish this unheard-of limit maneuver load factor, its pilot would fly the airplane from a prone position. Although unusual, prone piloting wasn't a new configuration; the Wright brothers' gliders employed this design, as did some of the first powered flyers.

The top speed of a production P-79 with its design 2,000-lbf propulsive system was projected to be about 540 mph at forty-thousand feet. With a primary mission of pursuit, interception, and destruction of enemy aircraft, it was to be armed with four M2 Colt-Browning .50-caliber heavy machine guns (two on either side of centerline) with 250 rounds of ammunition per gun. Small and light, the P-79 would be constructed of heavy-gauge magnesium alloy varying from ¾-inch thick at the leading edge of the wing to ⅛-inch thick at the trailing edge.

As construction of Northrop's three Model 12 gliders and three Model 14 XP-79s proceeded in concert, the development of Aerojet's rocket motor lagged

The one-off XP-79B (43-52437) sits on the ramp outside of Avion Corporation's facility near Burbank, California, shortly after its completion and prior to being transported to Muroc AAF. Powered by a pair of non-afterburning, 1,150-pound Westinghouse 19XB turbojet engines, it achieved an estimated top speed of about 550 mph. Since the XP-79B had two vertical tails, it was considered a "semi-flying wing." Its empty weight was 5,840 pounds; its loaded weight was 8,669 lbs. *NMUSAF*

The XP-79B as viewed from above. It spanned thirty-eight feet, was seven feet, six inches high, and fourteen feet long. It was the first aircraft to have twin vertical tails, a feature not employed again until the advent of the Lockheed YF-12A in 1963. *NMUSAF*

behind. In an effort to get in step with the program—that is, to meet its contract obligations and actually get a rocket-powered aircraft into the air—Aerojet built a smaller, interim 200-lbf rocket motor, the XCALR-200A. The installation of this motor meant that glider number one would now serve as a powered air vehicle for the P-79 program under the MX-324 program. Following the crash and complete destruction of glider one crashed during a test glide flight, though, the number two glider stepped up to become the powered MX-324.

Aerojet's development problems worsened, and a major delay in the rocket-powered XP-79 program was imminent. In spring of 1943, therefore, it was decided to modify XP-79 airframe number three for the installation of two 1,150-lbf Westinghouse J30 turbojet engines. Designated XP-79B (since the designation P-79A was reserved for production aircraft), this version was similar to the XP-79, except for its propulsion system and type of fuel. XP-79 airframes one and two were expected to fly with their intended 2,000-lbf rocket motors.

The XP-79 mockup inspection was held on June 3, 1943 (note that no mockup of the XP-79B version was required). The mockup evaluation group found the projected rocket-powered XP-79 version to be superior in climb rate, altitude, and maximum speed over the turbojet-powered XB-79B version. For this reason, the XP-79 version was deemed worthy of continued development. Nevertheless, the mockup inspection team was cautious of the XP-79's rocket propulsion issue, being understandably more comfortable with the turbojet-powered XP-79B version.

To meet its other aircraft production quotas (P-61 and so on), Northrop subcontracted its XP-79/-79B work to Avian, Inc. on June 29, 1943.

XP-79

The USAAF ordered three experimental XP-79 airplanes under contract number W535-AC-36997 (recall that USAAF AMC Secret Project number MX-365 was assigned to this highly classified program).

As proposed, in operation, the P-79 was to be a dedicated interceptor armed with four Colt-Browning M2 .50-caliber heavy machine guns and propelled

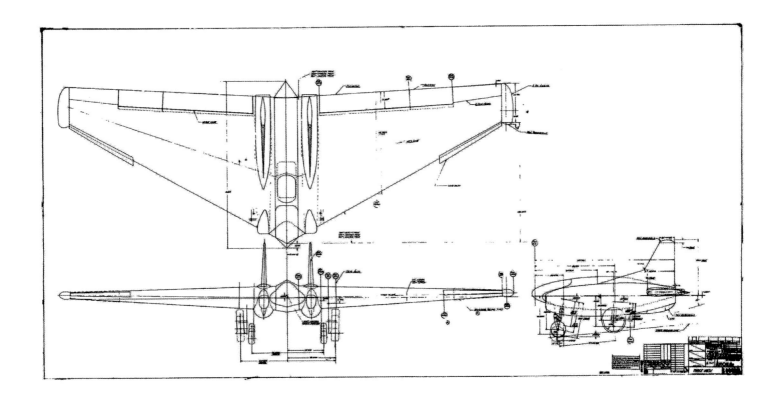

by a single-fuel chamber 2,000-lbf Aerojet General Model XCALR-2000A-1 "Rotojet" (XLR7-AJ-1 military designation) rocket motor.

XP-79B

On May 26, 1943, the US War Department approved USAAF-amended contract number W535-AC-36997 for a single XP-79B airplane. The preceding contract for three XP-79 airplanes was cancelled. The airframe used to create the XP-79B was formerly the number three XP-79, since it could be more easily converted to a turbojet-powered air vehicle rather than a rocket-powered air vehicle. The other two airframes were further along in construction and not so easily transformed. The one-of-a-kind XB-79B was issued USAAF serial number 43-52437. (It remains unclear as to what, if any, USAAF serial numbers were assigned to XP-79 numbers one and two.)

This XP-79B—Northrop Project Number N-14—was completed in early August 1945 and then secretly trucked to North Base on Muroc AAF. Once there, it underwent engine runs and systems tests followed by low-, medium-, and high-speed taxi tests. During these early taxi runs, several tires blew out and had to be replaced, delaying the upcoming flight test program by about a month. Subsequent taxi and engine runs found the airplane ready for its first flight.

Northrop chose test pilot Harry Crosby for the test. On the morning of September 12, 1945, Crosby taxied out and lined up for takeoff. He powered up

MX-324 and MX-334 Ground Run and Flight Dates

August 27–28, 1943....MX-334 glider two—ground runs . . . Crosby, Muroc Lake

September 14, 1943....MX-334 glider two—ground runs with short hop ending in a crash (repairs needed) . . . Crosby, Muroc Lake

September 30, 1943....MX-334 glider three—ground runs . . . Myers, Muroc Lake

October 2, 1943....MX-334 glider three—first flight . . . Myers, Muroc Lake

October 8, 1943....Cancelled Muroc Lake flight on MX-334 glider three; Myers had been injured in crash of Northrop XP-56 on same day so couldn't make the planned flight, while Crosby was on special duty in the South Pacific (no other glider pilots were available at this time)

November 9, 1943....MX-334 glider three—second flight . . . Crosby, Muroc Lake

November 10, 1943....MX-334 glider three—third flight . . . Crosby, Muroc Lake; glider went out of control and crashed, Crosby parachuted to safety

November 30, 1943....MX-324 glider one—first glide flight . . . Crosby, Muroc Lake

December 2, 1943....MX-324 glider one—second glide flight . . . Crosby, Muroc Lake

April 1944....MX-324 gilder one—ground runs . . . 1 Lt. Eisele, Roach Lake (mid-April; exact day unknown)

April 19, 1944....MX-324 gilder one third—glide flight . . . Crosby, Roach Lake

April 29, 1944....MX-324 gilder one fourth—glide flight . . . Crosby, Roach Lake

April 30, 1944....MX-324 gilder one—fifth, sixth, and seventh glide flights . . . Crosby, Roach Lake

May 5, 1944....MX-324 gilder one—eighth glide flight aborted due to towline failure . . . Crosby, Roach Lake

May 5, 1944....MX-324 gilder one—eighth glide flight (after towline repair) . . . Crosby, Roach Lake

May 11, 1944....MX-324 gilder one—ninth glide flight . . . Crosby, Harper Lake

May 19, 1944....MX-324 gilder one—tenth and eleventh glide flights . . . Crosby, Harper Lake

Northrop test pilot Harry Crosby lies prone in the cockpit of the unpowered MX-334 glider in this rare color photo. *WMOF*

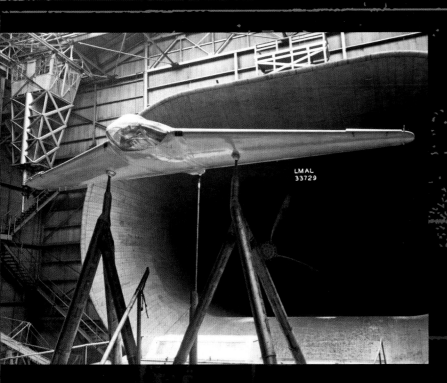

One of the three MX-334 gliders built shown in the NACA-Langley full-scale wind tunnel for aerodynamic evaluations, late 1943. This version had no vertical tail and featured a pure flying wing configuration. *NASA*

Harry Crosby enjoys a lighter moment standing up inside one of the MX-334 gliders. His upper body and head are outside the dorsal ingress/egress hatch while his legs and feet are outside the ventral emergency escape hatch. *WMOF*

June 23, 1944 ....MX-324 glider—two rocket-powered ground runs . . . Crosby, Harper Lake

(Note: After MX-334 glider three was lost, MX-334 glider two became the MX-324.)

July 1, 1944 ....MX-324 glider two—glide flight . . . Papana, Harper Lake

July 5, 1944 ....MX-324 glider two—glide flight . . . Crosby, Harper Lake

July 6, 1944 ....MX-324 glider two—first powered flight . . . Crosby, Harper Lake (most sources claim July 5, 1944, as the date of the first powered flight, but an official report claims July 6. This report, entitled History of Project MX-324, MX-334 TSEAL-2-4302-20-5 "Appendix-I History" and dated January 1, 1945, stated: "The next flight, on 5 July, with Crosby as pilot, indicated that the extreme nose heaviness experienced by Papana on the previous flight, was due entirely to the absence of the escape hatches. The release for these had been made more fool-proof, and no further trouble was experienced.")

July 11, 1944 ....MX-324 glider two—second powered flight . . . Crosby, Harper Lake

July 12, 1944 ....MX-324 glider two—third, fourth, and fifth powered flights . . . Crosby, Harper Lake

July 14, 1944 ....MX-324 glider two—glide flight . . . Crosby, Harper Lake

July 18, 1944 ....MX-324 glider two—two glide flights . . . Crosby, Harper Lake

July 19, 1944 ....MX-324 glider two—glide flight . . . Crosby, Harper Lake to Muroc Lake

July 20, 1944 ....MX-324—aborted takeoff, then a glide flight on second takeoff attempt . . . Crosby, Muroc Lake

July 25, 1944 ....MX-324 glider two—sixth powered flight . . . Crosby, Muroc Lake

July 26, 1944 ....MX-324 glider two—seventh powered flight followed by a glide flight . . . Crosby, Muroc Lake

July 27, 1944 ....MX-324 glider two—two glide flights ending the glider program . . . Crosby, Muroc Lake

August 1, 1944 ....Powered MX-324 and unpowered MX-334 glider programs officially end

MX-324 AND MX-334
PRODUCTION BREAKDOWN

MX-324:	1–none assigned	created from the number two unpowered MX-334 glider
MX-334:	3–none assigned	unpowered MX-334 glider number two was modified to serve as the powered MX-324

the plane's two Westinghouse Electric Corporation Model 19B engines to begin his takeoff roll and, at about 120 mph, rotated the nose landing gear wheel, and lifted off from the dry lakebed.

At first, the unique, semi-flying wing piloted by Harry H. Crosby performed well. Then, about fifteen minutes into the test, the twin turbojet-powered XP-79B went completely out of control. The craft entered a steep vertical spin and crashed on the ground where it burned completely, fueled by the highly flammable magnesium metal of its body. At the first sign of trouble, Crosby, who had been piloting from a prone position, struggled to escape. He managed to remove the escape hatch above his head, but the thirty-eight-year-old Crosby, most likely knocked unconscious by the wildly gyrating air vehicle, was bucked out of the cockpit. He fell to his death with his parachute still packed.

After the crash, Northrop's founder and president issued the following statement:

The takeoff for this flight was normal and for fifteen minutes the airplane was flown in a beautiful demonstration. The pilot mounted confidence by executing more and more maneuvers of a type that would not be expected unless he were thoroughly satisfied with the behavior of the airplane. After about fifteen minutes of flying, the airplane entered what appeared to be a normal slow roll, from which it did not recover. As the rotation about the longitudinal axis continued, the nose gradually dropped, and at the time of impact the airplane appeared to be in a steep vertical spin. The pilot endeavored to leave the ship but the speed was so high that he was unable to clear it successfully. Unfortunately, there was insufficient evidence to fully determine the cause of the disaster. However, in view of his prone position, a powerful electrically controlled trim tab had been installed in the lateral controls to relieve the pilot of excessive [g-]loads. It is believed that the deliberate slow roll may have been attempted (as the pilot had previously slow rolled and looped other flying wing aircraft developed by the company) and that during this maneuver something failed in the lateral controls in such a way that the pilot was overpowered by the electrical trim mechanism.

The USAAF accepted the number one XP-79B on a crash delivery basis in December 1945 and action was taken in January 1946 to terminate the remaining phases of the P-79 program, which had been an extremely advanced program for the era. The company scrapped the two incomplete XP-79 air vehicles (numbers one and two) and donated the lone MX-324 "Rocket Wing" (formerly MX-334 glider two) was donated to the National Air and Space Museum in Washington, D.C., where it remains on display. The remaining unpowered MX-334 (glider one) did not survive, and its final disposition remains unclear.

SUMMARY

The Northrop Aircraft Corporation XP-79 program led to the first piloted flight of a rocket-powered aircraft in America. The XP-79B was unique in its semi-flying wing design and the prone position of its pilot, as well as the four-point landing gear arrangement, twin vertical tails, and bellows-type flight control system attached to either tip of its wing. In addition, its body was constructed from magnesium alloy rather than aluminum alloy.

Unfortunately, America's first "Rocket Man," Harry H. Crosby, was lost during the XP-79B's first and only flight. With his tragic loss and the demise of the XP-79B, no replacement was ordered.

MX-324 AND MX-334

Northrop proceeded with its work on the three unpowered Model 12 MX-334 gliders, one of which would become the powered MX-324 air vehicle.

Construction on MX-334 glider number two—the first of the three to be finished—was completed in mid-August 1943 at Northrop's Hawthorne, California, facility. A few days later it was trucked to Muroc AAF, some 100 miles northeast of Hawthorne, where it was prepared for its first glide flight. The Northrop test pilot Harry Hume Crosby was chosen for this first flight. Born in Colorado, Crosby was an experienced commercial, general aviation, and race plane pilot when he joined Northrop in 1943. He soon jelled with the firm and earned his right to make first flights on several aircraft, including the MX-324, the first rocket-powered airplane to take wing in America.

During the tests of August 27–28, Crosby made a number of ground runs on glider number two, performed by towing the glider behind "souped-up" automobiles. On September 14, 1943, Crosby made additional ground runs on the glider, ending with a short hop that resulted in a crash.

In the meantime, glider number three had arrived on site and, on September 30, 1943, Northrop test pilot John Wescott Myers used it to make several ground runs.

XP-79 SPECIFICATIONS

Propulsive system
One 2,000-lbf Aerojet XCALR-2000A-1 "Rotojet" rocket motor

Length	13.22 feet
Height	4.73 feet
Wingspan	38.0 feet
Empty weight	4,348 pounds
Gross takeoff weight	13,500 pounds
Maximum speed	518 mph (projected)
Maximum range	454 miles (projected)

Armament
Four M2 Colt-Browning .50-caliber heavy machine guns

XP-79B SPECIFICATION

Crew
One (in the prone position)

Propulsive system
Two axial-flow, non-afterburning, 1,400-lbf Westinghouse Model 19B turbojet engines (J30 type), developed under USAAF AMC Secret Project MX-826

Length	14 feet, 0 inches
Height	7 feet, 6 inches
Wingspan	38 feet, 0 inches
Empty weight	5,840 pounds
Gross weight	8,669 pounds
Maximum speed	(estimated): 550 mph
Maximum range	(estimated): 993 miles
Ceiling	(estimated): 40,000 feet

Armament
Four Colt-Browning M2 .50-caliber heavy machine guns (never fitted)

Later, towed aloft by a Douglas C-47 transport, Myers made a successful maiden flight on the MX-334 unpowered glider (number three) on October 2, 1943.

On November 10, 1943, a third glide flight on glider number three spun out of control. Test pilot Crosby was able to parachute to safety, but the aircraft was a total loss. Despite the setback, the program proceeded satisfactorily through late 1943 and into early 1944 with the repaired glider number two.

At last it was time to install the interim 200-lbf rocket motor in glider number two. The airworthiness of the design had been proved, and Harry Crosby was appointed chief test pilot on the MX-324 program.

With the aid of Aerojet technicians, Northrop fitted glider number two (now dubbed the MX-324) with the XCALR-200A rocket motor in late spring 1944. The air vehicle was moved to Harper Lake, yet another dry lakebed in California, northwest of Hawthorne. Crosby initiated powered ground runs on June 23, followed by two glide flights that occurred on July 1 and July 5, respectively—one by Northrop test pilot Alexandru "Alex" Papana, and one by Crosby.

Early on the morning of July 6, 1944, Lockheed P-38 tow plane pilot USAAF Capt. Martin L. Smith towed Crosby and the MX-324 "Rocket Wing" up to an altitude of 8,000 feet. Crosby released the towline, flicked the motor's ignition switch to "on," and blasted forward and upward. Although rather mild from the 200-lbf, the instant kick was notable, though Crosby was underwhelmed, as he reported after the flight. He returned for a landing on the dry lakebed just four minutes after ignition.

The flight was hailed a complete success. That short flight marked the United States' entry into the Rocket Age—just twenty-one months after it had entered the Jet Age.

P-79
PRODUCTION BREAKDOWN

XP-79:	0-not assigned . . . three rocket-powered XP-79 airplanes were ordered, but only the third airframe was completed as the turbojet-powered XP-79B	airframes one and two were scrapped
XP-79B:	1-43-52437	created from the third XP-79 airframe

"Lulu-Belle," the one-of-a-kind XP-80 (44-83020), being prepared for its first flight on January 8, 1944. *LM via Denny Lombard*

The MX-324-powered glider made five more powered flights, all flown by Crosby, before the program was cancelled on August 1, 1944.

The rocket-equipped MX-324 made fifteen flights—nine glide flights without power-on, six flights with power-on—from July 1 to July 27, 1944—with the first powered flight occurring on July 6, 1944.

Three Northrop experimental test pilots and one USAAF test pilot—Crosby, Papana, and Myers—performed all of the ground run tests and test flights on MX-334 gliders one and three. USAAF 1st Lt. Arthur C. Eisele only performed a single ground run on MX-334 glider number one. Harry Crosby performed the bulk of the test runs and test flights and was the only pilot to fly glider number two, the powered MX-324.

LOCKHEED XP-80, XP-80A, YP-80A, XP-80B (P-80R), XFP-80 (XF-14), AND XFP-80A SHOOTING STAR

The Bell Aircraft Corporation chose not to proceed with the development of its Model 29, the single-engine version of its twin-jet Airacomet designated XP-59B. At this point, the USAAF went to the Lockheed Aircraft Corporation in Burbank, California, which had been trying to sell its in-house turbojet-powered pursuit design to the USAAF since March 30, 1942. This was Lockheed Temporary Design Number, or TDN L-133, coupled with its in-house turbojet engine design, the TDN L-1000.

With the Bell XP-59A Airacomet already in its advanced development stage, the USAAF passed on the Lockheed L-133 airframe and L-1000 powerplant option; they felt that the development of the particular combination would have been too time consuming, and they wanted a turbojet powered fighter as soon as possible.

In any event, Bell was out. If Lockheed wanted it, the contract for creating a single-engine pursuit powered by the British de Havilland Halford H.1B Goblin turbojet engine was in their hands. Lockheed immediately accepted the challenge and simultaneously shelved its private L-133/L-1000 airframe and powerplant combination.

Lockheed L-133

By 1940, Lockheed had established itself as one of the major airplane production firms in America. Like so many other aircraft manufacturers of that era, though, it was constantly looking for new business to survive and maintain its relatively small workforce. On order for fiscal year 1940 were sixty-five P-38 Lightnings and a single Lightning derivative, designated XP-49. These sixty-six aircraft, ordered by the then US Army Air Corps, represented a trifling number; Lockheed looked eagerly for additional orders. The next year would prove more lucrative, due to the expanding war in Europe.

While working at Lockheed on the XP-49 program as an engineer for aircraft engine supercharger systems, part-time inventor Nathan C. Price approached assistant chief engineer Clarence L. "Kelly" Johnson with his design for a gas turbine engine. Price had been working on the design since 1938, and Johnson was so impressed he introduced Price and his design to Lockheed Vice President and Chief Engineer Hall L. Hibbard.

Hibbard, Johnson, and aeronautical engineering genius Willis M. Hawkins had pushed the risky project up the chain by late 1941 to Robert E. Gross, head of Lockheed, who gave the go-ahead to develop a high-speed aircraft using Price's propulsive system. Price proceeded to finalize his design of an axial-flow gas turbine engine under Lockheed Temporary Design Number L-1000 (TDN L-1000). This gave rise to what became the Lockheed-built XJ37-1-LO turbojet engine program.

Lockheed TDN L-133, the first high-speed pursuit featuring the XJ37, was offered to the US Army Air Corps on March 30, 1942. As proposed, it was projected to be the first turbojet-powered airplane built and flown in the United States.

The L-133 project was Lockheed's initial attempt to build a turbojet-powered airplane. Lockheed considered the nascent turbojet propulsion technology to be a practical approach for attaining high speed, approaching seven-hundred-ten miles per hour—the speed of sound! The specific (albeit lofty) design parameters established by Lockheed included the following:

- High speed greater than any high altitude bomber or pursuit; assumed required speed of 600 mph
- Service ceiling of at least 40,000 feet
- Firepower sufficient to destroy a well-armored bomber or pursuit at high altitude; four 20mm cannon considered adequate
- Endurance sufficient for the aircraft to serve as a high-altitude patrol defender; three hours at fifty percent power to be sufficient
- Maneuverability sacrificed to obtain the desired high-speed performance
- Single-place aircraft, using jet propulsion units of Lockheed design
- Armor and bulletproofing minimal to keep the aircraft as small and light as possible; however, provisions for such items would be included (up to 160 pounds of pilot armor and normal fuel-tank bulletproofing)
- No provision for assisted takeoff that required special airfields or ground equipment

L-133 SPECIFICATIONS

Normal gross weight:	13.22 feet
Overload gross weight:	4.73 feet
Fuel capacity:	38.0 feet
High speed at sea level:	4,348 pounds
High speed at 20,000 feet:	13,500 pounds
High speed at 40,000 feet:	518 mph (projected)

Maximum climb rate at sea level:
3,740 feet per minute

Maximum climb rate at 20,000 feet:
5,670 feet per minute

Maximum climb rate at 40,000 feet:
6,350 feet per minute

Time-to-climb to 40,000 feet:
7.3 minutes

Takeoff distance over 50-foot object:
1,885 feet

Normal range at sea level:
320 miles

Normal range at 20,000 feet:
350 miles

Normal range at 40,000 feet:
390 miles

Terminal velocity at sea level:
710 mph

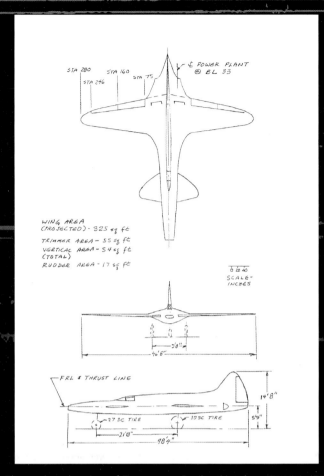

Three-view drawing of Temporary Design Number (TDN) L-133-02-01. *LM/Code One*

The L-133 was to be powered by two Lockheed L-1000 turbojet engines, and its simple design would make it adaptable to mass production. A midwing, tail-first airplane with a large-area vertical tail aft of the wing, its wings and forward-mounted tail plane or canard were to be full cantilever-tapered flying surfaces. The fuselage cross section was to be a flap ellipse with the major axis horizontal. The cockpit would have formed the leading edge or apex of a long dorsal fin, which was to fair into the vertical tail.

The L-1000 engines would be mounted aft of the pilot, one in each wing-to-fuselage fillet. The bifurcated engine air inlets for the turbojet engines would be in the extreme nose of the airplane. The cockpit, cannon armament, landing gear, fuel tanks, and so on would be contained in the fuselage between the engine air ducts. The ducts for engine air requirements would be D-shaped tunnels running aftward down either side of the fuselage. The four 20mm cannon would fire from the center of the engine air inlets, mounted in the extreme nose of the aircraft (gun-gas ingestion might have occurred and have been

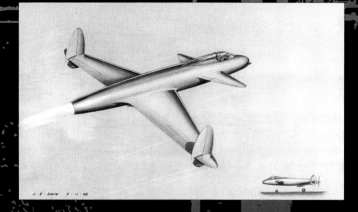

An alternate TDN L-133-02-05 design drawn on August 11, 1942, by J. E. Davis. *LM/Code One*

a problem with this particular configuration). The landing gear would be of the tricycle type, retracting into the fuselage. Estimated performance data, although high, for the proposed aircraft are as follows:

Had it been produced and met its performance estimates, the L-133 would have really turned some heads; for it would be ten more years before any jet fighter exceeded the performance goals of the L-133.

The proposed powerplant for the L-133, the L-1000 (or X-137, as it was subsequently designated), is a story in itself. Designed between 1940 and 1941, the L-1000 turbojet engine was estimated to provide a whopping 5,500-lbf static at sea level—totally unheard-of for its time. Lockheed estimated that two of these axial-flow turbojet engines would propel the L-133 to a top level attitude speed of about 620 mph at 20,000 feet. The XJ37 engine featured twin spools, a high compression ratio, and an afterburner section—all features that didn't find their way into turbojet engines until the late 1940s.

The XJ37 engine was to be built under license from Lockheed by the Menasco Manufacturing Company in Burbank, California, in 1944, under USAAF AMC Secret Project MX-411, and the patent for the engine were ultimately obtained by Nathan Price and the Curtiss-Wright Corporation in 1946. Several XJ37s were built, offering a disappointing maximum thrust of 2,200-lbf. What finally became of two of these engines is unclear. But one of the three examples is on display at the Planes of Fame Air Museum in Chino, California.

Nevertheless, Lockheed's early involvement with the L-133 and L-1000 airframe and powerplant projects clearly indicates why the USAAF went to this inventive firm for the design and development of what became a classic: the Lockheed F-80 Shooting Star.

Lockheed's XP-80 was completely restored for permanent display at the National Air and Space Museum in Washington, D.C., returned to the way it appeared upon its retirement from flight test duties. Its radio call sign, the large yellow number 78 on its nose, was designated by the flight test authorities based at Muroc AAF at the time.
LM via Denny Lombard

XP-80

On June 17, 1943, Clarence L. "Kelly" Johnson was at Eglin Army Air Field in Florida watching a flight demonstration of a new version of the Lockheed Lightning—a P-38J equipped with dive flaps—when he learned firsthand that turbojet-powered aircraft were flying in America.

USAAF Col. Marshall S. Roth, chief of the Air Materiel Command Propulsion Laboratory at Wright AAF, told Johnson of the XP-59A and YP-59A flight test programs at Muroc AAF. According to Johnson, the tests "had been disappointing because the General Electric turbojet engines made the aircraft slower than the P-38 they were watching," and Bell's Airacomet "would be no match for German turbojet-powered fighter aircraft."

"You wanted to build a jet for us once," said Colonel Roth. "Why don't you try your hand at putting a fighter airframe around the new de Havilland engine the English have promised us?"

Meanwhile, Royal Air Force Maj. Frank Bernard Halford had reconfigured a less complicated version of Frank Whittle's centrifugal-flow designs with the air intake on the front, including "straight-through" combustion chambers. Known initially as the Halford H.1B, the project was taken up by the de Havilland Aircraft Company, produced as the de Havilland Goblin. de Havilland eventually purchased Halford's company outright in 1944.

Johnson, filled with enthusiasm, told Roth, "Just give me the specifications of the engine." Johnson boarded an airliner and returned to Burbank. En route, he worked out some figures and preliminary design drawings of what he thought the airplane should look like. On his arrival, he immediately reported to Lockheed President Robert E. Gross and Chief Engineer Hall L. Hibbard.

"Wright Field wants us to submit a proposal for building an airplane around an English jet engine," Johnson told them. He added, "I've worked out some figures. I think we can promise them a 180-day delivery. What do you think?"

Heavy discussions began, and the group attacked drawing boards with vigor. Suddenly, top engineers found themselves up to their elbows in jet fighter plane designs. There was more to it than just a chance to build a new fighter, using a new type of engine for propulsion. It was a matter of pride, the culmination of efforts dating back to 1939 when Lockheed had begun planning for what it thought would be the first turbojet-powered fighter in America.

The deadline of 180 days was punishing. Gross, Hibbard, and Johnson knew that no company had ever designed and built a prototype airplane in less than a year, let alone in only six months. Moreover, a turbojet-powered airplane was a radical departure from contemporary piston-powered aircraft.

"OK, Kelly, it's your baby," Gross is reported to have said, and offering all the help they could. So began the saga of the Shooting Star.

One week after talking with Colonel Roth in Florida, on June 24, Johnson was at Wright Field showing Air Material Command officers and USAAF Commanding Gen. Henry H. "Hap" Arnold sketches of the proposed single-engine turbojet-powered pursuit, along with pages of detailed engineering data and specifications. Johnson, with his bosses' blessing, promised the USAAF: "We'll build it in 180 days!"

"Just when would those 180 days start?" General Arnold asked.

"Whenever you say," Johnson said. "Just as soon as we get the letter of intent," he added.

"Well you'd better get moving then," Arnold said, "This is day number one. We'll have your letter of intent ready this afternoon." And so it was: the letter of intent was dated June 24, 1943.

The USAAF AMC designated the program Secret Project MX-409. This added difficulty to the 180-day time limit, as every phase of the airplane's creation would be closely guarded. With the letter of intent, the USAAF promised to order on the forthcoming contract one XP-80 airplane (MX-409), two XP-80A airplanes (MX-409-A), twelve YP-80A airplanes (MX-409-B), and a single XF-14. USAAF serial number was assigned 44-83020 to the single XP-80; 44-83021 and 44-83022 to the two XP-80As; 44-83023, 44-83025 to 44-83035 was given to the 12 YP-80As, and 44-83024 to the lone XF-14 (formerly YP-80A number two).

Lockheed assigned Temporary Design Numbers (TDN) 140 (XP-80) and TDN-141 (XP-80A) to the program. USAAF contract number AC-40680 was approved on October 16, 1943, for a single experimental XP-80 airplane and two experimental XP-80A airplanes. Contract approval for the two XP-80As, twelve YP-80As, and one of XF-14 wasn't given until July 28, 1944.

Back in Burbank, Chief Engineer Hibbard and Assistant Chief Engineer Johnson assembled twenty-six aeronautical engineers and 105 shop workers to manufacture the prototype XP-80 airplane almost totally by hand.

Three view drawing of the XP-80 (TDN L-140) showing the craft as it appeared on its first flight date: January 8, 1944.
G. De Chiara © 2015

Donald "Don" Palmer and W. P. Ralston served as Johnson's assistant project engineers. Arthur M. "Art" Vierick, then chief of the Experimental Engineering Department, was put in charge of the shop workers. During one of the first group meetings with this team, Chief Project Engineer Johnson said: "The work at hand will be done on a six-day week, ten-hour day work schedule. No one will work on Sundays."

As the project moved forward, Johnson reminded his men of the strict enforcement of his no-Sundays rule. Some of the men had been getting sick, with up to 25% to 30% of the workforce out sick each day; this was unacceptable. Johnson told the men, "By coming back in here on Sunday, you're hurting the project. You don't get enough rest and you get sick. The next man I catch in here on Sunday goes back to the B-17s." He never caught anyone, but unexplained work continued to be done between Saturday evening and Monday morning.

In early November 1943, Lulu-Belle, as the XP-80 had been nicknamed, was removed from its assembly area during the night and secretly trucked to the North Base area of Muroc AAF. On November 12, 1943—day 139—the British Goblin turbojet engine roared to life. On day 143, the airplane was accepted by the USAAF as ready for flight. Lockheed chief engineering test pilot Milo Garrett Burcham flew her the next morning. Everything had gone well—too well.

Late in the evening of November 15, de Havilland engine expert Guy Bristow gave the H.1B Goblin a final run-up in preparation for its scheduled flight the

All twenty-eight of the aeronautical engineers involved in the design and development of the XP-80 signed this photograph of Lulu-Belle.
LM via Denny Lombard

next morning. As the engine roared at full power, both engine air intake ducts on either side of the fuselage sucked inward and collapsed. Before Bristow could shut down the engine, pieces of ducting metal were sucked into the Goblin's maw. A terrible grinding noise preceded engine stop. In this face of irreparable damage to this rare engine, a replacement engine would have to be delivered from Great Britain before Lulu-Belle could fly.

Engine air intake ducting repaired, a new Goblin engine was installed and tested. The XP-80 was now scheduled to fly during the early morning of January 8, 1944, fifty-four days later than had been originally scheduled.

Just before takeoff, Johnson gave Burcham a last-minute pep talk. "Just fly her, Milo," he said. "Find out if she's a lady or a witch. And if you have any trouble at all, bring her back. She's all yours from here. Treat her nice."

Lulu-Belle took off. Burcham initiated a climbing turn during which the wings began to wobble. He immediately nosed her down, came around, and landed on the dry lakebed. After he landed, Johnson asked Burcham, "What's the trouble?"

"Overcautious, maybe," Burcham answered. "She felt funny on the ailerons. Pretty touchy."

"You've got fifteen-to-one boost and a hot ship that's naturally sensitive—maybe you were over-controlling," Johnson said.

"Could be," Burcham agreed.

(Earlier that year, on an aircraft with turbojet propulsion used for familiarization purposes, Milo Burcham had checked out on a Bell YP-59A Airacomet. The XP-80 was a hot rod when compared with the woefully underpowered YP-59A.)

Head-on view of Lulu-Belle.
LM/Code One

Lulu-Belle's engine was restarted and Burcham lifted off again. This time she climbed up straight as an arrow. He made one low pass across the base, then climbed up and out of sight. Burcham came down from high altitude so fast that no one knew he was approaching until he passed overhead to the roar of the Goblin engine. Burcham then came back across the field doing full aileron rolls before he landed. Immediately after the flight, Burcham said, "You don't fly this airplane—you just hint to it where you want it to go. And it really goes."

The handful of USAAF officers who had watched the XP-80 demonstration flight were both surprised and delighted. They wanted production P-80s badly—and very soon.

British de Havilland H.1 Goblin engines, to be manufactured under license in the US by Allis-Chalmers as the J36, would not be available in quantity anytime soon. While this posed a serious problem for the USAAF and Lockheed, an answer to this dilemma was in hand.

General Electric's accelerated development and production of its new Model I-40 (J33) centrifugal-flow turbojet engine was rapidly making available this engine, larger than the Goblin and offering nearly 1,300-lbf more power. Unfortunately, the J33 would require a larger XP-80 airframe to accommodate it. The USAAF asked Lockheed's Johnson if it could be done without a great deal of difficulty. "Can do," Johnson said.

Thus the second and third airframes were enlarged to accommodate the General Electric J33-GE-11 turbojet engines. With their airframe redesign and new propulsive systems, they were given a new Lockheed TDN number, L-141, and redesignated XP-80A—under the same contract, amended.

XP-80A

Johnson's engineering and shop group finished the first XP-80A airplane in an incredible 132 days. Its first flight took place seven days later.

Lockheed test pilot Anthony William "Tony" LeVier made the first flight on the new XP-80A (44-83021) at Muroc AAF on June 10, 1944. This airplane was dubbed Gray Ghost because of its light gray paint scheme and for the way it seemed to vanish against the sky.

Although nearly two thousand pounds heavier, the XP-80A performed even more impressively than its predecessor, the XP-80. This was due in large part to the increase in propulsive energy: where the Goblin put out 2,460-lbf, the J33 developed 3,750-lbf, an increase of 1,290-lbf.

LeVier flew XP-80A number two (44-83022), nicknamed Silver Ghost because of its non-painted aluminum attire, on August 1, 1944. Its performance was equally outstanding.

The USAAF was convinced. In addition to its previous orders for the XP-80, XP-80A, YP-80A, and XF-14, it upped its order to include one thousand

ABOVE: Dubbed the Gray Ghost, XP-80A number one (44-83021) soars during its first flight on June 10, 1944, with Tony LeVier in the cockpit. *LM via Denny Lombard*

LEFT: XP-80A number two (44-83022, known as the Silver Ghost) after installation of its afterburning Westinghouse Electric XJ34-WE-15 turbojet engine for XF-90 propulsive system developmental tests. The dorsal spine held wiring and control cables for the engine and its Lockheed-designed-and-built afterburner section. *LM/Code One*

production P-80A aircraft. This contract, USAAF contract number AC-2527, had been approved earlier on April 4, 1944.

The XP-80A airplanes initially presented with a faulty cockpit pressurization valve, which caused unbearably high cockpit temperatures. Later iterations addressed and corrected this problem.

XP-80A number two was fitted with a second seat aft of the pilot's seat for a flight test engineer or observer. It was also the first P-80 type to be fitted with 165 gallon wingtip-mounted auxiliary drop-type fuel tanks.

The second seat came in handy when P-80 chief engineer and designer Johnson took a ride to pinpoint the airplane's unstable airflow through its engine air inlet ducts. Johnson found that this was due to boundary layer separation along the walls of the duct, remedied by adding a series of boundary layer bleed louvers along the upper edges of the air intake ducts. This fix was employed to subsequent production aircraft.

On March 20, 1945, while it was assigned to the 4144 Army Air Forces Base Unit of the ATSC at Muroc AAF, the first of the two XP-80A airplanes suffered a structural failure caused by an in-flight engine fire. Lockheed test pilot Tony LeVier managed to bail out and, though he was injured, he survived the ordeal, recuperated, and returned to work. The airplane crashed some six miles east of Rosamond, California, a total loss.

Later, as part of the upcoming XP/XF-90 Penetration Fighter flight test program, XP-80A number two was fitted with an axial-flow 4,100-lbf (with afterburning) Westinghouse Electric J34-WE-15 turbojet engine.

YP-80A

The YP-80A (Lockheed Model 080) was developed under USAAF AMC Secret Project MX-409-B, and was the immediate follow-on to the XP-80 and XP-80A programs. Thirteen YP-80As were ordered under USAAF contract number AC-2393, which was approved on July 1, 1944 (USAAF serial numbers 44-83023 to 44-83035). Of the thirteen YP-80As ordered, twelve were completed, while number two was completed under the designation XF-14 (discussed below).

The initial flight test of the first YP-80A (44-83023) took place on September 13, 1944, at Wright Field. It had been completed early, on July 1, 1944, and was then trucked from Burbank to Wright Field for static tests. After its static testing and several additional test flights at Wright Field, it was ferried back to Burbank, where it was specially instrumented and turned over to the NACA Ames Aeronautical Laboratory at Moffett Field in Santa Clara, California, for high-speed dive tests. On January 13, 1945, during a NACA test at Muroc AAF, pilot Lawrence A. Clousing crash-landed it on Rogers Dry Lake.

The YP-80A airplanes, closely related to the XP-80As, used the same propulsive system. They were employed extensively for service test and armament trials.

YP-80A number three (44-83025) crashed immediately after a takeoff from Lockheed Air Terminal at Burbank on October 20, 1944, killing Milo Burcham. It was assigned to USAAF ATSC 4035 Base Unit at that time.

Four YP-80As were sent abroad—two to England (44-83026 and 44-83027) and two to the Mediterranean (44-83028 and 44-83029). Of the two YP-80As sent to England, 44-83026 was fitted with a prototype Rolls-Royce Nene (Model B.41) for flight trials. It crashed to destruction at Bold near Widnes during its second flight out of Burtonwood, England, with the Nene engine, killing its test pilot, Maj. Frederick A. Borsodi. While still in the European Theater of Operations, the other YP-80A (44-83027) suffered an engine failure on November 14, 1945, and was destroyed in the crash.

Of the two YP-80As that went to the Mediterranean (based in Italy during Operation Extraversion), both flew operational sorties but never encountered enemy aircraft. Both were shipped back to the United States. One of them (44-83029) was lost in a crash on August 2, 1945, some two miles southeast of Brandonberg, Kentucky, killing its pilot, Ira B. Jones III. The other Mediterranean Theater of Operations veteran (44-83028) was modified and later served as a pilotless drone.

One YP-80A (44-83032), assigned to USAAF ATSC 4144 AAF Base Unit at Muroc AAF, was involved in two mishaps. On March 4, 1945, pilot Donald C. Craig was forced to make a gear-up belly landing six miles south of base due to engine failure. Then, on November 4, 1945, pilot Charles F. Hale also had to make a gear-up belly landing on Rogers Dry Lake due to a fuel system malfunction and engine flameout.

XF-14 (XFP-80) AND XFP-80A

The one-of-a-kind Lockheed XF-14-LO (Model 080) was a modified YP-80A (YP-80A number two, USAAF serial number 44-83024), produced as an unarmed experimental photographic reconnaissance and mapping airplane. The XF-14 was manufactured under USAAF contract number AC-2393 approved on July 1, 1944.

After its first flight at Burbank and USAAF acceptance, it was flown to Muroc AAF, where it was assigned to ATSC 4144 AAF Base Unit for a series of evaluations.

On December 6, 1944, it was involved in a nighttime mid-air collision with a B-25J Mitchell (USAAF serial number 44-29120) some seven miles south-southwest of Randsburg, California. Lockheed test pilot Perry Earnest "Ernie" Claypool Jr. and the four-man crew on the B-25J were all killed. If the one of XF-14 had survived, it would have been redesignated XFP-80.

No other XF-14 (XFP-80) airplanes were built to replace this craft. Instead, Lockheed manufactured numerous FP-80As that were redesignated as RF-80As in mid-June 1948.

The one-off P-80R *Racey* was used to establish a new (albeit short-lived) world speed record of 623.74 mph on June 19, 1947. *LM/Code One*

One XFP-80A was a modified P-80A-1-LO (44-85201) and was fitted with more photographic reconnaissance equipment than the XF-14 (XFP-80) in a modified nose. The modified nose, which hinged upward, allowed easier access to the cameras for offloading and reloading the films.

XP-80B (P-80R)

The XP-80B was a one-of-a-kind experimental version of the Shooting Star that was to demonstrate a thinner wing. It was built with a centrifugal-flow, 4,000-lbf, Allison-built J33-A-17 turbojet engine in an effort to improve overall performance. Created from the ninth production P-80A-1-LO (USAAF serial number 44-85200), it had first flown in 1945 as a P-80A. Its first flight as the XP-80B by Tony LeVier came in mid-1946 (the actual date remains unclear). After its flight-testing program ended, the XP-80B was placed in extended storage, only to reemerge as the P-80R.

On September 7, 1946, Royal Air Force Group Capt. Edward Mortlock "Teddy" Donaldson flew a modified RAF Gloster F Mk4 (F.4) Meteor to a new world speed record of 615.78 mph in Great Britain. In response, USAAF Commanding Gen. Henry H. "Hap" Arnold determined that he wanted to beat Donaldson's record and set a US world speed record. He looked to Lockheed and its P-80 to accomplish the feat.

The XP-80B seemed to be the logical choice for the mission. Lockheed was given the green light to modify the XP-80B with experimental NACA flush engine air inlets and a low-profile cockpit canopy; it was to retain its J33-A-17 turbojet engine. With these changes, it was redesignated P-80R (the R suffix for "racer") and nicknamed *Racey*.

(Note: Other references to this aircraft erroneously designate it as XP-80R. The correct designation is P-80R.)

In late October 1946, the P-80R was flown in an attempt to bring the United States a world speed record. The plane would have to average more than 615.9 mph during four passes over a three-kilometer course, which it failed to do. The airplane was returned to Lockheed in Burbank for further modifications.

The troublesome NACA flush intakes were replaced by large area engine air inlet openings, a special 4,600-lbf axial-flow non-afterburning Allison J33-A-23 turbojet engine (Allison Model 400) was installed, and an additional fuel tank was installed where the nose gun armament would have been carried.

The project's leader, USAAF Colonel Albert G. "Al" Boyd, elected himself to fly the speed record attempt mission over a measured three-kilometer course on the lakebed. He would have to engage the course four times to reach an average speed acceptable for an official new speed record. On June 19, 1947, Col. Boyd piloted the P-80R over the three-kilometer course and averaged 623.753 mph, which indeed established a new record.

General Arnold was pleased. The historic P-80R survived and can be seen at the National Museum of the United States Air Force (NMUSAF) in Dayton, Ohio.

SUMMARY

The F-80 Shooting Star holds the distinction of being the first operational turbojet-powered combat aircraft to operate in America.

The foresighted design of the F-80 lent itself to numerous versions and spinoffs, including the F-80 series itself, the USAF TP-80C (the later T-33A), the USN/USMC TO-1 (the later TV-1), the USN T2V-1 (the later T-1A SeaStar), and, finally, the USAF F-94 series of aircraft.

CONSOLIDATED VULTEE XP-81 AND YP-81 SILVER BULLET

The *Consolidated Vultee Aircraft* (Convair) XP-81 was designed in order to solve the limited range problem of early turbojet-powered fighters.

The projected duty of an operational fleet of P-81s was for long-range bomber escort missions. The concept was to use the turbojet engine only for takeoff and during combat, while the turbopropjet engine would be employed for cruising. In this way, the airplane would use less fuel and achieve the longer range required.

This composite-powered, push-me/pull-me concept for improved fuel economy seemed logical at the time, but this goal was never realized: the advent of in-flight refueling by tanker aircraft made this

XP-80 SPECIFICATIONS

Propulsive system
One centrifugal-flow non-afterburning 2,460-lbf de Havilland Halford H. 1B Goblin turbojet engine

Length	32 feet, 10 inches
Height	10 feet, 3 inches
Wingspan	37 feet, 0 inches
Empty weight	6,287 pounds
Gross takeoff weight	8,620 pounds
Maximum speed	502 mph
Ceiling	35,000 feet
Maximum range	500 miles
Armament	None installed

XP-80A SPECIFICATION

Propulsive system
One non-afterburning centrifugal-flow 3,750-lbf General Electric J33-GE-11 turbojet engine

Length	34 feet, 9 inches
Height	11 feet, 3 inches
Wingspan	39 feet, 0 inches
Empty weight	7,227 pounds
Gross takeoff weight	9,600 pounds
Maximum speed	561 mph
Maximum range	600 miles
Service Ceiling	40,000 feet
Armament	None installed

ABOVE: Initial test flight evaluations with the interim Packard-built V-12 Rolls-Royce V-1710-7 Merlin engine for XP-81 number one began in February 1945.

BELOW: Head-on view of XP-81 number one. *NMUSAFh*

design feature irrelevant. Nevertheless, two XP-81 airplanes were built and flight-tested at Muroc AAF over several years.

Convair initiated its Model 102 design on January 5, 1944. Two experimental XP-81 airplanes were ordered under USAAF contract number AC-1887 on February 11, 1943, USAAF AMC Project MX-480, with USAAF serial numbers 44-91000 and 44-91001 assigned. The contract was approved by the US War Department on August 3, 1944. (Later, while still under construction in San Diego, the long-range escort mission changed to the penetration fighter mission and its former MX-480 number was changed to MX-796.)

The first example, fitted with an interim Packard Motor Car Company–built Rolls-Royce V-1650-7 Merlin V-12 piston engine, made a successful first flight at Muroc AAF on February 11, 1945. Convair Chief Design Engineer and chief engineering test pilot Frank Wilbur Davis piloted it on its first flight; he who reported few squawks.

Convair retrofitted the first aircraft to house the design XT31-GE-1 turbopropjet engine (developed under USAAF AMC Project MX-464), which flew for the first time in this configuration on December 21, 1945 with Frank Davis at the controls.

The second XP-81 (44-91001) had made its first flight earlier in mid-1945. It too was flown from San Diego to Muroc AAF for ongoing flight tests.

Labels on cutaway diagram:

4 BLADE PROPELLER

REDUCTION GEARS

AIR INLET

PRESSURIZED CABIN

ENGINE AIR DUCTS

JET EXHAUST THROUGH TAIL

I-40 G-E JET ENGINE

JET EXHAUST

G-E PROPJET (GAS TURBINE)

COMBUSTION CHAMBER

AXIAL FLOW COMPRESSOR

ABOVE: Convair chief engineering test pilot Frank Davis pilots the first of two XP-81 airplanes (44-91000) on its December 21, 1945 (first flight date). This plane was mounted with the XT31 (Model TG-100) engine, designed by General Electric. *NMUSAF*

LEFT: This detailed cutaway of XP-81 reveals the XP-81 number one's main components. *USAF*

XP-81 SPECIFICATIONS

Propulsive system
One 3,750-lbf mid-fuselage-mounted non-afterburning centrifugal-flow General Electric J33-GE-5 (Model I-40) turbojet engine, one 2,300 horsepower nose-mounted General Electric XT31-GE-1 (Model TG-100) turboprojet engine turning a four-bladed Aeroproducts propeller

Length	44 feet, 10 inches
Height	14 feet, 0 inches
Wingspan	50 feet, 6 inches
Empty weight	12,755 pounds
Gross weight	24,650 pounds
Maximum speed	507 miles per hour (attained)
Ceiling	35,500 feet
Maximum range	2,500 miles (projected)

Armament
Six .50-caliber machine guns or six 20mm cannon (not installed); 2,000 pounds of bombs

YP-81

Thirteen service test YP-81 airplanes were to be produced along with the two XP-81s on contract number AC-1887, but these airplanes were soon terminated under Termination Order Number T1-2148. The information derived from *Model Designation Army Aircraft* (eleventh edition, published January 1945 by Commanding General, Army Air Forces) simply states that the YP-81s were to be basically the same as the XP-81s.

Only two experimental XP-81s served as test aircraft until May 9, 1947, when the program was cancelled outright and they were declared obsolete on September 29, 1947, placed in storage at Muroc.

Since the XP-81 flight test program ended before June 11, 1948, the airplanes weren't officially redesignated XF-81s. Some references still list them as XF-81s or ZXF-81s.

SUMMARY

As a composite powered airplane the performance with the turbopropjet engine was barely better than with the interim piston engine, and was disappointing overall. Thus no production order of these aircraft was forthcoming.

As a proposed combat aircraft, thought, it was to be armed with six .50-caliber heavy machine guns or six 20mm cannon—three on either wing. It also was to carry two thousand-pound bombs—one under either wing or auxiliary drop-type fuel tanks.

The two XP-81 airplanes remained at Edwards AFB for many years until it was decided to restore one of them for display at the Museum of the USAF in Dayton, Ohio, using parts from the other one.

BELL XP-83

With the advent of jet propulsion in the United States, aeronautical engineers were free to explore new horizons. As with all developers of a new technology, though, these engineers faced problems with no easy solutions.

Early turbojet engines were too heavy, too underpowered, and too thirsty. With a best-in-class engine-thrust-to-weight ratio in the 2.20:1 range (as of early 1943), a powerplant that was rated at four thousand lbf was inadequate for any but the smallest, lightest airframes. These small airframes had little room for fuel, which, combined with early turbojets' extreme thirst, mission times were limited to about a half hour.

The Bell Aircraft Corporation had found this out the hard way on its P-59 Airacomet program. Bell's hard-working design team, headed by its ace chief design engineer, Robert J. Woods, kept plugging away

ABOVE: Close-up nose view of the first XP-83 showing its gun arrangement. *NMUSAF*

LEFT: In-flight photo of 44-84990 taken on February 25, 1945, during the first flight out of Niagara Airport. *NMUSAF*

on these problems in an attempt to resolve them. While they struggled with the short-ranging jet-powered pursuit dilemma, Woods and his design team were able to design a viable long-range offensive pursuit airplane that could tempt the USAAF.

This design was first approved by the founder and president of Bell Aircraft Corporation, Lawrence Dale "Larry" Bell. Woods and project engineer Edgar P. "E. P." Rhodes took the Model 40 proposal to the AMC at Wright Field, arriving there on March 29, 1943.

The largest, heaviest turbojet-powered pursuit produced up to that time, this big twin-jet was configured like Bell's earlier P-59. The Model 40's projected long-range capability was derived from its voluminous internal fuel tanks, which held 732 US gallons or 4,880 pounds. Additionally, the design had provision for two externally mounted, 150-US-gallon droppable fuel tanks. This added up to a total fuel load of 1,032 U.S. gallons or 7,665 pounds. It had a projected range of 1,730 miles on internal fuel alone, and 2,050 miles with external fuel. This was twice the range of contemporary pursuit types, and the Model 40 would match their speeds.

Bell's Model 40 proposal was well received, especially at a time when endurance was the primary goal. Bell had found a way to combine speed and range. The USAAF gave its approval without hesitation.

A letter of intent served to guarantee continued development of the Model 40. On March 24, 1944, the USAAF placed an order for wind tunnel models, engineering mockups and data, and two experimental aircraft, now designated XP-83 under USAAF AMC Secret Project MX-511, USAAF contract number W33-AC-2425. USAAF serial numbers 44-84990 and 44-84991 were issued to the two XP-83 airplanes. The first XP-83 was expected to fly in one year.

The only known photograph of XP-83 number two (44-84991), dated July 10, 1946, at Wright Field. *NARA via hDennis R. Jenkins*

Bell selected Jack Valentine Woolams to serve as chief engineering test pilot on the XP-83 program. As we have seen, Woolams had flight-tested several XP-59A and YP-59A aircraft at Muroc AAF. Born in San Francisco on Valentine's Day, 1917—thus earning his middle name—Woolams had recently set a US altitude record of 47,600 feet on December 15, 1943, while piloting the thirteenth and final service test YP-59A (USAAF serial number 42-108783), nicknamed *Mystic Mistress*, at Muroc AAF.

The first of the two XP-83 airplanes (USAAF serial number 44-84990) was completed on February 8, 1945. On February 25, Woolams completed a successful test hop out of Niagara Airport adjacent to Niagara Falls, New York.

Later that year, on June 10, 1945, Maj. Russell E. "Russ" Schleeh—chief of fighter test, ATSC Flight Test Division of the Air Materiel Command at Wright Field—attempted a takeoff on XP-83 number one from Niagara Airport. An engine malfunction forced Schleeh to abort the takeoff and made a successful emergency landing. The airplane suffered minor damage to its landing gear, but was quickly repaired. (Major Schleeh went on to be recognized as one of the USAAF's best test pilots and became a colonel before he retired.)

The second XP-83 (USAAF serial number 44-84991) made its first flight out of Niagara Airport on October 19, 1945, and was subsequently ferried to Wright Field for armament tests. It featured a longer nose section to house six ninety-two-inch long, .60-caliber T17E3 machine guns manufactured by the US Army Corps of Engineers at its Rock Island Arsenal facility, Rock Island, Illinois.

Both of the XP-83 airplanes flew well during their respective test programs. Their weight proved a challenge, though, as their available propulsive systems delivered only limited power. For this reason, they became test-bed aircraft for various programs.

One such program was the evaluation of ramjet engines. XP-83 number one was fitted with two experimental RJ30 Marquardt Aircraft twenty-inch-diameter Model C-20 ramjet engines—one slung under either wing on short pylons. If it could reach a high enough speed for the ramjets to start up and continue running, the aircraft might sustain flight after its primary J33s were shut down—essentially flying on ramjet power alone. In addition to its ramjet engine installation, a compartment was installed behind the cockpit; this would allow a flight test engineer to monitor ramjet operations on instrumentation and by eye from out of the side windows. A hatchway beneath the airplane allowed ingress and egress, using a ladder. Bell test pilot Chalmers H. "Slick" Goodlin and Bell flight test engineer Charles L. "Charlie" Fay were assigned to this program.

The ramjet engine tests were to begin in late summer 1946. On September 4, Goodlin and Fay took off to perform the first of these scheduled tests. Just as the airplane was to enter into its preplanned flight test corridor, the right-hand (starboard) ramjet engine caught fire and began burning out of control. The wing structure on which the ramjet was attached was quickly engulfed, and control of the right wing aileron was lost. With the threat of a fuel explosion imminent, Goodlin barked, "Bail out, Charlie!" Both men parachuted to safety, and the plane crashed to destruction on a farm some four miles south of Amherst, New York.

No production contract for the P-83 was forthcoming. XP-83 number one was gone. In 1948, XP-83 number two had completed its tenure at Wright AFB and was removed from flight status.

The end of the P-83 program was also the end of Bell's fighter aircraft. In fact, Bell never built another fighter-type prototype. It came close to doing so in the late 1950s with its proposed D-188A (YF-109), but this program was cancelled before an airplane was built.

SUMMARY

The XP-83 program was but a steppingstone on the pathway to find an ideal long-range, jet-powered fighter aircraft. Such an aircraft would not appear for some time, though.

The surviving XP-83 was used at Wright Field to test the .60-caliber M17E3 very heavy machine gun throughout 1946. Two years later, it was declared surplus and salvaged. For this reason, no XP-83 airplane survives.

XP-83 SPECIFICATIONS

Propulsive system

Propulsive system: Two centrifugal-flow, non-afterburning 3,750-lbf General Electric J33-GE-5 (GE Model I-40) turbojet engines

Length	
44 feet, 10 inches (45 feet, 4 inches XP-83 number two)	
Height	13 feet, 9.5 inches
Wingspan	53 feet, 0 inches
Empty weight	12,214 pounds
Gross weight	18,300 pounds
Maximum speed	522 mph (attained)
Maximum range	2,050 miles

Armament

Six Colt-Browning .50-caliber M2 heavy machine guns or six Colt .60-caliber T17E3 very heavy machine guns; up to two thousand pounds of bombs

The first of two XP-84 Thunderjet airplanes (45-59475) rests at Muroc AAF. *NMUSAF*

REPUBLIC XP-84, XP-84A, AND YP-84A THUNDERJET, YF-84F (YF-96A) THUNDERSTREAK, YF-84J THUNDERSTREAK II, XF-84H THUNDERSCREECH, YRF-84F THUNDERFLASH

Throughout 1944, Republic Aircraft Corporation (RAC) chief engineer and noted aircraft designer Alexander Kartveli initiated design work on a turbojet-powered fighter. His original design for such an airplane, based on the P-47 Thunderbolt, looked like an open-nosed version of that aircraft with a tailpipe protruding out its rear end. This quickly evolved into a much sleeker design, engineered to accept a single axial-flow-type turbojet engine that was in development at GE. This straight-winged design was offered to the USAAF as the Republic Model AP-23.

With its development initiated earlier in May 1943 under USAAF AMC Secret Project MX-414, the General Electric Model TG-180 turbojet engine was first test-fired in April 1944. Designated the J35, this turbojet engine initially produced 3,750-lbf of power. By the standards of the day very powerful, this amount of turbojet thrust made speeds of 600-plus mph for fighter aircraft a real possibility. Although General Electric's own Model I-40 (J33) centrifugal-flow-type turbojet engine produced about the same amount of power, it had limited growth potential—unlike the J35. In fact, the J35's designers hoped it might be capable of producing well over 4,000-lbf power once fully developed. In reality, with the later addition of afterburning, the J35 produced a maximum power output of 7,400-lbf.

With the potential of 600-plus-mph fighter speeds, AMC planners released a general operational requirement (GOR) on September 11, 1944, that in fact called for the design and development of a new 600-mph pursuit interceptor using the J35 engine.

On January 5, 1946, the US War Department approved USAAF contract number W33-(038)-AC-6248 for three experimental XP-84 airplanes to be built by RAC in Farmingdale on Long Island, New York, under USAAC AMC Project MX-578. The USAAF assigned serial numbers 45-59475 to 45-59477 to these three airplanes. A fourth XP-84 example (45-59478) was ordered, but it was cancelled as a flyable airplane and its airframe was instead used for static structural test purposes.

As work on the three XP-84s progressed, development problems became apparent, in particular the aircraft's excessive weight. While XP-84s number one and two were too far along in the construction process for any significant changes to be made, XP-84 number three had just entered into assembly. With the corrective changes made during its assembly, this airplane was redesignated XP-84A.

XP-84, XP-84A, AND YP-84A THUNDERJET

The premier XP-84 was completed in late December 1945, after which it was partially disassembled and shipped to Muroc AAF via the Boeing XC-97 Stratofreighter for flight testing. After it arrived at North Base on Muroc AAF, the plane was reassembled and prepared for its first flight. Then, on February 28, 1946, USAAF Maj. Wallace A. "Wally" Lien took the controls of the first XP-84 aircraft and completed a successful first flight.

The second XP-84 flew on August 25, 1946, at Muroc AAF. This particular XP-84 went on to establish a new, unofficial US national speed record of 607.2 mph on September 7, 1946, at Muroc AAF.

(During the first flight tests of the Republic XP-84 airplane, it was discovered that the craft completely lacked a stall warning. A short series of development tests of a suitable stall-warning device for the airplane was made on a $\frac{1}{5}$-scale model in the Langley 300-mph seven-by-ten-foot wind tunnel. Two similar stall-warning devices were tested, each designed to produce early root stall and provide a buffet warning. It appeared that either device would give a satisfactory buffet warning in the flap-up configuration, at the cost of an increase of eight or ten miles per hour in minimum speed. Although neither device seemed to give a true buffet warning in the flaps-down configuration, it appeared that either device would improve the flaps-down stalling characteristics by lessening the severity of the stall and by maintaining better control at the stall. The flaps-down, minimum-speed increase caused by the devices was only one or two miles per hour.)

XP-84 numbers one and two soar during test flights. *NMUSAF*

During high-speed testing at fifteen-thousand and thirty-five-thousand feet, the second XP-84 reached a maximum level-attitude flight speed of 0.84 Mach number (Mn), or about 625 mph. Excessive control force appeared to be the limiting factor in attaining speeds higher than 0.84 Mn, especially at altitudes below thirty-five-thousand feet.

XP-84A THUNDERJET

The one-of-a-kind XP-84A (formerly XP-84 number three) remained in Farmingdale, where it made its first flight on December 31, 1946. The XP-84A was powered by the uprated Allison-built GE J35-A-15 rated at 4,000-lbf power, whereas the XP-84s both used the 3,750-lbf General Electric J35-GE-7 engines. The XP-84A was bailed to the National Advisory Committee for Aeronautics (NACA) at Langley Field, Virginia, in 1947 for further modifications, including the installation of a solid nose-cone fairing to cover its open nose and flush NACA-type engine air inlets placed on either side of its fuselage trunk between its nose and wing roots.

YP-84A THUNDERJET

Additionally, on the same contract approval date and on the same contract as the YP-84s, the USAAF acquired fifteen YP-84A aircraft: three YP-84A-1-REs, seven YP-84A-5-REs, and five YP-84A-10-RE airplanes, with serial numbers 45-59482 to 45-59484, 45-59485 to 45-59491, and 45-59492 to 45-59496. These were to be delivered with armament (six .50-caliber nose-mounted machine guns) to Muroc AAF and Patterson AAF in Dayton, Ohio, for their evaluations by USAAF test pilots. They arrived at their delivery points by mid-1947, all with GE J35-GE-7 engines.

YP-84A testing was in full swing when RAC began delivering to the USAAF. The first production Thunderjet airplanes were designated P-84B, F-84B after June 11, 1948; there were no production P-84As.

YF-84F (YF-96) THUNDERSTREAK

Swept-back flying surfaces promised to eliminate most of the compressibility problems suffered by aircraft with straight flying surfaces. To this end, in 1949 RAC proposed a swept-back wing version of its F-84 under RAC Model AP-23X. RAC's design was accepted by the USAF and the Department of Defense (DOD) approved USAF contract number AC-22053 on December 29, 1948, for a single experimental XF-96A-RE prototype (USAF serial number 49-2430). This airplane, created from the last production F-84E-15-RE, became RAC Model AP-23M.

The Model AP-23M featured sweep-back on all of its flying surfaces and propulsion from a single 5,300-lbf axial-flow, Allison-built GE J35-A-25 turbojet engine fitted with an afterburner section. This engine offered a glimpse of the kind of power needed to close in on a 700 mph speed in level attitude flight. On June 3, 1950, with Republic test pilot Oscar P. "Bud" Haas at the controls, the XF-96A made a successful first flight at Muroc AFB. On August 9, 1950, the designation XF-96A was changed to XF-84F. The best speed the XF-84F attained during flight testing was 693 mph in level flight—close to but less than the 700 mph goal.

YF-84F number one (49-2430) cruises above the landscape in this in-flight photo. *NMUSAF*

ABOVE: The third YF-84F (51-1345) and the YRF-84Fs (51-1828) fly in formation. *NMUSAF*

RIGHT: YF-84F number three and XF-91 number two fly in formation at Muroc AFB. *NMUSAF*

The need for better performance at high altitude led the USAF to seek a more appropriate propulsion system. Arranging the acquisition of the British Armstrong Siddeley turbojet engine (known as Sapphire), the Air Force tabbed the engine division of the Curtiss-Wright Corporation to manufacture this engine in the United States as the J65 Sapphire.

To evaluate this engine, the DOD approved USAF contract number AF-14803 on June 8, 1951, for two preproduction service test YF-84F-RE airplanes (USAF serial numbers 51-1344 and 51-1345). Two F-84G-25-RE airplanes were used to create these two aircraft: the first was powered by the 7,220-lbf Wright YJ65-W-1 Sapphire engine, while the second airplane received the more reliable YJ65-W-1A.

To accommodate the afterburner-equipped J65 engine, the two F-84G fuselages were lengthened slightly and deepened seven inches. In order to improve airflow to the face of the engine, the nose air inlet opening was enlarged slightly on the first YF-84F and reconfigured to an oval shape. The second YF-84F was built with a solid nose and employed wing root engine air intakes, which, unfortunately, caused a loss of power; this design was carried forward on production F-84F aircraft.

The first Sapphire powered YF-84F took to the air for the first time on February 14, 1951. The second YF-84F example, powered with the more dependable YJ65-W-1A, first flew in mid-1951.

XP-84F SPECIFICATIONS

Propulsive system
One axial-flow, afterburning, Allison 5,300-lbf (with afterburning) XJ35-A-25 turbojet engine (YF-84F number one); one non-afterburning axial-flow 7,200-lbf Wright YJ65-W-1 turbojet engine (YF-84F numbers two and three)

Length	43 feet, 1 inch
Height	15 feet, 2 inches
Wingspan	33 feet, 7 inches
Empty weight	12,150 pounds
Gross takeoff weight	23,230 pounds
Maximum speed	693 mph
Maximum range	1,700 miles
Combat Ceiling	40,000 feet

Armament
Six M2 .50-caliber heavy machine guns; provision for underwing stores

Workers construct the second of three YF-84Fs (51-1344) in the Republic factory. *COAM*

The needle-nosed one-off YF-84J Thunderstreak II (51-1708) soars in this rare photo. This aircraft was created from a modified F-84F-25-RE. Although the J variant could hit 700-plus mph, it was passed over in favor of the new and supersonic F-100. *NMUSAF*

The sweptwing F-84F Thunderstreak was ordered into production, initially with J65-W-1/-1A engines, but later with J65-W-3 engines.

YRF-84F THUNDERFLASH

On June 12, 1951, USAF contract number AF14810 was approved by the DOD for a single service test photographic reconnaissance airplane designated YRF-84F-RE, and built under RAC Model AP-23T (USAF serial number 51-1828). This airplane was created from a modified F-84F-30-RE. February 3, 1952, saw the one YRF-84F make a successful first flight at Edwards AFB with Republic test pilot Carl A. Bellinger at its controls.

The success of this one prototype prompted the USAF to order the RF-84F Thunderflash into production. It became the workhorse photographic reconnaissance airplane until the advent of McDonnell RF-101 Voodoo.

YF-84J THUNDERSTREAK II

Under USAF contract number AF14803, approved by the DOD on August 31, 1951, RAC produced a single service test YF-84J. Unofficially nicknamed *Thundersteak II*, the YF-84J airplane made its first flight at Edwards AFB with Republic test pilot Russell Morgan "Rusty" Roth in control on May 7, 1954. To create the YF-84J, Republic modified an F-84F-25-RE (USAF serial number 51-1708) by deepening its fuselage, enlarging and reshaping its nose-mounted engine air inlet to an oval rather than round shape, and installing an axial-flow, 8,920-lbf-thrust GE YJ73-GE-7 turbojet engine. With this engine, the YF-84J reached a maximum level flight speed of 1.09 Mn (about 810 mph) at thirty-five thousand feet in a subsequent test flight.

ABOVE: Scaffolding and other equipment surrounds the premier XF-84H (51-17059) during its construction phase. *COAM*

LEFT: The second of two XF-84H airplanes (51-17060) directly after takeoff at Edwards AFB. *NMUSAF*

A second F-84F-25-RE (USAF serial number 51-1709) was to be modified to make another service test YF-84J, but its modification was terminated in June 1954 and the entire F-84J program was cancelled soon after, on August 31, 1954.

As it happened, the up-and-coming service test YF-105A Thunderchief and other, more capable turbojet fighter prototypes were soon to appear. The need for a fleet of less capable F-84J aircraft became moot.

XF-84H THUNDERSCREECH

Two Model AP-46 airplanes designated XF-84H-RE were ordered from RAC under USAF contract number AF-20501, which was approved by the DOD on December 15, 1952. Two F-84F-35-RE airplanes were taken from the production line for this program and, for a brief time, these XF-84H airplanes were designated XF-106.

Known as the Supersonic Propeller Program, this evaluation was initiated to investigate high-speed flight by a non-piston-powered, propeller-driven aircraft propelled by a turboprop engine. It was to be operated by the USAF's Propeller Laboratory branch of the Wright Air Development Center at Wright-Patterson AFB, but the aircraft themselves would undergo their respective flight test evaluations at Edwards AFB.

The two aircraft were completed at RAC's Farmingdale facility, then partially disassembled for their journey to Edwards AFB via C-124 Globemaster transport aircraft. The first example (USAF serial number 51-17059) arrived in early June 1955, where it was reassembled and prepared for its initial flight. Republic test pilot Henry G. "Hank" Beaird Jr. launched the first hybrid XF-84H into the air on July 22, 1955, at Edwards AFB.

Both of the XF-84H airplanes were powered by a 5,850-horsepower (estimated shaft) Allison XT40-A-1/Model 500-B2-1 turboprop engine spinning a twelve-foot-diameter Aeroproducts three-blade propeller with squared-off, paddle-type blade tips. Engine exhaust provided additional power. The planes' expected top speed was to be 760 mph, or 1.0 Mn, with the ultimate goal of achieving supersonic flight with a propeller-driven airplane.

The XF-84H took on the nickname Thunderscreech because of the tremendous high decibel noise generated by its propeller. In fact, the tips of the propellers rotated at supersonic speeds and produced constant sonic booms—approximately 1.18 Mn (875 mph)—which explains the excessive noise of these aircraft in operation.

The first of two XF-84H airplanes (51-17059) sits on the tarmac at Republic's Farmingdale, Long Island, New York facility. *NMUSAF*

With Beaird at the controls, the first XF-84H made only eleven flights, with all but one ending in forced landings. Beaird was the only pilot to fly the number one XF-84H. The second XF-84H flew only once, and Republic test pilot Lindell C. "Lin" Hendrix refused to fly the airplane again; he claimed afterward, "I never flew it over 450 knots indicated [about 520 mph or 0.70 Mn], because at that speed, it developed an unhappy practice of 'snaking,' apparently losing longitudinal stability."

The XF-84H program was terminated in September 1956. The first example is on display at the National Museum of the United States Air Force at Wright-Patterson AFB in Dayton, Ohio. The second example was scrapped, but its engine was saved and transferred to the US Navy's Douglas XA2D-1 Skyshark program.

SUMMARY

In his interview with Oscar P. "Bud" Haas for the February 22, 1955, issue of *The Birmingham Press*, Pulitzer-prize-winning Associated Press writer Harold Vincent "Hal" Boyle reported that Haas was "a veteran test pilot who won his wings at 17." The article relayed the following details:

"In most airplanes today if you didn't have a speed indicator you wouldn't know you'd gone through the sound barrier." [Haas] said, "You usually don't even know when it happens."

Haas, director of flight operations at Republic Aviation Corp., supervises a crew of first class test pilots who tested the new

EXPERIMENTAL AND SERVICE TEST F-84
PRODUCTION BREAKDOWN

XP-84-RE:	2-45-59475 and 45-59476	third example to XP-84A
XP-84A-RE:	1-45-59477	modified third XP-84
YP-84A-1-RE:	3-45-59482 to 45-59484	53-2431 to 53-2436
YP-84A-5-RE:	7-45-59485 to 45-59491	53-2437 to 53-2446
YP-84A-10-RE:	5-45-59492 to 45-59596	54-149 and 54-150
XF-84F-RE:	1-49-2430	designated XF-96A until 8/9/50
YF-84F-RE:	2-51-1344 and 51-1345	preproduction service test airplanes
XRF-84F:	1-51-1828	preproduction service test airplane
YF-84J:	1-51-1708	a second YF-84J was cancelled before it was built
XF-84H-RE:	2-51-17059 and 51-17060	temporarily designated XF-106

F-84s. These planes can exceed 650 miles an hour and carry an atomic bomb.

[According to Haas], "most of them are former armed forces fliers with combat records."

. . . . The fliers check in at 7 a.m., and check out at 3 p.m.

There is also plenty of fresh oxygen and exercise on the job. The test pilot really starts his main chores at an altitude of 40,000 feet.

But no daredevils need apply. No manufacturer wants to put a hot 'rod flier in the cockpit of a $400,000 airplane. The test pilot is an intensely serious young man who must serve both as a scientist and a highly paid bookkeeper as he puts a plane through its final performance check before it is turned over to the Air Force.

"When I flew my first plane, an OX5 standard, a plane similar to the old Curtiss JN-4 Jenny," recalled Bud, "all I had to worry about was the ignition switch and three instruments. They showed the oil pressure, the water temperature and the engine speed. No compass. No radio.

"We had a dip stick to tell us how much gas we had left, and we filled her up with five-gallon cans from the nearest auto filling station. She burned four gallons an hour, and flew at 60 miles an hour.

"A jet fighter-bomber today burns more than 1,000 gallons of fuel an hour at sea level, and the cockpit has more than 200 switches and instruments."

After exhaustive ground checks, the completed plane is taken aloft and put through 40 different flight test points, ranging from steep climbs to supersonic power dives and "slow flying" maneuvers. Then another pilot takes the plane up and does it all over again. Before accepting delivery, the Air Force then puts the plane through the tests a third time with one of its own pilots.

What is the life expectancy and the future of a jet test pilot?

"There is a limit to how long a man wants to do it as a regular routine," said Haas, who still makes a couple of flights a week himself. "After 40 or 45, most men don't want to take the physical punishment of high altitude flying. But their work gives them an all-around knowledge of airplanes, and usually they take another job in the industry."

How dangerous is it? Jet test pilots have to pay an extra premium if they take out personal life insurance. But Hass has never had a serious injury.

"A couple of my men have had to bail out." he said, "but we've never lost a man in 120,000 takeoffs and landings in 10 years.

The dual turbojet-powered XB-43 Jetmaster aircraft were a direct development of the piston-powered XB-42 Mixmaster aircraft. The first example (44-61508), complete with its original "bug eye" cockpit canopies and its Lexan nosepiece, is shown here. Its first flight was on May 17, 1946, at Muroc AAF. *NMUSAF*

The mortality rate among all test pilots is probably lower than among service fliers on ordinary day-to-day tactical missions."

The success of the Thunderjet, Thunderstreak, and Thunderflash programs led to the production of 4,285 F-84s, 2,711 F-84Fs, and 715 RF-84Fs for the USAF, as well as a number produced for air forces of US allies and friends. No fewer than 7,711 examples of all types were manufactured by Republic and the General Motors Corporation.

DOUGLAS XB-43 AND YB-43 JETMASTER

The Douglas Aircraft Company's XB-43 Jetmaster holds the distinction of being the very first turbojet-powered bombardment airplane to be built and flown in the United States.

The XB-43 was essentially a turbojet-powered version of the piston-powered Douglas XB-42 Mixmaster, featuring side-by-side mounted turbojet engines installed in the general area where twin Allison inline piston engines had been located in the XB-42. The engines were installed aft of the twin "bug-eyed" cockpit canopies, fed air via "cheek-type" inlets on either side of the fuselage, with long, side-by-side exhaust pipes exiting the tail. To this end, for fear of

Special F-84 Projects TIP-TOW, FICON, and TOM-TOM

During the course of *Thunderjet*, *Thunderstreak*, and *Thunderflash* development and production, the USAF created several unique programs in order to make the most of these designs.

SPECIAL F-84 PROJECTS

Two F-84D-1-REs were modified to EF-84D (USAF serial numbers 48-641 and 48-661) with coupling devices; 48-641 for the starboard wing, and 48-661 for the port wing. This was for Project TIP-TOW under USAF AMC Project MX-1016, which called for aerial wing coupling maneuvers.

A B-29A was modified and redesignated EB-29A (USAF serial number 44-62093) with coupling devices on both of its wingtips. Because of the difference in landing gear lengths, the three aircraft took off separately to couple and uncouple while they were in flight. The pilot of 48-641 was USAF Maj. John M. Davis, while the pilot of 48-661 was three-time World War II P-51 fighter ace Maj. Clarence E. "Bud" Anderson.

Several coupling/uncoupling flights were flown by the three aircraft, with a number of successful cycles of attaching and detaching using one EF-84D at first and then both of the EF-84Ds. The pilots of the EF-84Ds maintained manual control when attached to the EB-29A, with roll axis maintained by elevator movement rather than aileron movement. The engines on the EF-84Ds were shut down in order to save fuel during the towing segment by the mother ship, after which the engines were restarted in flight.

The experiment ended in disaster on April 23, 1953, during the first attempt to provide automatic flight control of the EF-84Ds; the cause, apparently, was an electronics malfunction. The left-hand EF-84D disengaged but immediately rolled over and onto the left wing of the EB-29A while still partially connected. Both connected aircraft crashed, with the loss of all personnel on board the EB-29A, and Major Davis, the pilot of the EF-84D. Major Anderson had previously uncoupled, so he did not go down with the other two aircraft. This tragedy took place two miles south of Southampton, New York.

FICON

The *FI*ghter *CON*veyor, or "FICON" program, under USAF AMC Project MX-1602, began in the early 1950s. The USAF had decided to conduct experiments on the feasibility of B-36s carrying fighter aircraft that were suspended from their bellies, under USAF AMC Project MX-1602-B. This not only provided the bomber with its own fighter protection; it would also make carrying the fighter long distances to a combat zone possible. Upon reaching the edge of the enemy's territory, the fighter would be released to conduct reconnaissance or bombing missions on its own.

A production RB-36F-1-CF (USAF serial number 49-2707) was modified with a special trapeze-type mechanism within its bomb bay and redesignated GRB-36F, while a production F-84E-1-RE Thunderjet (USAF serial number 49-2115) was fitted with a retractable hook in the nose, at the front of the cockpit. The hook would link the fighter to the trapeze, which would hold the aircraft in the bomb bay during flight, lower it for deployment, and raise it back after the mission. Due to the size of the fighter, only the cockpit, the fuselage spine, and the tailfin actually fit inside the GRB-36F, which increased drag considerably and reduced the big bomber's range by five to ten percent. On a positive note, the fighter pilot was able to leave his aircraft while attached to the carrier, making the ten-hour flights to and from the target much more bearable.

The initial FICON trials were performed in 1952. The first hookup took place on January 9, 1952, with the first retrieval into the bomb bay on April 23, and the complete system's first flight from takeoff to landing on May 14. In 1953, the GRB-36/F-84E was sent to Eglin Air Force Base in Florida, where 170 airborne launches and retrievals were subsequently performed. In May 1953, the F-84E was replaced by the faster Republic F-84F *Thunderstreak*, with the original YF-84F (USAF serial number 49-2430) modified for the role, when it was briefly designated GRF-84F.

When the RF-84F *Thunderflash* tactical reconnaissance fighter began entering service, the FICON role was changed from attack to reconnaissance. As

YF-84F number one (49-2430) attaches to the trapeze assembly on the GRB-36F (49-2707) during the FICON program. *NMUSAF*

with the F-84F, the RF-84F was supposed to apply its smaller size and superior agility to overfly heavily defended targets and gather intelligence while the bomber loitered outside the range of enemy defenses. The scheme was found to be "tactically sound," and USAF ordered ten production RB-36Ds to be converted to GRB-36D carrier airplanes with a complement of 25 RF-84K tactical reconnaissance fighters. The RF-84K differed from RF-84F in its retractable hookup equipment and anhedral horizontal tail planes to better fit inside the GRB-36D. Since the RF-84K retained an armament of four .50-caliber machine guns, it could also act as an escort fighter. The RF-84K could be deployed at altitudes up to twenty-five-thousand feet.

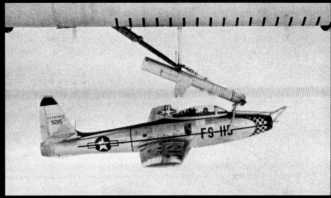

A close-up view of an EF-84E (49-2115) during FICON tests. *NMUSAF*

TOM-TOM

Project TOM-TOM was designed to demonstrate how fighter aircraft could accompany strategic bombardment aircraft, such as the Convair B-36, without burning their fuel. The USAF provided two RF-84F *Thunderflash* airplanes (USAF serial numbers 51-1848 and 51-1849) and a modified B-36 designated GRB-36F (USAF serial number 49-2707) under USAF ARDC Project MX-1602 to test the theory that the aircraft could be towed by bombers until needed to ward off attacking enemy fighter aircraft. This GRB-36F was also used in the earlier FICON Program. The aircraft were attached wing-tip-to-wing-tip with articulated arms and clamps. Although several successful hookups

were performed by Convair test pilots Arthur S. "Doc" Witchell, Beryl A. Erickson, and Raymond Fitzgerald in 1956, turbulence and vortices continued to present major problems.

On September 23, 1956, during test flight number eighteen, the RF-84F piloted by Beryl Erickson was torn away from the right wing-tip of the GRB-36F. All aircraft landed safely, but the concept was deemed too dangerous. At the time, developments in the area of in-flight refueling promised a much safer way to extend the range of fighters.

Project TOM-TOM was cancelled on March, 28, 1957, ending these dangerous in-flight hookups between "mother" and "daughter" aircraft for good.

The "bug eye" canopies covered the side-by-side seats within the cockpit in which the pilot and the copilot/navigator sat to operate the XB-43. The bombardier was seated in the nose section. Production B-43s were meant to carry up to eight thousand pounds of bombs. *NMUSAF*

dragging the tail pipes on runways during takeoff rotation, the XB-42-style ventral stability fin was eliminated from the XB-43 design. The resulting loss of stability would be countered by a taller vertical tail with increased area mounted on the XB-43 airplanes.

The turbojet engines used by the XB-43 were axial-flow turbojet engines provided by General Electric and designated J35-GE-3. These engines generated a maximum of four thousand lbf each.

XB-43

Two XB-43 Jetmaster airplanes were built and flown under US Army Air Forces Air Technical Service Command (ATSC) Secret Project MX-685. The two airplanes were ordered on USAAF contract number AC-40188, approved by the US War Department on March 31, 1944, as a high-priority program. USAAF serial numbers 44-61508 and 44-61509 were issued to these two airplanes.

If the experimental XB-43s proved successful, the USAAF planned to order an initial batch of fifty B-43 aircraft. Douglas planned to build up to two hundred B-43s per month, as long as further orders were forthcoming. Two versions of the B-43 were planned: (1) a bomber to carry up to eight thousand pounds of bombs and (2) an attack version armed with sixteen nose-mounted forward-firing .50-caliber machine guns and thirty-five five-inch rockets. For self-defense, both versions were to have twin .50-caliber machine guns mounted in a radar-directed, remote-control tail turret located between the engine exhaust pipes at the lower area of the vertical stabilizer just below the horizontal tail plane and rudder.

A delayed delivery of the General Electric Model TG-180 (J35) turbojet engines to Douglas for its Model 466 put the XB-43 program on hold for months. When they were finally delivered, V-E Day had already taken place, making the program less necessary.

Nevertheless, the first XB-43 was finally completed at Douglas's Santa Monica, California, facility in September 1945.

In October 1945, an engine run-up during the XB-43's preflight ground testing at Clover Field precipitated the failure of first-stage compressor blades inside the right-hand J35 engine. The engine was destroyed and disintegrated shards ripped through the top of the fuselage and its starboard side. The required engine replacement and systems and fuselage repairs caused seven months of additional delays to the program.

Finally fully repaired and ground tested, the first XB-43 (USAF serial number 44-61508) was trucked to Muroc AAF. The craft got airborne for the first time on May 17, 1946, with Douglas test pilot Robert P. "Bob" Brush in the left seat and Douglas test pilot Russell W. "Russ" Thaw in the right seat.

During one of its subsequent flight tests, the plane's transparent nosepiece cracked and had to be replaced. Since a spare Plexiglas nosepiece wasn't available, the replacement nosepiece was fashioned out of laminated plywood.

Numerous test flights were flown over and around Muroc for several years, with the airplane performing as well as could be expected for being the first of its kind.

On February 1, 1951, USAF test pilot Col. Gust "Gus" Askounis took off in the first XB-43 at Naval Air Station (NAS) Miramar. An accident due to engine failure ensued, and the airplane was badly damaged. What was left of it was subsequently returned to Edwards AFB via tow truck to be used for spares on the second airplane.

YB-43

The number two XB-43 (USAF serial number 44-61509) was first fitted with twin "bug-eye" canopies, but these were soon replaced with a side-by-side cockpit and canopy, according to the plans for production B-43s under USAAF contract number AC-11417. This aircraft made its first flight on May 15, 1947, as one of the YB-43s out of Clover Field; after its first test hop, it landed at Muroc AAF.

Douglas XB-42A

Although the Douglas XB-42 *Mixmaster* was originally built as an experimental piston-powered, propeller-driven bombardment airplane, it was decided to install a pair of Westinghouse Electric turbojet engines onto the first of the two XB-42s built (USAAF serial number 42-50224). After this, the airplane was redesignated XB-42A and added to the ranks of the turbojet-powered bombers discussed above, albeit as a composite-powered, bombardment-type airplane.

In order to serve as a turbojet engine test-bed, the first of the two XB-42 airplanes was flown from Muroc AAF to Santa Monica, California, for modifications. These included the replacement of its two Allison V-1710-25 engines with uprated, 1,375-horsepower V-1710-133 engines. To make them composite-powered aircraft under USAAF AMC Project MX-880, two podded turbojet engines—US Navy-supplied, 1,600-lbf Westinghouse Model 19XB-2A (J30-WE-2A) axial-flow turbojets—were installed, one under either wing.

After it had been trucked back to Muroc AAF in its modified form under USAAF contract number AC-40188, the Douglas XB-42A-DO (Douglas Model 740) was first flown on May 27, 1947. It was piloted by Douglas test pilot Russ Thaw and copiloted by Douglas test pilot Bob Brush. Douglas flight test crews flew it another twenty-one times over a total of seventeen hours. During these twenty-two flights, the airplane reached a maximum level attitude flight speed of 488 mph.

The one-off XB-42A (43-50224)—created from XB-42 number one—made its first flight on May 27, 1947. It was used primarily as a J30 turbojet engine testbed. With an all-out speed of 488 mph, both its piston and jet engines turning and burning, it made twenty-two test flights. On its twenty-second flight it made a hard landing, damaging its ventral stabilizer/rudder assembly; the damage was repaired, but the XB-42A never flew again. *NMUSAF*

XB-43 number two (44-61509)—later redesignated YB-43—shown on an early test flight near the Tehachapi mountain range west of Muroc AAF. Its first flight was on May 15, 1947—almost exactly one year after XB-43 number one. *NMUSAF*

Two more advanced and newer four-jet experimental bombers had taken wing by this time (the XB-45 on March 17 and XB-46 on April 2); for all intents and purposes, these new bombers made the XB-43 obsolete. For this reason, and to serve as an engine testbed, the second XB-43 became a service test airplane for the then new General Electric J47 turbojet engine. Redesignated YB-43, it was aptly nicknamed *Versatile II* because of its numerous and varied flight test duties.

Versatile II flew until late 1953, when it was retired and donated to the Smithsonian Institution the following year. Later transported to the National Museum of the United States Air Force (NMUSAF, formerly the United States Air Force Museum) at Wright-Patterson AFB, six miles northeast of Dayton, Ohio, it awaits restoration for public display.

SUMMARY

The Douglas Aircraft Company had hoped that its experimental and service test B-43 program would bear fruit and grow into a production program. By the time the second XB-43 had entered into flight test activities, though, there was no further need for the type, largely because its payload wasn't significant enough so satisfy USAAF planners. Moreover, it was underpowered and too small to carry an adequate fuel supply.

Although its story ends there, it remains significant in US aviation history as America's first turbojet-powered bomber.

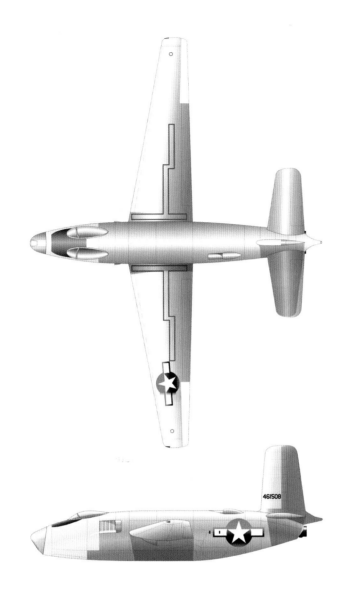

XB-43
SPECIFICATIONS

Propulsive system
Two non-afterburning, 4,000-lbf General Electric J35-GE-3 axial-flow turbojet engines

Crew
Three (pilot, copilot/gunner, navigator/bombardier)

Length	51 feet, 5 inches
Height	24 feet, 3 inches
Wingspan	71 feet, 2 inches
Gross takeoff weight	40,000 pounds
Maximum speed	503 mph (attained)
Range	Approximately 2,500 miles
Service Ceiling	38,500 feet

XB-43 Jetmaster number one in three-view. *G. De Chiara © 2015*

TWO

TRIALS AND TRIBULATIONS

Never, ever, ever, ever give up even when, and especially when, there is no chance of winning.

— Ann Curry

Progress on turbojet engine programs in the United States remained limited and slow during and immediately after World War II. A number of the aircraft produced in this time didn't come close to living up to the expectations accompanying them from their planning and construction to their field tests and— for the lucky aircraft—first flights. Nevertheless, several other turbojet engines developed in this period held much more promise, and even excelled. It was time to generate newer and better bomber and fighter airframes, driven by improved powerplants.

The Boeing B-47A Stratojet makes a full assisted takeoff (ATO) demonstration on December 31, 1948, with all eighteen of its 1,000-lbf Aerojet Model 14AS1000 ATO units firing; each unit had a fourteen-second burn duration. *AFTC/HO*

First flown on St. Patrick's Day in 1947, the XB-45 (NAA Charge Number NA-130) became America's first operational jet-powered bomber. Its tail-gunner compartment—like its bombardier compartment—had fake, painted-on windows. *NMUSAF*

XB-45 SPECIFICATIONS

Propulsive system
Four non-afterburning, axial-flow, 3,750-lbf Allison-built GE J35-A-7/-9 turbojet engines

Length	74 feet, 0 inches
Height	25 feet, 2 inches
Wingspan	89 feet, 6 inches
Gross weight	66,820 pounds
Maximum speed	500 mph
Maximum Range (unrefueled)	2,920 miles
Service Ceiling	38,500 feet
Armament	None installed
Payload	N/A

As the first embers of the Cold War were being fanned following the end of World War II, the Soviet Union, was as much a beneficiary of British and German jet engine development as the United States. Indeed, the USSR was busy moving forward on its own designs for jet technology, pushing hard to field its own fleets of turbojet-powered bomber and fighter aircraft.

The USAAF had only two operational turbojet-powered fighters and no jet-powered bombers in operation, which led to its decision to concentrate on turbojet-powered bombers in tandem with its continuing jet fighter development programs. While desperate to acquire turbojet-powered bombers, this only added to the pressure on teams as they suffered the trials and tribulations of developing a completely new aviation technology.

NORTH AMERICAN XB-45 AND B-Y5A TORNADO

The North American Aviation, Inc. B-45 Tornado program spawned the first operational turbojet-powered bomber to be operated in America. Developed as a medium-class, conventional bomber, this plane evolved into two different types: a nuclear-capable bomber—the B-45C—and a dedicated reconnaissance bomber—the RB-45C.

At first the B-45 was optimized to carry and deliver conventional, general-purpose, high-explosive bombs—up to 16,500-pound bombs. Later, under the Back Breaker program, it could carry a single 12,000-pound Tall Boy bomb or a single 22,000-pound Grand Slam bomb in its modified bomb bay. Further refinements allowed it to carry various nuclear devices, and it became the first turbojet-powered bomber in the world to drop a nuclear weapon.

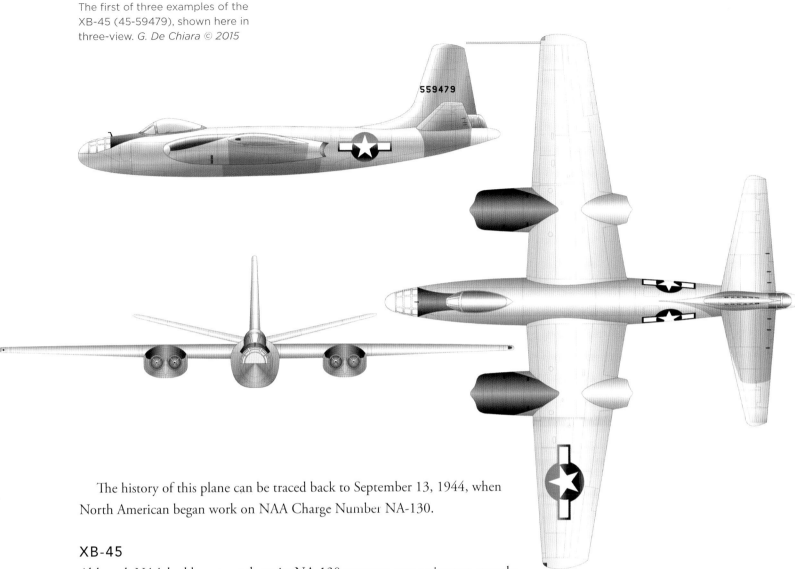

The first of three examples of the XB-45 (45-59479), shown here in three-view. *G. De Chiara © 2015*

The history of this plane can be traced back to September 13, 1944, when North American began work on NAA Charge Number NA-130.

XB-45

Although NAA had begun work on its NA-130 program some nineteen months earlier, USAAF contract number AC-5126 wasn't approved until May 2, 1946, for the three NA-130 airplanes that had been ordered under the XB-45 designation. These airplanes were issued USAAF serial numbers 45-59479 to 45-59481, and they were developed under USAAF AMC Secret Project MX-553.

The B-45 design was optimized to be propelled by four axial-flow, non-afterburning, 5,230-lbf (with water injection) General Electric Model TG-190 (two XJ47-GE-1 and two J47-GE-3) turbojet engines. Early on, though, these engines were unavailable, and the three experimental XB-45 airplanes and the initial batch of 22 B-45A-1-NA airplanes were powered instead by four non-afterburning, axial-flow, 3,750-lbf General Electric Model TG-180 (two J35-GE-7 and two J35-GE-9) turbojet engines.

Of the several new turbojet-powered bombers that made their first flights in the year 1947, the XB-45 was the first. On March 17, 1947, the first of three XB-45 airplanes made its initial flight.

Right-hand (starboard) side profile view of XB-45 number one on the dry lakebed at Muroc AAF, showing its fake bombardier nose application. There was no transparent Lucite available for the airplane at the time, and the greenhouse-type bombardier nose section windows were painted on. *NMUSAF*

With NAA test pilot George W. Krebs in command and Paul Brewer as flight test engineer, XB-45 number one (USAF serial number 45-59479) made a forty-minute first flight at Muroc AAF. Since its landing gear doors wouldn't fully close, the crew flew under restricted speed in a show of appropriate caution.

The first XB-45 was accepted by the USAF on July 30, 1948, the second XB-45 on August 31, 1948, and the third and last XB-45 in early 1949.

On June 28, 1949, the first of the three XB-45s built (USAF serial number 45-59479) crash-landed at Wright-Patterson AFB in Ohio. As famed USAF test pilot Col. Albert A. "Al" Boyd flight-tested the XB-45 that day, its landing gear got stuck during retraction shortly after takeoff; while the aircraft circled as Colonel Boyd attempted to extend the gear again, multiple failures occurred in the pressurization cooling system, the electrical system, and the number four engine. The pilot subsequently executed a crash landing on the runway, with the plane skidding for a mile before coming to rest on the edge of a big ditch. The plane was written off due to the extensive damage it had suffered, but Colonel Boyd and his crew survived without injury.

Nothing further is known about the accomplishments of XB-45 number two (45-59480), nor is its final disposition documented.

The sole surviving XB-45 (USAF serial number 45-59481)—the third of the three built—was transferred to ATC on May 15, 1950, now serving as a training aid for future USAF airframe and powerplant (A&P) mechanics set to work on the Tornado fleet. Its final disposition is unknown, but it was most likely recycled along with the first two articles.

XB-45
PRODUCTION BREAKDOWN

XB-45-NA:	3-45-59479 to 45-59481

none of these historic aircraft survived

B-45A-1-NA
PRODUCTION BREAKDOWN

B-45A-1-NA:	22-47-001 to 47-022

used for service test and combat crew training

B-45A

No service test Y prefix appeared in the designation of early production B-45A-1-NA (NA-147) airplanes that were used as service test and training aircraft, although a few were designated TB-45A-1-NA; in fact, twenty-two were built on USAF contract number AC-15569. The first B-45A made its first flight on February 24, 1948. These aircraft, like the XB-45s, were still powered by interim J35 engines. The B-45A-5-NA airplanes were the first Tornados to receive design J47 engines.

All three XB-45s were powered by four non-afterburning, 3,750-lbf, Allison-built GE J35-GE-4 turbojet engines that required jet assisted takeoff (JATO) during maximum payload tests. *NMUSAF*

On September 20, 1948, during a NAA flight test on B-45A-1-NA number one (USAF serial number 47-001), disaster again struck the Tornado program. The flying crew that day included NAA chief engineering test pilot George W. Krebs—who had made the first flight on the premier XB-45—and NAA flight test engineer Nicholas G. Picard. Flying out of Muroc AFB, Krebs put the plane into a dive to test its design load factor. An engine in a starboard (right-hand) nacelle exploded, which tore off several cowling panels. These hit and damaged the horizontal stabilizer, at which point the B-45A pitched up and both wings separated due to the excessive g-load. With no ejection seats, both Picard and Krebs were killed. The airplane impacted the ground 1.5 miles south of Alpaugh, California, northwest of Muroc AFB.

SUMMARY

The B-45 Tornado was the first operational turbojet-powered bomber in the US arsenal. More important: it was the first of the many steppingstones that led the USAF to groundbreaking operational bombardment aircraft.

Although it got off to a rocky start (with several fatal crashes), the B-45's design package coalesced into a very good weapon system. It became an important reconnaissance bomber that gathered critical intelligence and served to destroy significant targets during the Korean War.

In addition to the three XB-45s, North American produced ninety-six B-45As, ten B-45Cs, and forty-three RB-45Cs, all of which first served with the USAF Strategic Air Command before they were retired by the end of 1960. In total, 152 B-45s were produced and put into service before the B/RB-45s mission was overtaken by the B/RB-58A Hustler.

Overall, the success of the three experimental XB-45 aircraft led to the production of America's first operational turbojet-powered bombardment airplane: the North American B-45 Tornado.

XB-45A-1-NA SPECIFICATIONS

Propulsive system
Four axial-flow, non-afterburning, General Electric J35-GE-1/-3 turbojet engines

Length	75.3 feet
Height	25.2 feet
Wingspan	89.0 feet
Empty weight	45,208 pounds
Gross takeoff weight	88,400 pounds
Maximum speed	580 mph
Maximum Range	1,656 miles
Combat Ceiling	42,100 feet
Payload	22,000 pounds of bombs

Armament
Two tail-mounted .50-caliber M2 heavy machine guns with six hundred rounds of ammunition each

ABOVE: The XB-46 was initially powered by four Chevrolet-built, GE 3,750-lbf, J35-C-3 engines, later changed to Allison-built, 3,750-lbf, J35-A-5 engines. Both firms built these engines to GE drawings and specifications. The aircraft was unofficially nicknamed Needle due to its long, sleek fuselage. Here, the plane soars during its first flight, on April 2, 1947, from San Diego to Muroc AAF. *NMUSAF*

RIGHT: A rare color view of the XB-46 preparing for takeoff on its first flight out of Lindbergh Field in San Diego, California. *AFTC/HO*

The number one XB-48 shows off its left (port side) profile during its third test hop out of NAS Pax River. Note the retracted landing gear. *NMUSAF*

CONVAIR XB-46 NEEDLE

On February 27, 1945, the US Army Air Corps ordered three Convair (*Con*solidated-*V*ultee *Air*craft) Model 109 airplanes under USAAC Air Material Command Project MX-583. It called for the manufacture of three experimental, medium-class bombardment airplanes designated XB-46; USAAC serial numbers 45-59582/-59584 were assigned.

By design, these aircraft would be powered by four 3,820-lbf, General Electric-designed, Chevrolet-built, J35-C-3 axial-flow Model TG-180 turbojet engines. By June 1946, however, two of these aircraft had been cancelled, and only 45-59582 would be built and flown. USAAF contract number AC-7674 was approved and the airplane accepted on June 7, 1946, though its first flight was still almost ten months off. (Funding for the cancelled XB-46s was transferred to the Convair XA-44 program, which ultimately became the short-lived XB-53 program in December 1953.)

The one-of-a-kind XB-46, nicknamed Needle, took off for the first time on April 2, 1947, from Lindbergh Field in San Diego, California. Its landing

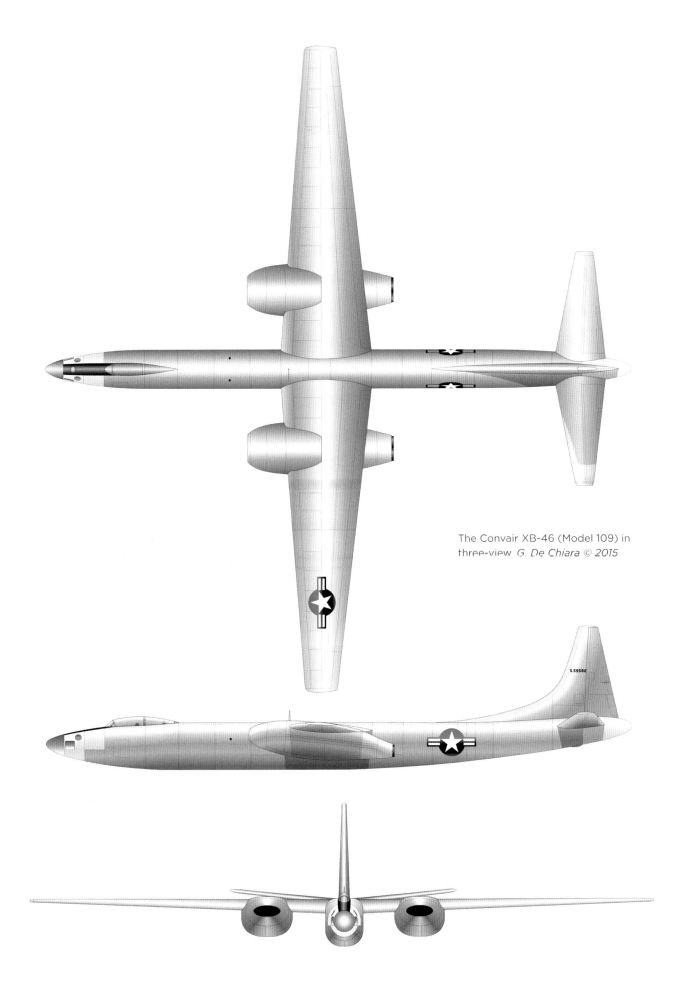

The Convair XB-46 (Model 109) in three-view *G. De Chiara © 2015*

XB-46
SPECIFICATIONS

Propulsive system
Four non-afterburning, axial-flow, 3,820-lbf Chevrolet J35-C-3 turbojet engines (initially); later refit with four non-afterburning, axial-flow, 4,000-lbf Allison J35-A-3 turbojet engines

Length	105 feet, 9 inches
Height	27 feet, 11 inches
Wingspan	113 feet, 0 inches
Empty weight	48,018 pounds
Gross takeoff weight	95,600 pounds
Maximum speed	545 mph at 15,000 feet
Maximum Range	2,870 miles
Combat Ceiling	40,000 feet
Payload	22,000 pounds (projected)

Armament
Two tail-mounted M2 .50-caliber heavy machine guns (not installed)

XB-46
PRODUCTION BREAKDOWN

XB-45-NA:	3-45-59479 to 45-59481

none of these historic aircraft survived

XB-48
PRODUCTION BREAKDOWN

XB-48	2-45-59585 and 45-59586

not available

gear was retracted and, after its initial test hop, it landed at Muroc AAF. The XB-46 featured an inline, tandem-seat cockpit with a long, narrow, bubble-type canopy. In the front seat was chief engineering test pilot Ellis D. "Sam" Shannon, while experimental test pilot William J. "Bill" Martin took the back seat.

Convair/USAAF flight-testing at Muroc ended after only fourteen flights. The airplane—accepted by the USAF on November 7, 1947, was refit with GE-designed, 4,000-lbf, Allison-built J35-A-3 turbojet engines before it was flown to Wright Field near Dayton, Ohio, for further evaluations.

On August 12, 1948, while at Wright-Patterson AFB under an assignment to 4000 Base Unit, the airplane suffered a minor landing accident. Its pilot at the time was USAF 1st Lt. Paul M. Butman.

From Wright-Patterson AFB (formerly Wright Field), the XB-46 flew to West Palm Beach AFB, Florida, for a series of tests performed between August 1948 and August 1949. These included tail section vibration tests and noise measurements.

The airplane sat static for about a year until July 1950, when the XB-46 was flown to Eglin AFB in Florida for climate testing. After these tests were completed, the USAF ordered the airplane scrapped. During the recycling process in February 1952, its nose section was saved and sent to the USAF Museum in Dayton, Ohio. Unfortunately, this is all that remains of the *Needle*.

SUMMARY

Considered by many to be one of the most elegant bombardment-type aircraft ever built and flown, the XB-46 could not match the performance of its main rival, the Boeing XB-47 Stratojet, which entered into production around the same time as the XB-46, along with another of its rivals, the North American XB-45 Tornado. Its last rival, the Martin XB-48, was likewise cancelled.

Streamlined to near perfection, this so-called "flying cigar" could only muster a top speed of 545 miles per hour, meeting the original requirement but not adequate in competition against the six-engine, sweptwing XB-47, which easily surpassed 600 mph.

Only the nose section of the *Needle* survives; it looks more like a thimble than a piece of the elegant airplane.

MARTIN XB-48

The Glenn L. Martin Company of Baltimore, Maryland, produced thousands of piston-powered bombardment aircraft during World War II. The company became one of the first US airframe contractors to build and fly a turbojet-powered airplane.

On November 17, 1944, a call went out for a medium-class, turbojet-powered bombardment airplane, and Martin responded with its proposed Model 223. The design was offered to the USAAF less than a month later, on December 9, 1944.

The ATSC awarded Martin a letter of intent contract—USAAF number W33-038-AC-7675—to produce a full-scale engineering mockup of its Model 223 for inspection. One year later, on December 9, 1945, the Model 223 was officially designated XB-48 and USAAF AMC Secret Project MX-598 was assigned to the XB-48 program.

Design and development of the XB-48 proceeded, and on December 13, 1946, the US War Department approved USAAF contract number W33-038-AC-13492 for the manufacture of two experimental XB-48 bombardment airplanes. USAAF serial numbers 45-59585 and 45-59586 were assigned to the pair of XB-48 airplanes.

The first XB-48 was completed on April 11, 1947. With Martin test pilots Orville Edwin "Pat" Tibbs (left-hand seat) and Melvin Relvin "Dutch" Gelvin on board, the premier XB-48 made its first flight on June 22, 1947, a thirty-seven-minute test hop from the Glenn L. Martin Airport in Baltimore to Naval Air Station Patuxent River, also in Maryland. As the airplane touched down, however, all four tires blew out on its fore-to-aft bicycle-type landing gear; Tibbs managed to bring the craft to safe stop without injury to himself or copilot Dutch Gelvin.

The second XB-48 made its first flight on October 16, 1948—one month after the program was cancelled.

On July 18, 1950, as it was flown by Ralph Stutsman of the Flight Test Division (FTD) of USAF ARDC, the number two XB-48 was involved in a ground accident at Wright-Patterson AFB due to an onboard engine fire.

Pat Tibbs (right) and Dutch Gelvin review flight test parameters with their flight test engineer. *NMUSAF*

SUMMARY

The two XB-48 airplanes were be not much more than historic footnotes due to their lack of performance. In fact, they were obsolete even before they ever took off.

That said, the XB-48 can lay claim to being the first of this kind of aircraft built with bicycle-type landing gear assemblies.

The first XB-48 was grounded in the fall of 1949, becoming a source of parts to keep the second XB-48 in the air for ongoing tests, such as thermal de-icing systems. Then, in September 1951, it was delivered to the Aberdeen Proving Grounds in Maryland for static structural testing, where it was ultimately tested to destruction.

XB-48 number one taxis out for its first takeoff on June 22, 1947, with test pilot Pat Tibbs at the controls. *NMUSAF*

NORTH AMERICAN XP/XF-86, YF-86A, YP-86C (YF-93A), YF-86D (YF-95A), YF-86H, AND YF-86K SABRE JET

Produced for many years and in large quantity by North American Aviation, Inc. in Inglewood, California, and Columbus, Ohio, the F-86 Sabre Jet doesn't

XB-48 number one was a truly handsome aircraft. Or was it? *AFTC/HO*

XB-48 SPECIFICATIONS

Propulsive system
Six non-afterburning, axial-flow, 3,750-lbf General Electric J35-GE-7 (Model TG-180B-1) turbojet engines

Length	85.8 feet
Height	27.5 feet
Wingspan	108.3 feet
Empty weight	63,603 pounds
Gross weight	102,600 pounds
Maximum speed	542 miles per hour at 13,000 feet (maximum power)
Maximum Range	1,330 miles
Ceiling	42,500 feet

Armament
Two tail-mounted, remote-control 20mm cannon (not installed); an operable dummy tail turret was installed without cannon

Payload (proposed)
One 12,000-pound M-121 (formerly T-10) Tall Boy bomb; one 22,000-pound T-14 Grand Slam bomb

require an introduction in the traditional sense, as it was the forerunner for all turbojet-powered fighter types that have served and continue to serve in the USAF. The F-86 was the first operational turbojet-powered aircraft manufactured and flown in the United States with design sweptback flying surfaces to combat the effects of compressibility. It proved to be one of the most significant and successful fighter aircraft ever produced.

Not only was the F-86 the leader in sweptback flying surfaces technology and the leader in close-in, gun-firing aerial jet fighter combat: it made every American flyer a jet ace during the Korean War. In addition, it served as a go-to fighter for numerous Allied nations for many years.

As the first sweptback-winged fighter aircraft employed by the USAF, the F-86 scored consistent aerial victories over Russian-built MiG-15s during countless dogfights in the skies above North Korea, accounting for a final kill ratio of 10.5 to 1 and creating forty jet aces in the process, with at least five kills each.

F-86As, F-86Es, and F-86Fs were the standout fighters of the Korean War, and they were the only turbojet-powered fighters that could handle the MiG-15s. The phrase "you're 86'd!" (meaning "you're out of here!" in USAF jargon) was coined during this form of aerial melee.

The history of the F-86 can be traced back to specific US Navy Bureau of Aeronautics (BuAer) requirement calling for a turbojet-engine powered, aircraft-carrier-based, daytime fighter. This emerged as an NAA design, NAA Charge Number NA-134. Three examples, designated XFJ-1 and named Fury, were ordered by the USN on December 27, 1944. As designed, the NA-134 was a single-seat, single-engine prototype powered by the General Electric Model TG-180, a non-afterburning, axial-flow turbojet engine designated J35.

North American prudently co-developed a land-based version of the sea-based Fury, NAA Charge Number NA-140, which was offered to the USAAF. The proposed design was accepted, and, on May 23, 1945,

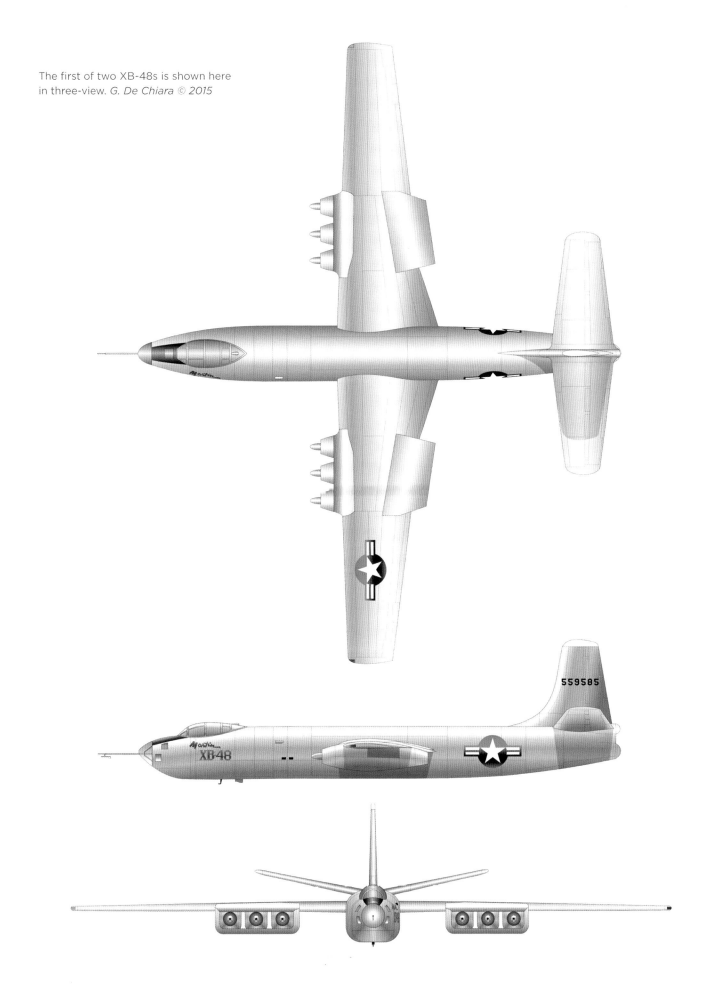

The first of two XB-48s is shown here in three-view. *G. De Chiara © 2015*

The sweptback-winged XP-86 (NAA Charge Number NA-140) began life as a straight-winged USAF version of the US Navy XFJ-1 Fury (NA-134). German aeronautical engineering data on the advantages of sweptback flying surfaces fell into US hands, and NAA asked for—and received—permission to apply this new technology to its XP-86. *USAF*

a trio of new X-fighters, designated XP-86, was ordered on USAAF contract number AC-11114 under USAAF AMC Secret Project MX-673.

Materially, and in their looks, both the USN NA-134 and USAAF NA-140 were quite similar. In fact, both designs featured straight laminar-flow wings; straight tail planes, tricycle landing gear (relatively new at the time); aftward-sliding bubble-top cockpit canopies; nose-type engine air inlets with straight-through ducting (downward and under the cockpit floor, then upward) to the face of the J35 engine mounted amidships; and six nose-mounted .50-caliber machine guns (three on either side of the fuselage). The USAAF design, however, sported thinner wings and a higher fuselage fineness ratio, which gave the XP-86 a sleeker look.

As development of the XP-86 proceeded through mockup stage, NAA engineers studied captured German aeronautical information on sweptback wings; the data was found to be sound. NAA asked for and received permission from the USAAF to build its NA-140 with thirty-five-degree sweptback flying surfaces. This change proved to be invaluable for the program.

Another boost to the program was the advent of the General Electric Model TG-190 axial-flow, afterburning turbojet engine being developed under USAAF AMC Secret Project MX-827. The TG-190 was designated J47, and would be employed by the prototypes as soon as it became available to them.

In the interim, though, the experimental F-86s would use the non-afterburning, axial-flow General Electric Model TG-180 turbojet engine, designated J35, which had been developed earlier under USAAF AMC Secret Project MX-414.

In September 1945, the USAAF approved NAA's XP-86 proposal and gave its go-ahead to produce all three XP-86s on order with sweptback flying surfaces. The previously allotted USAAF serial numbers remained the same: 45-59597 to 45-59599.

In the meantime, the first of three straight-winged XFJ-1 h airplanes made its first and subsequent test flights. It could only attain a best speed of 533 mph while flying straight and level, which was expected. Could the XP-86, propelled by the same engine type (with a different dash number), with its sweptback flying surfaces, actually top the Fury's maximum speed by one hundred mph, as advertised?

XP/XF-86

The first XP-86, unofficially dubbed Silver Charger, was completed on August 8, 1947. It was trucked to the North Base area of Muroc AAF on September 11, where it underwent a number of system checks, engine run-ups, and taxi tests. And on September 30, America's first sweptwinged airplane was ready for flight.

The next day, October 1, 1947, NAA chief engineering test pilot George Schwartz "Wheaties" Welch rotated and lifted off the dry lakebed. After some thirty minutes of trouble-free flight, Welch lowered the landing gear for recovery, but the nose landing gear refused to extend. Undaunted, Welch brought her in for a slow, nose-high landing. The vibrations from the main gear touchdown shook the nose gear enough to force its extension and complete a normal landing.

The only real squawk Welch reported after the flight was the lack of power produced by the 3,750-lbf Chevrolet-built General Electric J35-C-3 (Model 7E-TG-180-A5) turbojet engine. The airplane later used an Allison-built General Electric J35-A-5 (Model TG-180-A7) turbojet engine, but this provided the same propulsive force.

When the design J47 engine finally became available, the first of three XF-86 airplanes was fitted with an afterburning XJ47-GE-1 (Model 7E-TG-190-A1) turbojet engine. With afterburning, this engine produced 5,820-lbf and vaulted the XP-86 to speeds over 650 mph during straight and level flights.

Since this XP-86 had a new propulsive system, it was redesignated YP-86A for ongoing service test purposes. Its performance excelled. In a flight test piloted by Welch on April 25, 1948, the YP-86A unexpectedly broke the sound barrier during a dive.

Two of the three XF-86s suffered mishaps, all at Edwards AFB while they were assigned to Headquarters Squadron, 3077th Evaluation Group. A mechanical failure was the cause of the accident on February 25, 1949, which took place at Muroc AFB during a flight test of the XF-86 number three (45-59599) piloted by USAF Maj. Charles E. "Chuck" Yeager. On June 26, 1950, USAF test pilot Edward L. Johnson experienced a taxiing accident due to a fire, again on the XF-86 number three (45-59599). The third accident, on June 7, 1951, saw USAF test pilot David Sharp suffering a mechanical failure on the XF-86 number one (45-59597).

Finally, in September 1952, the first XF-86 (now YF-86A), which had launched the Sabre Jet, crashed to destruction after logging 241 flying hours.

Numbers two and three flew until April 1953, when they were retired from flight. Only the third XF-86 was ever armed for gun-firing evaluations.

Extensive flight-testing of one, two, and even three times per day proved that the XP-86 was more than just an average turbojet-powered fighter aircraft.

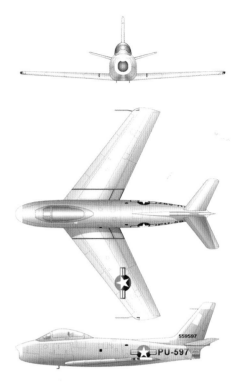

Three-view of the XP/XF-86. Engineering on what became the XP-86 began on June 3, 1945, and ended on January 20, 1949. The final engineering hours totaled 801,386, 340,594 for drafting alone. When the XP-86 made its first flight, 413,086 engineering hours had already gone into the program.
G. De Chiara © 2015

Long before the first XP-86 had flown, the USAAF ordered thirteen service test examples (without the Y prefix) and twenty-one production P-86A-1-NA airplanes on USAAF contract number AC-16013. NAA began work on these on November 20, 1946, applying NAA Charge Number NA-151 to these aircraft. This contract was later amended to include another 188 P-86A-5-NA airplanes under NA-151.

F-86A

There were no service test YF-86A aircraft per se, but the first thirteen of the J47-powered F-86A-1-NAs rolling off the production line were used for extensive service test activities at Edwards AFB. The NAA eventually produced 521 F-86A airplanes.

P-86B

The second batch of 188 P-86A aircraft was originally to be built as P-86B aircraft. Although the USAAF had ordered 190 P-86B aircraft under NA-152 on September 24, 1947, the proposed B version of the P-86 possessed beefier landing gear and larger wheels/tires for rough field operations—and little else. In light of this, the P-86B (NA-152) program was cancelled and the USAAF reordered 188 of the 190 cancelled P-86Bs as P-86A-5-NA airplanes, which were then transferred to the NA-151 program.

YP-86C

The last two of the original two hundred P-86A-5-NAs (NA-151?) on order were cancelled, since they would be used to create the two YP-86C Penetration Fighter service test airplanes (discussed below). This was the temporary designation later assigned to the YF-93A program (NA-157).

All three XF-86s fly in formation.
AFTC/HO

YF-86D

Originally designated YF-95A, the pair of YF-86D airplanes (USAF serial numbers) were built under NAA Charge Number NA-164 to evaluate an all-rocket-armed, all-weather interceptor version of the Sabre under Weapon System 206A. These were ordered under contract number AF-9211, which was approved on June 2, 1950. The USAF assigned serial numbers 50-577 and 50-578 to the two YF-86Ds.

The first YF-86D, powered by a single afterburning 6,650-lbf (with afterburning General Electric J47-GE-17 turbojet engine), made its first flight on December 22, 1949, at Muroc AFB with NAA experimental aircraft test pilot Joseph Arthur "Joe" Lynch Jr. at the controls (most references name George Welch, but Lynch was in fact the pilot).

YF-86H

Two service test YF-86H-1-NA Sabre Jet airplanes (USAF serial numbers 52-1975 and 52-1976) were ordered for USAF evaluation under NAA Charge Number NA-187. With USAF contract number AF-27681 in hand, NAA initiated work on its F-86H program on March 16, 1951. The contract was approved by the DOD on November 3, 1952.

On April 30, 1953, NAA test pilot Joe Lynch made a successful first flight on the first of the two prototype YF-86Hs at Edwards AFB.

The F-86H was a dedicated fighter-bomber version of the NA-172 (F-86E/F) Sabre with a higher g-load factor and more powerful afterburning GE J73-GE-3/-3A turbojet engines. The first two airplanes and a single static test article were built in Inglewood, California, while the remaining F-86Hs (173 F-86H-5-NH airplanes) were built in Columbus, Ohio.

NAA Columbus subsequently built another three hundred F-86Hs under NA-203 as F-86H-10-NHs. These were equipped with 8,920-lbf (with afterburning) GE J73-3D/-3E turbojet engines.

F-86J

The F-86J was a one-of-a-kind service test airplane built for evaluation by the Royal Canadian Air Force (RCAF), featuring a single 6,500-lbf Avro Orenda turbojet engine for propulsion. It was created from an F-86A-5-NA airplane (NAA Charge Number NA-161, USAF serial number 49-1069) on USAF contract number AC-21671.

The success of this program led to a fleet of Canadair-built CL-13 Sabre jets for the RCAF.

YF-86K

In mid-1952, the allied NATO air forces in Europe urgently requested a high-performance, all-weather interceptor from America, with the purpose of warding off the growing threat of attack by Soviet bombers. The F-86D was selected because it was available and considered to be the highest performance interceptor at the time.

The United States' allies did not know that the F-86D was equipped with the Hughes E-4 fire control system, which was fraught with issues and would likely have burdened allied NATO air forces with its less-than-ideal functionality. Additionally, the system was too secret to send out of the country, in particular because the Americans didn't want it to fall into Russian hands.

North American suggested that it create a simplified version of its F-86D to meet the requirements. With its E-4 fire control system and rocket armament replaced by the MG-4 fire control system, North American's revised version would operate four nose-mounted M-24A1 cannon, each with 132 rounds of 20mm ammunition. This plane would retain its nose-mounted APG-37 radar system and employ a modified AC Type A-4 gun sight.

The USAF accepted this plan, but it requested a two-man crew. NAA responded that a two-seat version of the F-86D would be much too costly and time-consuming to design, proposing instead the retention of a single-seat configuration of the F-86D. The USAF finally agreed, and NAA applied Charge Number NA-205 to the project. This new version of the F-86D was designated F-86K.

Two service test YF-86K airplanes were created from two modified F-86D-40-NA airplanes (USAF serial numbers 52-3630 and 52-3804) under NAA Charge Number NA-205. The airplanes had been bailed back to NAA for the modification program.

NAA test pilot Raymond S. "Ray" (a.k.a. "Silky") Morris flew the first of the two YF-86K airplanes (52-3630) on July 15, 1954. The second YF-86K

Rare color photo previews a special armament test for F-86D in Day-Glo paint scheme at Edwards AFB. Ordnance technicians are loading Mighty Mouse 2.75-inch FFARs into its retractable rocket tray. *NMUSAF*

XF-86 number three soars with a load-out of five-inch diameter HVARs under its wings. *AFTC/HO*

(52-3804) followed shortly thereafter, and, following, their respective flight test programs, they were shipped to the Fiat aircraft production facility in Italy.

F-86L

Although no experimental XF-86L or service test YF-86L airplanes were built, the F-86L was an important development to the Sabre family.

The first F-86L made its first flight on December 27, 1955. All F-86Ls were powered by single axial-flow, afterburning 7,650-lbf (with afterburning) General Electric J47-GE-33 turbojet engines.

The F-86L version of the Sabre was created from modified F-86D airplanes featuring modernized electronic equipment. It operated with the Semi-Automatic Ground Environment (SAGE) datalink system, incorporation of the 6-3 slotted wing leading edge flaps, and the addition of twelve inches to the wingtips for improved performance and handling qualities; 981 F-86Ls were built under NAA Charge Numbers NA-190 and NA-201.

TF-86F

Two service test (without the Y prefix) TF-86F Transonic Trainer airplanes were built at separate times with the idea of garnering additional orders for supersonic-capable jet pilot trainer and transition aircraft.

The first TF-86F was created under NAA Charge Number NA-204 by modifying an F-86F-30-NA under USAF contract number AF-6517.

Piloted by NAA test pilot Ray Morris, the first of two TF-86F airplanes (USAF serial number 52-5016) made its first flight on December 14, 1953; NAA Charge Number NA-204.

Tragedy struck the TF-86F program on March 17, 1954, when NAA test pilot Joe Lynch was killed at Nellis AFB, Nevada, when the first of two TF-86F airplanes crashed to destruction on takeoff. The unfortunate accident was attributed to a

The F-86 Sabre created forty aces (with at least five kills each) during the Korean War. Thirty-nine triple-, double-, and single- "MiG Maulers" are shown here. A fortieth F-86 ace with five kills, 1st Lt. Charles G. "Chick" Cleveland, was so named on January 15, 2008, when he was officially added to this elite group of combat pilots. *NMUSAF*

stall caused by an intended slow roll performed by the pilot of the airplane during takeoff, a move that had been successfully performed by Lynch on previous flights. In this event, the nose of the airplane appeared to be too high during his intended maneuver, resulting in an unrecoverable stall. Lynch had no time to escape.

A second TF-86F (USAF serial number 53-2138) was built to replace the lost example; it made its first flight on August 5, 1954, NAA Charge Number NA-216, USAF contract number AF-22304. Created from the last production F-86F-35-NA off the production line, this aircraft featured a small ventral fin (installed to address the problem associated with the first example's crash), two nose-mounted .50-caliber machine guns, and underwing attachment points (one on either wing), for either external drop-type fuel tanks or bombs.

This second TF-86F was delivered to Nellis AFB, Nevada, on January 31, 1955, as a combat trainer. On February 7, 1956, though, the USAF informed NAA that the TF-86F would be transferred to Edwards AFB and that no orders for TF-86Fs would be forthcoming. Its legacy lies in its use over several years at Edwards AFB as a chase/camera aircraft for recording the flight tests of other aircraft.

The overall performance of the TF-86F was acceptable for its day, but NAA had come up with a better option for the USAF: its own TF-100C Super Sabre, which eventually went into production (see below).

SUMMARY

The North American F-86 Sabre series of fighter aircraft were the springboard that launched the USAF into the modern era of turbojet-powered combat aircraft, especially fighter aircraft. It set the standard for all of the sweptback winged fighters to come.

The F-86 Sabre Jet became a fighter pilot's fighter, and was America's first sweptwing fighter, as well as the first to feature an all-movable horizontal tail called a "stabilator" (a combined horizontal stabilizer and elevator).

North American Aviation eventually manufactured four daytime (non-all-weather) versions of the Sabre: the F-86A, F-86E, F-86F, and F-86H. It also manufactured or remanufactured three nighttime (all-weather) variants: the F-86D, F-86K, and F-86L (the latter remanufactured as F-86Ds).

NORTHROP YB-49 AND YRB-49A

John Knudsen "Jack" Northrop, the founder and president of Northrop Aircraft Corporation, was a true visionary. He foresaw the potential for flight by all-wing aircraft, predicting less drag, more lift, and longer range. His fight to produce such aircraft yielded results with the design and manufacture of numerous flying-wing aircraft types, including those discussed above.

Unfortunately, none of Northrop's designs were ever put into full production for the civilian, military, or private aircraft markets. It wasn't until the advent of the B-2 stealth bomber that a flying-wing aircraft actually went into production.

The two types of flying-wing aircraft came on the heels of eight flying-wing-type aircraft designed, built, and flown before them. These included the 1929 Flying Wing, the MX-334 rocket-powered flying wing, the MX-324 unpowered flying wing glider, the JB-1 Jet Bomb, the experimental XB-35, and service test YB-35 Flying Wing bombers, the XP-56 semi-flying wing, and the XP-79B.

Although Northrop never saw his designs go into production, he did live long enough to witness a model of the Advanced Technology Bomber or ATB (B-2) in April 1980.

YB-49

On December 17, 1942, the US War Department approved USAAF contract number AC-33920 for the manufacture and flight-test of two turbojet-powered service test YB-49 Flying Wing bomber aircraft (USAAF serial numbers 42-102367 and 42-102368). These would be created from two modified

The first example (52-5016) of two prototype TF-86F Transonic Trainer airplanes built with the idea of generating additional orders for the type. It crashed on March 17, 1954. A second TF-86F example (53-1228) was built to replace it, but, on February 7, 1955, the USAF opted for two-seat supersonic F-100Fs instead. Both examples were created from modified F-86F aircraft and built under NAA Charge Numbers NA-204 and NA-216. *NMUSAF*

The authorization to convert two YB-35s to the YB-49 configuration was given in June 1945. The first YB-49 was accepted in June 1948, the second in May 1949. Operational B-49s were designed to carry up to 16,000 pounds of bombs. One payload configuration was two four-thousand-pound general-purpose (G.P.) bombs; another was ten 1,600-pound, all-purpose (A.P.) bombs. *NMUSAF*

piston-powered YB-35 airplanes, which were unfinished and still on the outdoor assembly line in Hawthorne, California, at the time the contract was ordered.

The first YB-49 (USAF serial number 42-102367) was completed in September 1947 and thoroughly checked out prior to its first flight on October 21, 1947. Its maiden flight took place from Hawthorne to Muroc AAF by Northrop test pilot Max R. Stanley, with Northrop test pilot Fred C. Bretcher as copilot.

On January 13, 1948, the second of the two YB-49s (USAF serial number 42-102368) made its first flight from Hawthorne to Muroc AAF with Stanley and Bretcher again at the helm.

YRB-49A

The one turbojet-powered YRB-49A was created from the third of five piston-powered YB-35A aircraft ordered on amended USAAF contract number AC-33920 (USAF serial number 42-102376). It differed from the YB-49s in that it was powered by six GE-designed, 4,900-lbf, Allison-built J35-A-19 non-afterburning, axial-flow turbojet engines, instead of the eight Allison-built J35-A-15 turbojet engines employed by the YB-49s. In this case, four of its engines were located inside the wing, while the other two were housed in underwing pods on either side of centerline.

On May 4, 1950, Northrop test pilot Bretcher, along with Northrop copilot John J. "J. J." Quinn, controlled the first flight on the YRB-49A from Hawthorne to Edwards AFB.

Even with its six jet engines at maximum power, the YRB-49A's top speed was only 438.4 mph at 35,332 feet. The eight-jet YB-49 wasn't much better, reaching only 492.5 mph at 20,000 feet for its maximum speed.

SUMMARY

Neither of the two YB-49s nor the single YRB-49A survive. None of the XB-35s, YB-35s, or YB-35As from which they came survive, either, as all were ordered to the scrap yard by Headquarters USAF, "to be scrapped as soon as possible" after their respective development and flight test programs had ended.

This marked the end of the Northrop Flying Wing bomber—until the late 1970s. At that time, the USAF initiated its ATB program, which led to the Northrop (now Northrop Grumman) B-2A Spirit, a pure flying-wing bomber.

The maximum takeoff weight of an YB-49 was 193,938 pounds with capacity for up to 14,542 gallons of JP-1 fuel. Its combat range with maximum fuel onboard was projected to be 2,254 miles. *NMUSAF*

BOEING XB-47, XB-47D, YB-47C (YB-56), YB-47F, YB-47J, YDB-47B, AND YDB-47E STRATOJET

The Boeing Airplane Company B-47 Stratojet was the first sweptwing bombardment airplane to enter service anywhere in the world, under contract with USAF Strategic Air Command (SAC). More than two thousand of these medium-class strategic bombardment aircraft were built, serving with USAF SAC until February 11, 1966, when the last two B-47s were retired from service.

The B-47 was also the first aircraft to receive a weapon system (WS) classification. The all-bombardment versions became WS-100A, while the reconnaissance bombardment versions became WS-100L.

XB-47

The experimental XB-47 aircraft's first designs were initiated in December 1945, when Boeing received a letter of intent to purchase two flying examples. These went through a number of configuration studies and tentative propulsive system applications before the final configuration, the Model 450-3-3, was frozen.

The design phase was comprised of four different designs, including Boeing models 424, 432, 448, and 450-1-1 (similar to the 450-3-3 but with the two outboard engines attached to the wingtips).

The most significant design features of the XB-47 were its sweptback flying surfaces (wings and tail planes), which set this aircraft apart from its competition and helped make it a resounding success.

In April 1946, the USAAF Air Materiel Command held its 689 engineering board inspection of the full-scale engineering XB-47 mockup, which passed the examination. Boeing was given a final go-ahead to build two XB-47 airplanes under USAAF AMC Secret Project MX-584. USAAF contract number W33 (038) AC-8429 was approved for these on July 10, 1947, and the aircraft were issued USAAF serial numbers 46-065 and 46-066.

YB-49 stands a lonely vigil until its next flight test. Note the sharp lines and the sleek profile of the aircraft's shadow. *NMUSAF*

TOP, LEFT: The YB-49 prepares for another test flight. *NMUSAF*

TOP, RIGHT: For nighttime photographic reconnaissance work, an operational RB-49A was designed to carry six 188-pound T-89 flash bombs to illuminate the terrain below. *NMUSAF*

ABOVE: The YRB-49A was powered by six Allison-built J35-A-19 turbojet engines rated at 4,900-lbf each (maximum sea level static), for a total output of 29,400-lbf—just 600-lbf less than the YB-49 but with two fewer engines. *NMUSAF*

The first XB-47 rolled out on September 12, 1947, at Boeing Plant 2 in Seattle, Washington. It was towed across State Highway 99 to Boeing Field, where it underwent a volley of ground checks and engine tests before making its first flight.

Boeing chief engineering test pilot Robert M. "Bob" Robbins and copilot Edward S. "Scott" Osler took XB-47 number one (46-065) aloft for the first time on December 17, 1947, from Boeing Field. They landed at Moses Lake in eastern Washington for subsequent test flights.

The number two XB-47 (46-066) first flew on July 21, 1948, and it too landed at Moses Lake. It was subsequently flown to Eglin AFB in Florida for weapons tests.

XB-47 number one was accepted on November 29, 1948, while XB-47 number two was accepted on December 18, 1948.

Both examples of the XB-47 were originally propelled by six non-afterburning General Electric J35 turbojet engines. Both were refitted with six non-afterburning General Electric J47s, and the first flight with the J47s occurred on October 8, 1949.

Earlier, on February 8, 1949, the first XB-47 was flown from Moses Lake to Andrews AFB near Washington, D.C., in record time. Still powered by six J35s, it flew the 2,289-mile distance in three hours, twenty-six minutes, to average 602.2 mph; this established an unofficial transcontinental speed record. The mission was flown by USAF Maj. (later Col.) Russell E. "Russ" Schleeh (pilot) and USAF Maj. (later Col.) Joseph W. "Joe" Howell (copilot).

Only XB-47 number two (46-066) survives today. It is on display in Rantoul, Illinois, at the Chanute Air Museum, located on the former Chanute AFB (closed in 1993).

XB-47D

In the late 1940s and early 1950s, a number of turbopropjet engines were in development. In the simplest terms, a turbopropjet engine is a gas turbine engine that uses its exhaust for thrust but also turns a propeller to provide high cruise speeds with better fuel economy.

Seeking an increased unrefueled range at a high cruising speed of nearly 600 mph, the Boeing Airplane Company was authorized to proceed with its plans in April 1951. The company built two composite-powered versions of its six-jet Stratojet, designated XB-47D under USAF Air Materiel Command Secret Project MX 1637; Boeing model numbers 450-162-48 and 450-162-49. As composite-powered demonstrators, the two four-engine XB-47D airplanes were each powered by two 8,500-shaft horsepower Wright YT49-W-1 turboprop engines and two non-afterburning 5,900-lbf General Electric J47-GE-23 turbojet engines. The Wright YT49-W-1 turboprop engines turned reversible fifteen-foot-diameter Curtiss Electric four-bladed propellers. The first XB-47D was a modified B-47B-30-BW (51-2103), while the second was a modified B-47B-25-BW (51-2046).

The Wright YT49-W-1 was a turboprop version of the British license-built Armstrong Siddeley Sapphire turbojet engine, designated J65. The program was

ABOVE: Both XB-47s were re-equipped with six 5,200-lbf J47-GE-11 turbojet engines. XB-47 number one made its first flight with this revised propulsive system on October 8, 1949. XB-47 number two was also refit with J47 engines. With this change, their top speed increased from about 580 mph to approximately 630 mph. *NMUSAF*

BELOW: The first of two XB-47 airplanes (45-065 and 46-066), initially powered by four 3,750-lbf J35-GE-7 (inboard) and two non-afterburning, axial-flow, 4,000-lbf J35-GE-9 (outboard) turbojet engines. The USAF accepted these airplanes on November 29, 1948, and December 18, 1948, respectively. *NMUSAF*

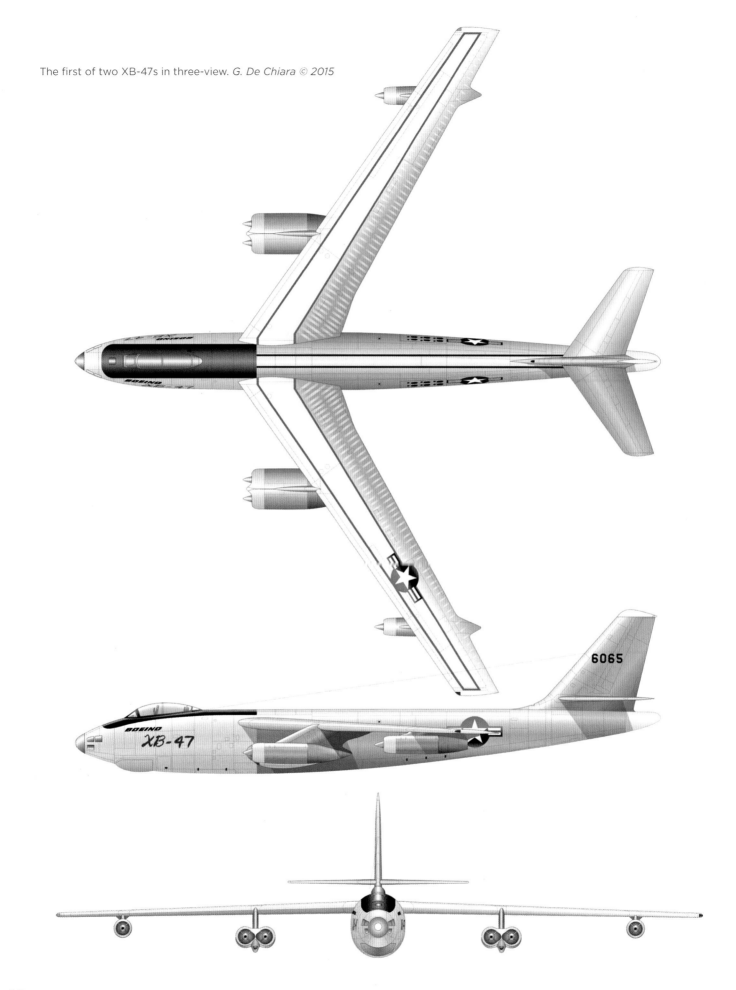

6065

BOEING XB-47

delayed by problems with the T49 engine, which failed to pass its fifty-hour qualification run. Additional problems with the engine-propeller combination and shortages of government-furnished equipment delayed progress still further. It was not until late 1955 that the aircraft were ready for their first flights.

Boeing test pilot Samuel Lewis "Lew" Wallick Jr. piloted the first flight of the number one XB-47D on August 26, 1955. The second XB-47D made its first flight on February 15, 1956, again with Wallick at the controls. Both first flights were made at Boeing Field in Seattle, Washington. Maximum speed reached during flight test was 597 mph at 13,500 feet, making it the fastest combined turboprop- and turbojet-powered airplane to fly, though this was accomplished at full-power settings with excessive fuel consumption. Finally, as demonstrated, the XB-47D achieved a best cruise speed of 462 miles per hour at 42,400 feet altitude. Since Boeing's all-jet-powered B-47 Stratojet counterpart was deemed preferable, the XB-47D project was abandoned.

The XB-47D airplane was 106 feet, 8 inches long, 27 feet, 9 inches high, with a Wingspan of 116 feet. Its maximum gross takeoff weight was 184,428 pounds, and its maximum unrefueled range was 5,759 nautical miles. It was to have a three man crew: pilot, copilot/gunner, and navigator/bombardier.

YB-47C (YB-56)

The Boeing Airplane Company was to create a single YB-56 for USAF evaluation. YB-47C was the original designation for this design project, initiated in January 1950 under Boeing Model Number 450-19-10, with the purpose of redesigning the B-47B aircraft for four Allison-built General Electric J35-A-23 turbojets in place of the six General Electric J47s. The proposed B-47C model configuration was nearly identical to the B-47B, with the exception of the inboard engine nacelles; one J35 was to be mounted in place of the two J47s on the production B model. One B-47B-20-BW (USAF serial number 50-082) was bailed back to Boeing and scheduled for modification, and the first flight was projected for April 1951.

Because of delays and problems with the engines, a number of alternatives were proposed. The original Allison J35-A-23s turbojets had estimated performance of 9,700-lbf each at maximum power (with afterburners) and a normal power rating of 8,200-lbf each. The actual J35 performance was nowhere near the estimates (about 7,400-lbf maximum power and 5,600-lbf normal), so alternative engines were proposed.

XB-47 SPECIFICATIONS

Propulsive system
(1) Six non-afterburning, axial-flow General Electric J35 (Model TG-180) turbojet engines: four 3,750-lbf J35-GE-7 (inboard) and two 4,000-lbf J35-GE-9 (outboard);
(2) both airplanes were refitted with six Model TG-190 non-afterburning, axial-flow 5,200-lbf J47-GE-3 turbojet engines

Length	107 feet, 5 inches
Height	27 feet, 7 inches
Wingspan	116 feet, 0 inches
Empty weight	74,623 pounds

Gross weight
109,000 pounds (combat);
162,500 pounds (maximum)

Maximum design speed
578 mph (with J35s);
over 600 mph with J47s

Maximum Range	4,000 miles
Service Ceiling	37,500 feet

Crew
Three (pilot, copilot, navigator)

Armament
Two tail-mounted .50-caliber Browning M2 heavy machine guns with 1,200 rounds of ammunition (600 rounds per gun)

Payload	22,000 pounds (maximum)

The XB-47s and early production B-47s featured eighteen Aerojet 1,000-lbf ATO rocket motors (fourteen seconds duration) to help get them airborne, since their gross takeoff weight (162,500 pounds) far exceeded the power produced by their propulsive system (23,000-lbf). *NMUSAF*

The Allison axial-flow J71-A-5 turbojet engine was first considered for replacing the J35-A-23. The J71 had maximum thrust above 10,000-lbf at military power, but other problems and delays forced it to be dropped from the B-47C project as well. By this time, the YB-47C project designation was dropped and redesignated YB-56. The Pratt & Whitney J57 was evaluated for the B-56 project, but it was still in development and delivery delays proved unacceptable. The Boeing B-52 project was well underway by this time and had higher priority for the J57s when they became available.

The initial prototype conversion of a B-47B was first designated YB-47C under USAAF ARDC Project MX-584-A. The YB-47C was ordered on USAF contract number AF-22413 approved on November 14, 1949.

The YB-47C was redesignated YB-56, and the intended production version of the aircraft was designated B-56A, with a photo reconnaissance version designated RB-56A (this was changed to RB-47C under Boeing Model Number 450-24-26).

Features incorporated in the B-56A preliminary design included cabin pressurization, thermal anti-icing, bombing-navigation radar, gun-laying radar, anti-skid brakes, bailout spoiler door, internal fuel purging, camera provisions, braking parachute, single-point refueling, air-refueling provisions, and a maximum fuel capacity of 17,350 US gallons.

By the time the program was cancelled, these aircraft had been redesignated B-47C and RB-47C.

Boeing and the USAF were intrigued by the idea of a four-engine version of the six-engine B-47, offering near-equal performance (speed) and less fuel consumption (more range), though the concept never came to completion.

Ultimately, the B-56 program was cancelled in December 1952 before conversion work on the B-47B set aside for the project had begun. An appropriate turbojet engine for the project was never determined, and the early development problems with the J47 engine had been corrected. For these reasons, the four-engine B-56 airplane was no longer needed.

Boeing built two XB-47D airplanes to evaluate the use of turbopropjet engines. Each XB-47D employed two axial-flow, 8,500 shaft horsepower, Wright YT49-W-1 turboprop engines spinning Curtiss Electric four-blade, fifteen-foot diameter propellers. *NMUSAF*

YB-47F

One B-47B (USAF serial number 50-069) was fitted with a probe for in-flight refueling via the US Navy–type probe and drogue aerial refueling system and it was designated YB-47F. The tests were flown with a combined piston/turbojet-powered Boeing KC-97G tanker cargo airplane fitted with an in-flight refueling hose. The tests proved less than successful for B-47 operations, and the standard Boeing designed and developed Flying Boom method prevailed.

YB-47J

To evaluate the MB-2 radar bombing/navigation system, a single service test YB-47J was created from a modified B-47E. The MB-2 low-altitude bombing system (LABS) was slated for use by the RB-52B.

YDB-47B

One B-47B (USAF serial number 51-2186) was created in 1953 to carry a single Bell GAM-63 Rascal for service tests. The follow-on YDB-47E aircraft completed these evaluations.

YDB-47E

Two B-47E airplanes (USAF serial numbers 51-5219 and 51-6220), built in Wichita, Kansas, were completed as specially modified YDB-47E aircraft. Their sole purpose was to fly service tests with the Bell GAM-63 Rascal air-to-surface guided missile, which carried a W-27 nuclear warhead. A single GAM-63 (previously designated B-63) was carried on the starboard side on each YDB-47E, behind the trailing edge of their right wings.

   In practice, after a missile was launched, it was guided to its target via remote control by an operator aboard the aircraft. The two YDB-47E airplanes bore Boeing Model 450-167-50 and had an unrefueled combat radius of 1,230 miles with an 18,000-pound payload flying at 490 mph.

YB-56 SPECIFICATIONS
(WITH J35-A-23 ENGINES)

Propulsive system
Four afterburning, Allison 9,700-lbf J35-A-23 turbojet engines; four afterburning 10,700-lbf Allison J71-A-5 turbojet engines; or four 10,000-plus lbf (with water injection) Pratt & Whitney J57-P-1W turbojet engines (no definite **propulsive system** was ever finalized)

Length	106 feet, 7 inches
Height	27 feet, 9 inches
Wingspan	116 feet, 0 inches
Empty weight	74,623 pounds

Rocket assisted takeoff (RATO)
One Aerojet YLR45-AJ-1 or one M. W. Kellogg YLR47-K-1 with four chambers; 5,000-lbf each to total 20,000-lbf

Maximum speed
600 mph at best altitude (estimated)

Maximum range (unrefueled)
3,500 miles (estimated)

Service ceiling	42,200 feet (estimated)
Crew	Three

Armament
Two radar-directed, remote-control 20mm cannon with 600 rounds ammunition each

SUMMARY

The Boeing B-47 Stratojet series of bombardment aircraft proved to be a prodigious asset to the USAF's Strategic Air Command. As a dedicated strategic nuclear bombardment platform, its medium-class nomenclature was a misnomer. It served with SAC from June 1951 until its retirement in 1966.

Eventually, Boeing, Douglas, and Lockheed produced six versions of the Stratojet: B-47A, B-47B, B-47E, B-47E-II (modified B-47Bs), RB-47E, and RB-47H.

CURTISS–WRIGHT XP/XF-87 (XA-43) BLACKHAWK

The Curtiss-Wright Corporation's XP/XF-87 Blackhawk was first proposed as an attack aircraft, designated XA-43 (Model 29). The XA-43 was to be powered by four General Electric J35 turbojet engines, with dive recovery flaps, fitted fuselage-to-engine-nacelle under the wing. On November 21, 1945, the USAAF cancelled the XA-43 program and all of its current and future funding; its USAAF serial numbers were transferred to the XP-87 (Model 29A) program.

On October 3, 1947, the DOD approved USAAF contract number AC-6266 for a single four-engine XP-87 (USAAF serial number 45-59600). A second XP-87 (45-59601) was cancelled and reordered as the XP-87A (USAAF serial number 46-522) on the same contract (amended), approved on the same date.

The one-off XP-87A was to be completed as a preproduction prototype with two axial-flow, 6,000-lbf, General Electric J47-GE-7 turbojet engines—one mounted to either wing.

In the interim, the XP-87 was nearing its final assembly phase; work on it proceeded on schedule. It was completed and rolled out of the factory in mid-January 1948, after which it underwent a series of ground checks to test its systems. It was then partially disassembled for transport to Muroc AAF, where it would be thoroughly flight-tested. It arrived at Muroc in mid-February and was reassembled and ground tested again.

Curtiss test pilot B. Lee Miller performed number low-, medium-, and high-speed taxi tests through March 4, 1948, before he dared to make its first flight. Satisfied, Miller said he would fly her on the next day. On March 5, the one-off XP-87 made its first flight. Successful for the most part, Miller reported her sluggish performance, which had been expected: the cumbersome XP-87 (weighing nearly forty-thousand pounds during its initial flights) was woefully underpowered with its four non-afterburning axial-flow Westinghouse XJ34-WE-7 turbojet engines, which together only provided 12,000-lbf (3,000-lbf each).

On June 10, 1948, an order was placed for fifty-seven P-87A and thirty RP-87A airplanes. The next day, the "P for pursuit" prefix became an "F for fighter"—the XP-87 became XF-87, the P-87A and RP-87A became F-87A and RF-87A, respectively.

The Curtiss-Wright Model Number 29A was assigned to the XF-87.

The prototype XP-87 (45-59600) was finally ready in early 1948. It was trucked out to Muroc Dry Lake from Columbus, Ohio, in early February 1948 for its first flight. It took to the air for the first time on March 5, 1948, where it was piloted again by Miller.

Initial performance of the XP-87 (redesignated XF-87 three months later, in June 1948) was promising, although the top speed was some 12 percent slower than promised and there were some problems with buffeting at high speeds.

Since the XP-87 had proven to be "gutless," the production F-87As were to have been powered by two 6,000-lbf General Electric J47-GE-7 engines. Plans were made for the second XP-87 prototype (46-522) to be fitted with a pair of these J47 turbojets as a production prototype, one under either wing in a nacelle. It was redesignated XP-87A.

If the XF-87 had been flight-tested with its intended J47 propulsive system, it most likely would have fared much better against its competition. This projection was never realized: the entire F-87 program was cancelled October 10, 1948, including the hoped for follow on F-87A and RF-87A production airplanes. This was a major blow to the Curtiss-Wright Corporation, and it proved almost fatal.

A Curtiss-Wright artist's conception of the proposed XA-43 (Model CW 29) four-turbojet engine-powered airplane emerging from the clouds; early 1945. The XA-43 program was cancelled on November 21, 1945, at which time its funding was transferred to the XP-87 (CW 29A) program. *CW*

SUMMARY

After the XF-87 Blackhawk program was cancelled, the Curtiss-Wright Corporation closed the doors to its aircraft manufacturing facility in Columbus, Ohio, and sold it to North American Aviation, Inc.

The *Blackhawk* was the last Curtiss-built, combat-type airplane from a long line of successful aircraft that had preceded it. (Curtiss-Wright did manufacture an X-19 VTOL research aircraft (Model X-200) in Caldwell, New Jersey, in 1963; its first hover flight was accomplished on November 25 of that year.)

The Curtiss-Wright Corporation remains in business today, but it no longer produces aircraft. Instead, it produces a long line of electronic equipment and other materials for number of industries including the aerospace industry.

NORTHROP XP/XF-89, YF-89A, YF-89D, YF-89E, YF-89F, AND YF-89H SCORPION

The Northrop Aircraft Corporation's F-89 Scorpion series aircraft were designed as dedicated, nighttime, all-weather fighters, fighter bombers, and fighter interceptors. The program to acquire these aircraft began on August 28, 1945, when the USAAF issued requirements calling at first for a piston-powered, propeller-driven, night (all-weather) fighter-bomber to replace the Northrop P-61

YF-87 SPECIFICATIONS

Propulsive system
Four non-afterburning, axial-flow, Westinghouse 3,000-lbf J34-WE-7 turbojet engines

Length	62 feet, 0 inches
Height	20 feet, 4 inches
Wingspan	60 feet, 0 inches
Gross weight	37,350 pounds
Maximum speed	520 mph
Maximum range	1,000 miles
Service ceiling	41,000 feet

Crew
Two (pilot, copilot/navigator/gunner)

Black Widow. The requirements included an unrealistic top speed of 525 mph for a piston engine at thirty-five thousand feet, 550 mph at sea level, a combat radius of six hundred miles, and a time-to-climb to thirty-five thousand feet in twelve minutes. Four months later, although relatively new at the time, all of the competing airframe contractors were encouraged to consider a turbojet propulsive system as well.

In addition to Northrop, five other airframe contractors submitted proposals. These included Bell, Convair, Douglas, Goodyear, and Curtiss-Wright.

XP/XF-89

According to official USAF documentation dated February 14, 1958, the USAAF ordered a single XF-89 and a single YF-89A on USAAF contract number AC-14541. This contract was approved on May 21, 1947. The two airplanes were to be built and delivered under USAAF AMC Secret Project MX-808.

According to an earlier plan, these two airplanes were to be produced as a pair of XP-89s and built under Northrop Model N-24. The second example, however, was to be built with a different propulsive system, with other improvements. To this end, it was redesignated YF-89A under Northrop Model N-49.

Northrop test pilot Fred Charles Bretcher successfully flight-tested the XF-89 (45-678) at Muroc AFB on August 16, 1948.

On February 22, 1950, during its 102nd flight, the one-off XF-89 suffered a structural failure of its vertical and horizontal tail planes after a high-speed pass over Hawthorne Field (a.k.a. Northrop Field). The aircraft crashed to destruction two miles northeast of Manhattan Beach, California. It had been on a demonstration flight for attending USAF officials. Northrop test pilot Charles "Chuck" Tucker was thrown clear of the disintegrating airplane and survived his parachute descent and landing; for some reason, Northrop flight test engineer Arthur A. "Art" Turton wasn't able to parachute out of the airplane and was killed in the crash.

The one-off XP-87 rolled out on August 22, 1947, at Curtiss-Wright's St. Louis, Missouri, facility. This rare color image of XF-87 (redesignated after June 11, 1948) shows its right side in profile. *AFTC/HO*

The second XP/XF-87A was to be armed with a movable up-and-down, side-to-side nose turret housing four 20mm cannons, with 270 rounds of 20mm ammunition each; it was never built. *NMUSAF*

Front view showing the wingtip auxiliary fuel tanks. *NMUSAF*

YF-89A

Work to convert XF-89 number two (46-679) to the YF-89A began on November 15, 1949.

On June 27, 1950, the YF-89A made a successful first flight with Northrop test pilot John J. "J. J." Quinn at the controls.

YF-89D

A modified F-89B Scorpion served as the lone service test YF-89D (USAF serial number 49-2463). Under Northrop Model N-67, the YF-89D made its first flight on October 23, 1951.

The one-off YF-89D was lost on October 20, 1953, when it crashed to destruction six miles northeast of Edwards AFB. Its Northrop test pilot, Walter P. Jones, was killed along with Northrop radar operator Jack Collingsworth.

YF-89E

One F-89C-10-NO (USAF serial number 50-762) was modified to accept two axial-flow, 9,500-lbf Allison YJ71-A-3 turbojet engines, and was evaluated under Northrop Model N-71. The lone YF-89E made its first flight on June 10, 1954. Subsequent tests found little improvement over the F-89B/C, and the F-89E Scorpion program was terminated. The YF-89E continued to fly at Edwards AFB for armament test purposes.

ABOVE, LEFT: The XF-89 cuts a dark silhouette during its first flight on August 16, 1948. *NMUSAF*

ABOVE, RIGHT: The YF-89A (46-679) soars during its maiden flight on November 15, 1949. *NMUSAF*

YF-89F

The YF-89F began life as the so-called "Advanced F-89D Scorpion" under Northrop Model N-82 on December 3, 1951.

This program was cancelled in August 1952.

YF-89H

Three F-89Ds were modified to serve as service test F-89H airplanes, with USAF serial numbers 52-1830 (an F-89D-35-NO), 52-1939 (an F-89D-45-NO), and 52-2149 (F-89D-45-NO). The only one to receive YF-89H designation, though, was 52-1939. (Some sources claim that 53-249 was the third F-89D to be converted, but this serial number was actually issued to a Boeing KC-97G.)

The YF-89H (Northrop Model N-138) was primarily used to test the feasibility of carrying the radar-guided and infrared (heat)-guided Hughes Falcon guided aerial missiles (GAR).

The production F-89H Scorpion carried forty-two 2¾-inch-diameter folding-fin aerial rockets (FFAR) and six GAR-1 or GAR-2 Falcon air-to-air guided aerial rockets in all. Each wingtip had a pod carrying twenty-one 2¾-inch rockets and onto which three GAR-1s or three GAR-2s were attached. Northrop produced 156 F-89Hs.

The GAR-1 (changed later to AIM-4A) was radar-guided, while and the GAR-2 (changed later to AIM-4B) was infrared (heat)-guided. The acronym AIM stands for "air intercept missile."

The three service test F-89Hs were later modified into F-89Js.

SUMMARY

The F-89 Scorpion series of aircraft played a significant role in the defense of America during the Cold War. Armed with cannon, missiles, and/or rockets, these aircraft defended US borders from mid-1951 until their retirement in late 1959.

The Northrop Aircraft Corporation produced 1,052 F-89s in all.

McDONNELL XP/XF-85 GOBLIN

When the gargantuan piston-powered, propeller-driven Convair XB-36 lifted off terra firma and flew for the first time on August 8, 1946, it became the largest and heaviest bomber airplane ever to do so. It was followed into the air by the YB-36 on December 4, 1947. One year later, the USAF SAC had twenty-two production B-36A intercontinental bombers on tap. Unofficially dubbed the Peacemaker, the B-36 was the prized trophy atop SAC's mantelpiece until the operational debut of the eight-jet Boeing B-52 in June 1955.

Operational B-36s could carry a ten-thousand-pound bomb load five thousand miles, deliver it, and return to base flying nonstop without in-flight refueling. At the time, it served as the epitome of heavy bombardment aircraft.

To ensure that B-36s got through to their targets—and make it home—they carried an impressive self-defense armament. The USAF wanted even more protection for its B-36, though, and so sent out a request for a very-long-range bomber escort fighter.

But what fighter—in service or on the drawing board—could begin to match the long-range capability of the B-36?

There simply was no fighter available that could match the range of the B-36. With this knowledge, an assortment of bomber escort fighter ideas came to light, most bordering on the ridiculous. For example, one plan was to tow a fighter behind a B-36 until the convoy neared its target, at which point the fighter would be released from the towing mechanism (much like an unpowered glider) to fend off the enemy. Then, somehow, the fighter would be retrieved by the bomber (if it had survived) for the trip home. This retrieval part of the plan was never figured out, not to mention the problems associated with the bomber's drag increase from towing a fighter behind it, and making the fighter behave aerodynamically in the B-36's slipstream.

The answer to this problem came in an unexpected form.

The B-36's bomb bay was long and wide, roomy enough to house its bomb load—and, possibly, its own escort fighter within it. But could such a small, light, turbojet-powered fighter airplane with adequate armament and performance—and capable of being launched and retrieved—be developed? If so, it was determined, it would have to be a parasite. Thus, the USAAF Parasite Fighter program was set in motion under USAAF AMC Secret Project MX-472, for exclusive protection of B-36s in enemy airspace. Project MX-472 specifically called for preliminary design studies for an unconventional fighter.

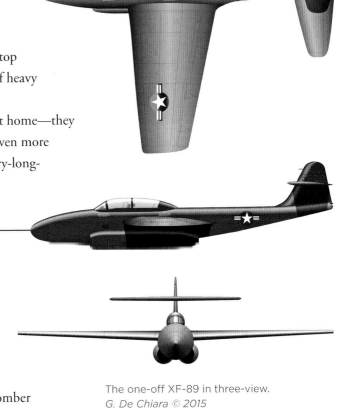

The one-off XF-89 in three-view.
G. De Chiara © 2015

In mid-1945, the USAAF issued its requirements for a Parasite Fighter to US airframe contractors, which included the exact measurements of the B-36 bomb bay. These specifications, based upon many unknowns, included:

1. Length not to exceed fifteen feet; height not to exceed nine feet; wingspan not to exceed five feet (It was recommended that the airplane have folding wings for an adequate aspect ratio, though this could not exceed twenty-two feet when fully extended.)
2. Gross weight not to exceed six thousand pounds
3. Armament comprised of four .50-caliber heavy machine guns
4. Maximum speed of six hundred miles per hour; service ceiling forty-thousand feet; one-hour mission time, engine start to engine stop
5. Aircraft would be single-place, single-engine

Every airframe contractor that received the requirements balked at the project, except for the McDonnell Aircraft Corporation of St. Louis, Missouri. McDonnell obliged the USAF with a good proposal.

The preliminary design team at McDonnell, orchestrated by Herman D. Barkey, began with the search for a suitable turbojet engine for what it called its Model 27. With an overall length of 112 inches, a diameter of 27 inches, a dry weight of 1,207 pounds, and a static sea-level thrust rating of 3,000-lbf, the engine of choice was the Westinghouse Model 24C-4B axial-flow, non-afterburning J34-WE-22 turbojet engine. The J34 was a development of the earlier Westinghouse J30 turbojet under USAAF AMC Secret Project MX-825.

Aeronautical engineering genius Barkey and his design team proceeded to wrap a mini-fighter airframe around this powerplant. What they developed was the smallest and lightest fighter airplane in the world, ironically, to protect the largest and heaviest bomber in the world.

McDonnell offered its initial Model 27 Parasite Fighter design for USAAF scrutiny in September 1945. A bit too heavy and too large, it was therefore unacceptable; but refinements could still be made.

McDonnell received the go-ahead to proceed with further development. The engineers ran through a series of designs powered by either the J30 or J34 engine—Models 27A through 27D—before finally settling on Model 27E. McDonnell built a mockup and in June 1946 it was inspected and approved. The USAAF bought it and it would be built.

Under USAAF contract number AC-13496, the manufacture of two XP-85 Parasite Fighter airplanes assigned USAAF serial numbers 46-523 and 46-524 was authorized, and the contract was approved on February 5, 1947. These would be built under a new directive—USAAF AMC Secret Project MX-667 (Project MX-472 was shelved).

Thus, this particular "odd couple," the gigantic B-36 and the diminutive P-85, would be mated. Or would they?

The P-85 was not only small, it was unique. It had an oversize engine nacelle for its fuselage, upward-folding wings (hinged just outboard of the roots), and a small cockpit that sat atop the engine, creating the illusion that its pilot was straddling the engine. It had no landing gear: for emergency landings, it incorporated a heavy-plate landing skid. In its final Model 27E configuration, the P-85 employed five tail planes in a cruciform configuration. On centerline, just forward of the cockpit canopy windscreen, it mounted the retractable hook that attached to the B-36's bomb bay trapeze mechanism during transport and docking; after release from the trapeze, the hook retracted, as did the bomber's trapeze for aerodynamic efficiency for both aircraft. The .50-caliber machine guns on production P-85s were to be placed on either side of the fuselage (two per side) with three hundred rounds of .50-caliber ammunition per gun. No external stores, fuel, or ordnance were considered.

As previously mentioned, the P-85 was to be carried, launched, and retrieved by the B-36 exclusively. As XP-85 flight testing neared, though, no B-36s were either modified with the trapeze mechanism or made available for modification, as B-36 development was running some two years behind schedule. It was therefore decided to modify a Boeing B-29 Superfortress for the XP-85's flight-test phases. This particular B-29, designated EB-29B and nicknamed *Monstro*, received a trapeze mechanism similar to the type to be installed in B-36s for future Parasite Fighter operations; it had to be modified to fit Monstro.

The first of the two XP-85 airplanes (46-523) was completed, less its powerplant, in October 1947. For wind tunnel testing, it was transported to the NACA eighty-by-forty-foot Ames Aeronautical Laboratory wind tunnel at

Ed Schoch practices with the trapeze hookup in the XP/XF-85 mockup cockpit. *NMUSAF*

ABOVE: The complicated trapeze assembly is clearly visible as the XF-85 dangles under Monstro. *NMUSAF*

TOP: Monstro prepares to lift off with XF-85. *NMUSAF*

Naval Air Station (NAS) Moffett Field, south of San Francisco, California. It arrived there on November 9, 1947. As the end of the wind tunnel tests neared, a malfunction of the trapeze bridle, holding the XP-85 hook, caused the plane to fall some forty feet. Although it landed relatively upright, it was damaged enough to need major repair. By this time, though, the number two XP-85 (46-524) had been finished, and it was decided to proceed with flight-testing that plane first, while number one underwent repairs.

The phase would be captive tests, designed to test the trapeze (extension and retraction maneuvers in flight), and the interaction between the two very different aircraft. On June 1, 1948, the number two XP-85 arrived at Muroc AFB via a Boeing C-97 transport airplane.

The USAF aircraft redesignation system put into effect on June 11, 1948, caught up with the XP-85 program, and it became the XF-85.

Meanwhile, McDonnell had selected a former US Navy combat pilot, Edwin F. "Ed" Schoch, to serve as chief engineering test pilot on the XF-85 program. Schoch, a former US Navy Lieutenant Commander and F6F Hellcat combat pilot with four confirmed kills, had joined McDonnell after his enlistment ended. Long before the XF-85's first flight, its future pilot had become heavily involved in the program and even helped with the design of its cockpit and other system functionalities.

Prior to the first free-flight of XF-85 number two, a series of five captive flights were flown. The first was on July 22, 1948, with subsequent flights taking place on July 30 and August 2, 10, and 18. It was now time for phase two: free-flight testing.

Nicknamed Bumble Bee by those closely associated with it, and officially named Goblin by McDonnell, the number two XF-85 was uploaded into *Monstro* on the evening of August 22. The next morning, *Monstro* took off and climbed to twenty-thousand feet over Muroc Lake. After being lowered for release from its trapeze—at a speed of two hundred miles per hour—Schoch started up the J34 beneath him, brought it up to necessary thrust, then pulled the release lever. Separation was smooth and clean. He flew Goblin two away from the mother plane to investigate its basic flying qualities.

Satisfied his ride was flight-worthy, he headed back toward *Monstro* for capture. He failed to dock on his first attempt, and as he closed in on his second attempt, the trapeze's bridle hit the upper part of the cockpit canopy. The canopy shattered, and the bridle knocked Schoch's helmet and oxygen mask off, shocking him temporarily. Before Schoch regained control of the falling plane, it had dropped several thousand feet. He decided to land on the skid rather than attempt another hook-on maneuver; moreover, fuel was becoming critical. He therefore made a good skid-landing on the dry lakebed, touching down at an estimated 210 mph.

Repairs to Goblin two were made and two additional captive flights were flown on October 11 and 13, 1948. On October 14, the second free flight took place. Two more successful retrievals were accomplished on October 15. The fifth free flight and hook-on was attempted on October 22; however, the hook-on failed, and Schoch made another emergency landing.

The number two Goblin was returned to McDonnell's St. Louis facility, where it joined number one, now repaired, for some required aerodynamic modifications and skyhook changes.

By this time, however, the USAF was losing interest in the XF-85 program because it had developed another Parasite Fighter plan (discussed in the previous chapter), which mated the original Republic YF-84F (formerly YF-96A) Thunderstreak (49-2430) to another special trapeze within the belly of a B-36. That program was largely successful and, in fact, the system became operational.

Meanwhile, both Goblins were returned to Muroc AFB for additional testing.

The eighth and ninth captive flights with *Monstro* for XF-85 number two were accomplished on March 8 and 9, 1949. Then, on March 14, XF-85 number one made a true captive-carry flight. All of the improvements made to the Goblin air vehicles in St. Louis appeared to be adequate, and a free flight of XF-85 number two, its sixth, was performed on March 19, 1949.

Unfortunately, due to a bad separation maneuver between the Goblin and *Monstro's* trapeze, Schoch was forced to make his third skid-landing on Muroc Lake. The B-29's trapeze was seriously damaged, and it took three weeks to repair it.

On April 8, 1949, XF-85 number one made its first free flight. Once again, though, the aircraft was unable to hook up with the trapeze for in-flight recovery, leading to yet another skid-landing in the desert.

At this point, the Parasite Fighter program was cancelled. Counting both Goblin aircraft, the program's total free-flight time added up to only two hours and nineteen minutes.

The USAF's revised Parasite Fighter program, called FICON (*FI*ghter *CON*veyor), faired a great deal better, though it, too, was short-lived.

Goblin climbs back up to attach to the trapeze under *Monstro*. NMUSAF

SUMMARY

The McDonnell Aircraft Corporation XP/XF-85 Goblin program was unique, though it was also plagued by developmental problems.

The XF-85 was essentially a flying engine nacelle with wings and horizontal and vertical stabilizers. Its pilot, seated about midship, seemed to ride atop the engine, which was directly beneath him.

The need for air-launched, internally carried, wingtip-towed bomber escort aircraft became moot with the advent of aerial refueling for fighter aircraft, and these almost ludicrous tag-along-with ideas of bomber protection thankfully fell by the wayside.

Finally, since the contractor flight-test program was abbreviated due to malfunctions and other issues, Ed Schoch was the only pilot to fly both Bumble Bee aircraft.

CONVAIR XP-92 AND XP/XF-92A DART

On August 6 and 9, 1945, two Boeing B-29 Superfortress bombers, nicknamed *Enola Gay* and *Bockscar*, each dropped a single atomic bomb on Hiroshima and Nagasaki, two cities in Japan, bringing about V-J Day and the end of World War II. These bombs, *Little Boy* and *Fat Man*, devastated the cities in large nuclear blasts. Just as suddenly as these two cities had been obliterated, the horror of the so-called Nuclear Age had descended upon mankind. Worse, the devastating power of nuclear weapons had become an operational reality in military planning.

Clearly, foreign powers would seek to develop their own nuclear arsenals, and America immediately put plans into action to save its own cities from nuclear attack. In an effort to defend US cities, the USAAF announced a competition for the development of a near-supersonic, point-defense, interceptor-pursuit in late August 1945. One response was Consolidated Vultee Aircraft Corporation's XP-92 Dart program, begun in response to the USAAF's call for an advanced daytime interceptor capable of reaching an altitude of fifty-thousand feet (nine-and-a-half miles) in four minutes, with a maximum speed of seven-hundred miles per hour, or just below the speed of sound (Mn 0.94). This goal was optimistic in the extreme, when we consider that it had been only twenty-six months since man's first supersonic flight in October 1947!

Nevertheless, several airframe contractors pursued the project with the idea of getting a contract to build what promised to be a most radical interceptor aircraft.

In May 1946, two finalists, Convair and Republic, won the right to proceed with their respective offerings. Both firms offered designs that would employ two propulsive systems: turbojet engines and rocket motors.

At Convair's Downey, California, facility alongside Vultee Field, the creation of this unique interceptor was turned over to Chief Engineer Charles R. "Jack" Irvine, Assistant Chief Engineer Frank W. Davis, Chief of Design Adolph Burstein,

and Chief of Aerodynamics Ralph H. Shick. Their work produced a proposal that won a letter of intent to purchase on May 2, 1946, for three experimental airplanes designated XP-92 under USAAF AMC Secret Project MX-813. The USAAF issued serial numbers 46-682 through 46-684 to the three XP-92s, on USAAF contract number AC-14547, dated June 28, 1946.

The Convair XP-92 design of May 2, 1946, Consolidated Vultee Model 115, featured thirty-five-degree sweptback flying surfaces, a V-tail, and the use of a single, non-afterburning, axial-flow 1,365-lbf Westinghouse Electric XJ30-WE-1 (Model X19B) turbojet engine and liquid-fueled rocket motors, with sixteen fuel-injected, helium-pressurized, petroleum-fuel/liquid-oxygen rockets (rated at 50-lbf each) and four helium-pressurized, petroleum fuel/liquid oxygen, 1,500-lbf rocket chambers housed in a Reaction Motors XLR11 rocket motor. The turbojet engine was to be used for takeoffs, cruising, and landings, while the rocket motors were to be used purely for boosted takeoffs and rapid ascensions during the interception of enemy bomber aircraft. The propulsive system for the XP-92 was radical for the era, but it was well received.

XP-92

At this time, it became apparent that two major problems must be isolated and corrected if this program was to move forward: the development of a near-supersonic airframe and of a tri-power propulsive system. In November 1946, the XP-92 contract was amended to address these two problems. One led to extensive laboratory and wind tunnel experimentation in aerodynamics, the other to extensive investigation of ramjet engines and/or petroleum/liquid oxygen-fueled rocket motors. The former program led to the development of Convair's delta wing configuration. In 1947, a full-scale engineering mockup of the proposed XP-92, depicting air vehicle number two (46-683) was unveiled to the USAAF.

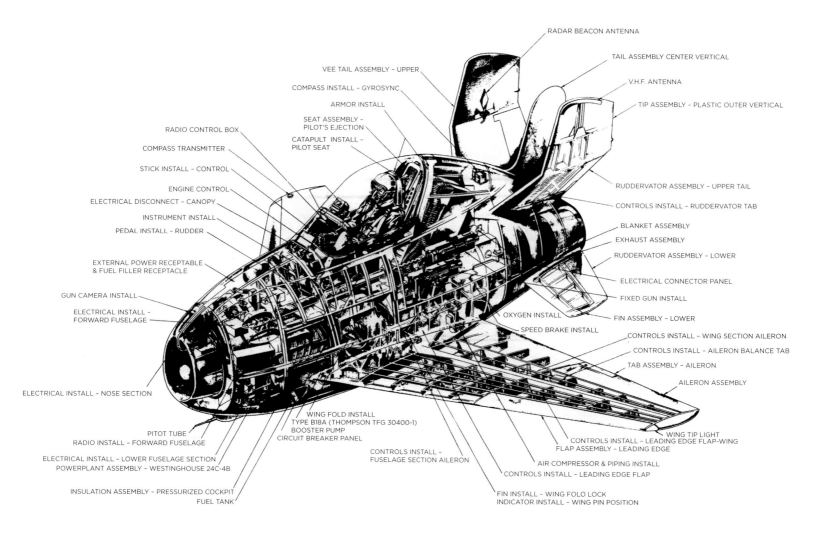

RADAR BEACON ANTENNA

TAIL ASSEMBLY CENTER VERTICAL

VEE TAIL ASSEMBLY – UPPER

V.H.F. ANTENNA

COMPASS INSTALL – GYROSYNC

TIP ASSEMBLY – PLASTIC OUTER VERTICAL

ARMOR INSTALL

SEAT ASSEMBLY –
PILOT'S EJECTION

RADIO CONTROL BOX

CATAPULT INSTALL –
PILOT SEAT

COMPASS TRANSMITTER

RUDDERVATOR ASSEMBLY – UPPER TAIL

STICK INSTALL – CONTROL

CONTROLS INSTALL – RUDDERVATOR TAB

ENGINE CONTROL

BLANKET ASSEMBLY

ELECTRICAL DISCONNECT – CANOPY

EXHAUST ASSEMBLY

INSTRUMENT INSTALL

RUDDERVATOR ASSEMBLY – LOWER

PEDAL INSTALL – RUDDER

ELECTRICAL CONNECTOR PANEL

EXTERNAL POWER RECEPTABLE
& FUEL FILLER RECEPTACLE

FIXED GUN INSTALL

GUN CAMERA INSTALL

OXYGEN INSTALL

FIN ASSEMBLY – LOWER

ELECTRICAL INSTALL –
FORWARD FUSELAGE

SPEED BRAKE INSTALL

CONTROLS INSTALL – WING SECTION AILERON

CONTROLS INSTALL – AILERON BALANCE TAB

TAB ASSEMBLY – AILERON

AILERON ASSEMBLY

ELECTRICAL INSTALL – NOSE SECTION

WING FOLD INSTALL
TYPE B18A (THOMPSON TFG 30400-1)
BOOSTER PUMP
CIRCUIT BREAKER PANEL

WING TIP LIGHT

PITOT TUBE

CONTROLS INSTALL – LEADING EDGE FLAP-WING

RADIO INSTALL – FORWARD FUSELAGE

FLAP ASSEMBLY – LEADING EDGE

ELECTRICAL INSTALL – LOWER FUSELAGE SECTION

CONTROLS INSTALL –
FUSELAGE SECTION AILERON

POWERPLANT ASSEMBLY – WESTINGHOUSE 24C-4B

AIR COMPRESSOR & PIPING INSTALL

CONTROLS INSTALL – LEADING EDGE FLAP

INSULATION ASSEMBLY – PRESSURIZED COCKPIT

FUEL TANK

FIN INSTALL – WING FOLO LOCK
INDICATOR INSTALL – WING PIN POSITION

Detailed cutaway revealing the inner workings of the XF-85 Goblin. *USAF*

During earlier development, still in May 1946, two so-called desk models of the original XP-92 configuration were built and shipped to Wright Field, where the entries were judged by USAAF AMC officers. Afterward, one sweptback-winged model was placed inside Convair's four-foot-long wind tunnel at Vultee Field to obtain rough preliminary aerodynamic data. The results of this test were instantly disappointing: an early report stated: "Tuft tests indicate that wingtip stall begins at a five-degree angle of attack." Lateral control difficulties were found as well, and Convair's engineering staff suspected that an entirely new wing-and-tail group configuration would be needed to meet performance goals. On July 5, 1946, this historic note appeared in one of the test reports: "A sixty-degree sweep-back wing of a delta-shape will be tested this week."

The concept of triangular flying surfaces, especially wings, was not new in mid-1946. Prewar studies by NACA had shown the theoretical high-speed advantages of various wing shapes with extremely low aspect ratios; a delta wing platform provides a very low aspect ratio. However, because propulsive systems for such speed (in the transonic regime) did not exist before World War II, delta wing development in the United States was shelved for possible future applications.

During World War II, the father of Germany's rocket-powered Messerschmitt Me 163 interceptor, Dr. Alexander Martin Lippisch, experimented with a number of delta-shaped flying surfaces. He built a delta wing glider that proved unsuccessful (it was brought to America after the war for USAAF evaluations), and he had conceived the Lippisch P-13a, which evolved into an unpowered development glider, the DM-1. The P-13a design combined a delta wing with ramjet engine power for a theoretical top speed of 1,500 mph—2.0-plus Mn! This airplane was not built by Germany during the war, and its entire data envelope fell into Russian hands after V-E Day.

Convair chief aerodynamicist Adolph Burstein first advocated giving the delta wing a try. Fuselage-to-wing filets were shaped to fill internal angles aft of the original thirty-five-degree sweptback wing on Model 115, creating a triangular wing planform of sorts. Wind tunnel performance of this wing were recorded by NACA Ames on July 12, 1946, as the "best to date."

After learning that Lippisch himself was at Wright Field, Convair's aerodynamics chief Shick arranged a conference with him in late July 1946. Convinced that his engineering group was on the right track, Shick returned to Downey with newfound vigor. Intensive investigation followed on a number of delta-shaped wing planforms, with leading-edge sweepbacks ranging from forty-five to seventy degrees. More than five thousand wind tunnel hours were devoted to the testing of these delta wing configurations.

When Lippisch's delta wing designs were revealed to have been too thick, one of the most outstanding advantages of the delta wing became clear: if thin enough, the wing's relatively low parasite drag allowed for transonic speed. Airfoils show a sharp rise in drag beginning at about 0.80 Mn (600 mph), peaking just above 1.20 Mn (800 mph), then tapering off as the so-called drag hump is left behind. Prepared for the USAAF in November 1946, Convair's studies showed a peak drag coefficient (Cd) of nearly 0.048 on a sixty-degree delta wing, compared with 0.072 Cd on a forty-five-degree sweptback wing of equal gross area. This was a difference of 0.024 Cd that, in aerodynamicist's circles, remains outstanding.

These wind tunnel tests and subsequent studies also proved that the triangular wing was exceptionally stable in thin air (high altitude) and possessed good low-speed handling characteristics, due in part to the absence of a distinct wingtip stall point. The triangular wing's delta shape is inherently strong, thereby permitting a thin structure that is relatively light and rigid. In addition, the area of a delta wing, which is larger than that of a comparable straight or aftward-sweptwing,

XF-85 SPECIFICATIONS

Propulsive system
One axial-flow, non-afterburning, 3,000-lbf Westinghouse J34-WE-22 turbojet engine

Length	14 feet, 10.5 inches
Height	8 feet, 3 inches
Wingspan	21 feet, 1.5 inches
Empty weight	3,740 pounds
Gross weight	5,600 pounds
Maximum speed	600 mph (estimated)
Climb rate	3,000 feet per minute (estimated)
Range	500 miles (estimated)

Armament
Four Colt-Browning .50-caliber M2 heavy machine guns with three hundred rounds per gun

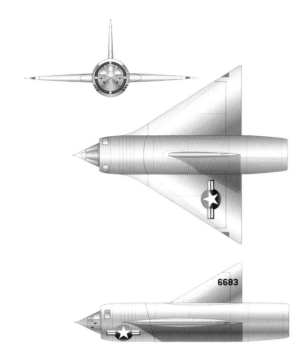

Three-view rendering of one of the original XP-92 configurations highlights its dart-like contours. *G. De Chiara © 2015*

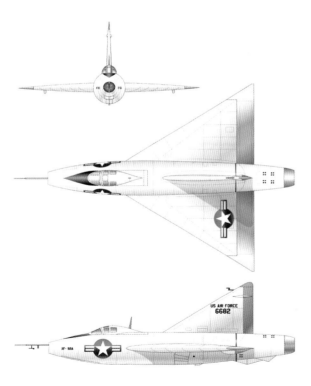

The more refined XF-92A in its final white livery. *G. De Chiara © 2015*

creates greater internal volume for increased fuel capacity. The only real drawback of the delta wing was its poor lift-to-drag ratio at low speed—the opposite effect experienced at high speed.

Generally satisfied with Convair's findings, the approval to construct one experimental sixty-degree delta wing airplane on the XP-92 contract came down from the USAAF in November 1946, using the first of the three airframes on order (46-682). As then stipulated, however, it was to be used for delta wing research only and powered by a single, standard turbojet engine for its evaluation of the delta wing configuration. To save development time and conserve research funds, the amended contract specified its completion "[i]n the shortest time possible using available materials . . . no attempt shall be made to meet existing XP-92 specifications, and existing parts from other aircraft shall be used where possible." This action boosted Republic's XP-91 effort and, essentially, doomed Convair's XP-92 program. In fact, XP-92 airframe numbers two and three were soon cancelled. Only later was this recognized as a blessing in disguise.

To distinguish the one-of-a-kind experimental research airplane from the original XP-92 program, the airplane was designed XP-92A. It was also known as 7-002 (nicknamed *Seven-Balls Two*) because this was the work order number used by Convair's accounting department when the program was initiated.

XP/XF-92A

As it set about constructing the one-off XP-92A, Convair managed to incorporate a nose landing gear assembly from a Bell P-63 King Cobra, the main landing gear assemblies from a North American FJ-1 Fury, an emergency ejection seat assembly from one of its own XP-81 airplanes, and various other hand-me-downs. Full construction on the XP-92A was proceeding well at Vultee Field in the summer of 1947, until Convair ran into financial woes and was forced to close its Downy facility. Convair subsequently moved its entire operation to San Diego, California, alongside Lindbergh Field. The move included the unfinished XP-92A airplane, whose unique airframe, minus the powerplant, was completed in the fall of 1947.

The original sweptwing XP-92 had evolved into a delta wing airplane with an open nose to feed air to its turbojet engine via bifurcated air ducts. The delta wing's root cord extended through almost two-thirds the length of its circular fuselage, which required a radical departure from conventional tail group configurations of the day. The craft had no horizontal tail plane whatsoever, and the vertical tail, like the wing, was delta-shaped, with a rudder on its trailing edge running nearly the fin's full height and above its tip. The functions of the wing ailerons and horizontal stabilizer-mounted elevators were combined in wing-mounted elevons (combined elevators and ailerons), attached to the trailing edges of the wings. The airplane was supported by a tricycle landing gear arrangement.

For full-scale wind tunnel evaluation, the XP-92A was shipped up the West Coast to San Francisco aboard a US Navy cargo ship, then trucked south to the NACA-Ames Research Facility at Naval Air Station Moffett Field in Santa Clara, where it arrived in December 1947. The wind tunnel tests confirmed the results of previous small-scale model tests, and the XP-92A was returned to San Diego aboard the USS Boxer, a USN aircraft carrier.

Unfortunately, during the immediate postwar years, turbojet powerplant development didn't keep pace with airframe development. In early 1947, Convair had anticipated that the Pratt & Whiney J42 (Model JT6)—actually the British Rolls-Royce Nene, built in the United States under license—or the General Electric J33 (Model I-40) would provide 7,000 to 8,000-lbf by the time the XP-92A was ready to fly. However, as the time neared for it to spread its wings, these particular centrifugal-flow turbojets, which the airplane had been designed to use, were producing only nominal power ratings of 4,000 to 5,000-lbf. This was a major concern in light of anticipated performance goals. Instead of being fitted with 7,000-to-8,000-lbf engines, the airplane now had to be fitted with an interim non-afterburning, centrifugal-flow, Allison J33-A-21 turbojet engine, with just 4,600-lbf total power.

Once the powerplant had been installed, the XP-92A was trucked to Muroc AAF on April 1, 1948. The J33-A-21, with its 4,600-lbf total power, didn't even offer enough power to allow for flight-testing and limited the airplane to low-, medium-, and high-speed taxi tests only. It wasn't until September 1948 that the improved 5,200-lbf J33-A-23 was installed for flight-test activities.

Finally, on September18, with Convair Chief Engineering test pilot Ellis D. "Sam" Shannon at the controls, the XF-92A rose up and away from the dry lakebed and made its first flight: an eighteen-minute test hop. Except for lag in hydraulic system response to Shannon's stick and rudder inputs, he reported that its performance was normal. As expected, though, the airplane was woefully underpowered. (The XP-92A was redesignated XF-92A on June 11, 1948.)

Nevertheless, the first piloted delta-winged airplane had flown in the United States. It was another historic event in US aviation's ever-growing chronology of firsts.

Up until the maiden flight of the XF-92A, only a small number of Convair engineers had jumped onto the delta wing bandwagon. Its radical configuration generated little enthusiasm within the ranks of the USAF, but, as its flight test

Color photo revealing the one-of-a-kind XF-92A in its original livery taken in late June 1948, shortly after its completion in San Diego, California. *LM/Code One*

The one-off XF-92A (46-682) lifts off for its maiden flight on September 18, 1948. *AFTC/HO*

Historical Note

The National Advisory Committee for Aeronautics (NACA) ceased to exist in its own right on September 30, 1958, becoming the foundation for the National Aeronautics and Space Administration (NASA), which opened for business the next day.

program proceeded, the Dart was able to attract more and more attention. Test pilot William J. "Bill" Martin was the second and only other Convair pilot to fly it.

On May 16, 1949, USAF contract number AC-14547 was finally approved by the DOD for the single XF-92A.

Original planning called for the retirement of the XF-92A after only fifty hours of flying time by Convair pilots alone. Instead, the USAF subsequently took over to perform a thirty-hour evaluation of its own. Then NACA became interested and appointed A. Scott Crossfield as its XF-92A project pilot.

After the installation of the afterburner-equipped, Allison J33-A-29 turbojet engine (8,200-lbf) in May 1951, Convair, the USAF, and NACA continued flight-testing what had become a heavily instrumented testbed airplane.

Shannon and Martin flew it on a rotating basis for Convair. Test pilots Maj. Charles E. "Chuck" Yeager and Maj. Frank Kendall "Pete" Everest Jr. did most of the early USAF flying on the airplane, while Crossfield performed all of the NACA flights. These five pilots found the XF-92A easy to maneuver and land. More important, they found it to be exceptionally stable in the transonic speed regime—mostly around 0.90 Mn (about 667 mph). The airplane lacked sufficient power to exceed the speed of sound in level-attitude flight, and is known to have exceeded the speed of sound only once, when Maj. Yeager flipped it over onto its back and pulled four g's in a full ninety-degree split-S dive to register 1.10 Mn (about 815 mph) before his pullout.

On December 5, 1950, the XF-92A was being tested by Headquarters Squadron, 3077th Evaluation Group, at Edwards AFB. With Major Yeager seated in the cockpit, the XF-92A crashed, making a belly landing.

Major Yeager was okay, but the XF-92A suffered some damage and had to be repaired before returning to flight.

After it was repaired, the USAF continued testing the XF-92A before it was eventually handed over to NACA (specifically, its High-Speed Research Station at Edwards AFB), for further flight testing by NACA test pilot Crossfield.

During its flying career, the XF-82A was flown 119 times to total some sixty-two flight hours.

SUMMARY

Flight-testing of the XF-92A accumulated 119 flights in five years of flying: forty seven by Convair, forty seven by the USAF, and twenty five by NACA. Thus, from its first flight on September 18, 1948, it racked up some sixty-two hours of flight time. Its last two flights were flown on October 14, 1953; upon landing the second time, the nose landing gear collapsed and the airplane tipped over slightly onto its right side. NACA test pilot Crossfield emerged unharmed, but the XF-92A never flew again, eventually finding its way into the National Museum of the United States Air Force collection, where it can be seen today.

The XF-92A proved the practical application of delta-wing aircraft. So practical were these planes, in fact, that they spawned the F-102 Delta Dagger, F-106 Delta Dart, YF-7A (formerly YF2Y-1) Sea Dart, and the B-58 Hustler (all built by Convair), and others such as the Lockheed YF-12A and the North American XB-70A Valkyrie.

XF-92A SPECIFICATIONS

Propulsive system
One centrifugal-flow, afterburning 8,200-lbf (with afterburning) Allison J33-A-29 turbojet engine (final configuration)

Length	42 feet, 5 inches
Height	17 feet, 8 inches
Wingspan	31 feet, 3 inches
Empty weight	9,100 pounds
Gross takeoff weight	14,600 pounds

Maximum speed
1.10 Mn (815 mph), attained in a dive; 0.90-plus Mn (667-plus mph), attained in level-attitude flight

Maximum Range	600 miles
Maximum Ceiling	45,000 feet
Armament	None

THREE

"MORE'S THE PITY."

But the fact that some geniuses were laughed at does not imply that all who are laughed at are geniuses. They laughed at Columbus, they laughed at Fulton, and they laughed at the Wright brothers. But they also laughed at Bozo the Clown.

— Carl E. Sagan

In 1945 and 1946, the USAAF Air Materiel Command (AMC) spawned many new turbojet- or turboprop-jet-powered bomber and fighter programs in its never-ending quest to field superior combat aircraft. The expanding fleet of USAAF combat aircraft was getting quite expensive for the American taxpayers, but the USAAF itself was not the body paying the enormous bills. As it was then, the American taxpayer still pays for these planes.

The McDonnell XF-88A number two
(46-526) soars above the St. Louis cityscape.
David J. Menard Collection

TABLE 3-1: REDESIGNATION SYSTEM OF USAF AIRCRAFT

Appendix B of Technical Order–(T.O.) 1-1-81, dated June 11, 1948, orders that the prefix "P for pursuit" be changed to "F for fighter." This change affected all the pursuit (now fighter) aircraft in current and future service. This order called for several other changes, as well including PP for Photographic Pursuit became RF for Reconnaissance Fighter, and TP for Trainer Pursuit became TF for Trainer Fighter. This table shows the designation changes associated with the experimental (X) and prototype/service test (Y) aircraft within this reference.

Previous Designation	Current Designation	Comment
XP-59A	not changed	out of service
YP-59A	not changed	out of service
XP-59B	not changed	not continued
XP-79	not changed	not continued
XP-79B	not changed	out of service
XP-80	not changed	out of service
XP-80A	not changed	out of service
XP-80B (P-80R)	not changed	out of service
XFP-80 (XF-14)	not changed	out of service
XFP-80A	not changed	out of service
XP-81	not changed	out of service
YP-81	not changed	not continued
YP-83	XF-83	
XP-84	not changed	out of service
XP-84A	not changed	out of service
YP-84A	not changed	out of service
XP-85	XF-85	
XP-86	XF-86	
YP-86A	YF-86A	
YP-86C	YF-93A	
XP-87 (XA-43)	XF-87	
XP-88	XF-88	
XP-89	XF-89	
XP-90	XF-90	
XP-91	XF-91	
XP-92	not changed	not continued
XP-92A	XF-92A	

McDONNELL XP/XF-88, XF-88A, AND XF-88B VOODOO

By the mid-1940s, turbojet engine development had advanced to the point where powerplant contractors could boast of a lower engine dry weight and specific fuel consumption, higher sea level static thrust ratings, and engine-thrust-to-weight ratios. These claims created a great deal of optimism or, for the first time, optimism that overshadowed an earlier pessimism.

It was with this new belief in progress that prompted the USAAF to sire several new and important turbojet-propelled fighter programs. One program was its revised long-range bomber escort and strike fighter program, which it called the Penetration Fighter program. As discussed in the previous chapters, the original long-range bomber escort and strike fighter program had spawned the Convair XP-81 and Bell XP-83 projects; it had become apparent by late 1945 that neither type would ever succeed.

The USAAF's Penetration Fighter program was created with the idea that an airframe contractor could, with the correct powerplant and design, develop a turbojet-powered, long-range bomber escort and strike fighter that would succeed where the XP-81 and XP-83 had failed. It was to supplement and ultimately replace World War II Lockheed P-38 Lightnings, North American P-51 Mustangs and Republic P-47 Thunderbolts, all of which were piston-powered and propeller-driven.

This head-on view of XF-88 number one reveals its fearful symmetry. *NMUSAF*

The Penetration Fighter program began officially on August 28, 1945, when the USAAF AMC released an invitation to bid (ITB) to the industry. With the ITB came a stringent list of specific operational requirements:

1. Single-place cockpit; bubble-type cockpit canopy
2. Twin-engine arrangement
3. Sweptback flying surfaces
4. Four or six 20mm cannon armament and provision for external ordnance and/or fuel
5. 600 mph maximum speed
6. Ten-minute time-to-climb to 35,000 feet; 40,000-foot service ceiling
7. 900-mile combat range with full combat load

But even before the ink had dried on this list of requirements, the authors of this ITB added more: combat range with full load of ordnance was increased to 1,500 miles; combat range with full load of ordnance was then reduced to six hundred miles; service ceiling would be fifty thousand feet; and time-to-climb was now five minutes, but to fifty thousand feet.

With these fluctuating requirements in mind, McDonnell Aircraft (McAir) began its work on its part of the Penetration Fighter program on April 1, 1946. It assigned Model 36 to the project and began by looking for a suitable powerplant, deciding on a pair of Westinghouse Model 24C (J34 turbojet) engines; the planners then proceeded to design the airframe to house them. It had also investigated the use of wingtip-mounted engines (Model 36A), and a version similar to its US Navy XF2H-1 Banshee (Model 36B), but decided against them.

McAir Chief Engineer Kendall Perkins selected Edward M. "Bud" Flesh to serve as project engineer and David S. "Dave" Lewis as chief of aerodynamics.

On October 13, 1945—less than two months after it had received the ITB—McAir submitted its proposed Model 36C to the AMC at Wright AAF for evaluation. McAir's proposed Penetration Fighter was quite large for a single-seat fighter plane, but its size had been dictated by internal fuselage volume for fuel. The internal fuselage volume was also to provide space for two axial-flow, 3,000-lbf-class Model 24C-4 Westinghouse J34-WE-13 turbojet engines, and room for 1,400 US gallons of fuel. Adequate space was also needed for the AN/APG-30 ranging radar system and six (as was decided) 20mm cannon with their associated ammunition cans, feeds, and mechanisms, not to mention cockpit area, radio equipment, and so on.

The Westinghouse J34 turbojet engine, based on earlier J30 and J32 designs, held a great deal of promise in late 1945, and a number of airframe contractors designed their aircraft around it. Moreover, two more powerful versions of the J34 were also under development—the Model 24C-8 (J40) and the Model 24C-10 (J46). McAir designed its Model 36 to accept any pair of these turbojet engines, but this called for even more internal volume due to the larger sizes of the upcoming J40 and J46 turbojet engines.

After a somewhat lengthy evaluation period, lasting a little over eight months, McAir was awarded a letter of contract on June 20, 1946, for engineering data, wind tunnel models, a full-scale engineering mockup, a static test article, and two flyable airplanes. The USAAF AMC assigned Secret Project MX-811 to McAir's offering and designated it XP-88. McAir received USAAF contract number W33-038-AC-14582, and the airplanes received USAAF serial numbers 45-525 and 46-526.

Of all the Penetration Fighter proposals submitted, only two of them, McAir's and Lockheed's, held enough promise for USAAF recognition. At the same time, Lockheed received a similar letter contract for a pair of XP-90 airplanes (discussed below).

McDonnell's XP-88 mockup, first inspected August 21–23, 1946, was approved. As a result, it received formal contract approval for two XP-88s on February 14, 1947. In the meantime, project engineer Flesh went forward with the afterburning J34 turbojet engine program. He asked Westinghouse and several other powerplant contractors if they were going to be able to provide a fifty-two-inch-long (required for aircraft rotation and takeoff clearance parameters), maximum-length, non-liquid injection-type afterburner section for application on the J34 engine, which would give the second airplane (now designated XF-88A, Model 36D) better takeoff, climb, and high-speed performance than the non-afterburner-equipped XP-88. Flesh never got an adequate response.

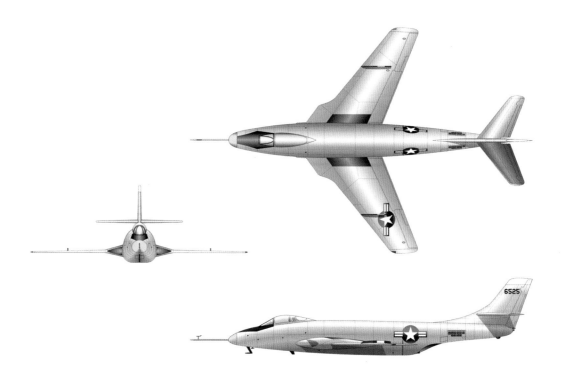

McAir was forced to undertake its own efforts at short-section J34 afterburner development. This created what was known in-house as the MAC Short Afterburner—a device only thirty inches long!

Lockheed was forced to do the same, since a J34 without afterburning would produce about 3,200 lbf. With afterburning, though, the J34's output would increase to 4,200-lbf; that is, 8,400-lbf via two afterburning J34 turbojet engines.

On September 18, 1947, and June 11, 1948, respectively, the USAAF became the USAF (US Air Force) and the "P for pursuit" prefix was changed to "F for fighter," changing the XP-88 to the XF-88.

On June 16–19, 1948, the AMC's 689 Engineering Board of Inspection of the XF-88 was held. The USAF was pleased with McAir's XF-88 airplane, especially with its overall effort in showing its many F-88 configurations (including a two-seat all-weather fighter interceptor) and numerous others. McDonnell showed how its basic Model 36 airframe could easily adapt to most any other mission requirement.

By this time, in keeping with the McAir tradition of naming its combat aircraft after spooky entities, the XF-88 was named Voodoo.

XP/XF-88

Now complete, the first XF-88 rolled out on August 11, 1948, at McAir's Lambert Field facility in St. Louis, Missouri. It was subsequently delivered to the High Speed Test Center (HSTC) at Muroc AFB, where a series of preflight tests were conducted.

Under the guidance of McDonnell test pilot Robert M. "Bob" Edholm, the maiden flight of the XF-88 Voodoo occurred on October 20, 1948. As had been expected, with only a maximum of 6,400-lbf of non-afterburning power from its two J34-WE-13 turbojet engines, Edholm's biggest squawk was its lack of performance. Further, he noted, even with the projected addition of 2,000-lbf via the MAC Short Afterburner sections, the top speed of the aircraft was not much above 1.0 Mn. Nevertheless, flight testing continued.

After McDonnell had completed its phase one flight test obligation, the USAF took over for phase two. It proceeded to fly XF-88 number one seventeen times between March 15 and March 26, 1949, accumulating seventeen hours and fifty seven minutes of flight time, or about one hour per flight. Still, the XF-88 was underpowered, although it was deemed easy to fly and quite maneuverable.

XF-88A

The number two airplane, designated XF-88A with its pair of J34-WE-15 turbojet engines (without the afterburner sections being installed as yet) and other changes rolled out on April 1, 1949. It made its first flight on April 26,

After installation of the MAC Short afterburner sections onto its two J34 engines, the XF-88 became the XF-88A (McAir Model 36D). Propulsive system engineer Richard G. Laucher, a McDonnell employee, designed and patented the MAC Short afterburner section. His patent filing was entitled *Afterburner for Turbojet Engines and the Like*; filed on April 9, 1948, it was issued on May 12, 1953. Here, McAir technicians install one of the two J34-WE-15s with a MAC Short Afterburner attached to it. *SDASM Archives*

1949, out of Lambert Field, with Edholm at the controls. Among its several refinements, the number two Voodoo featured a variable-geometry horizontal tail plane and was armed with six M39 20mm cannon (number one was never armed). Both aircraft performed almost trouble-free throughout their respective flight-test programs.

During the seventeenth test hop on XF-88 number one on May 12, 1949, Edholm took it up to an altitude of forty-two thousand feet and leveled off prior to an all-out assault on supersonic speed. Powered by its non-afterburning J34-WE-13 engines, Edholm entered into a straight-down ninety-degree split-S dive. As he passed through thirty-two thousand feet, he recorded a top speed of 1.18 Mn. Then, after pulling out of his dive at seventeen thousand feet, he landed safely, and became an early member of the then-exclusive "Mach One Club." Edholm noted soon after the flight that, when the airplane entered into the transonic speed regime, through the supersonic speed regime, and back to subsonic speed, the plane did not buffet. The flight had been smooth throughout, and the plane's aerodynamics were exceptional.

With a MAC Short Afterburner section installed on the left-hand (port) engine with only a fixed exhaust nozzle, the first in-flight afterburner operation with the XF-88A occurred on June 9, 1949. That test was a success, and an afterburner section was installed on the right-hand (starboard) engine as well.

Meanwhile, negotiations had been underway for installation of a Model 501F-1 Allison 2,500, estimated-shaft horsepower (eshp) XT38-A-5 turbopropjet engine in the nose of the number one Voodoo for a series of supersonic propeller demonstration flights. Thus, on July 15, 1949, McAir was awarded a USAF contract (AF-7442) to make the conversion to the XF-88B Trimotor airplane under USAF Project MX-1100. McDonnell Aircraft applied Model 36J to the XF-88B project. For this conversion, the number one Voodoo was removed from flight status after it had flown ninety times. Flight test evaluations on the XF-88A continued.

On November 9, 1949, the XF-88A made a crash landing on the Muroc AFB dry lakebed. It was returned to St. Louis for repairs, after which it flew again on March 27, 1950, out of Lambert Field. It subsequently returned to Edwards AFB to prepare for the upcoming Penetration Fighter fly-off competition. But on June 16, 1950, XF-88A number two (piloted by USAF Maj. Frank K. "Pete" Everest Jr.) made a second emergency gear-up belly landing, due to an engine fire at Edwards AFB. To meet the fly-off competition, Voodoo number one was substituted.

While in storage in St. Louis and awaiting its conversion into the XF-88B, MAC Short Afterburner sections had been installed on its new J34-WE-15 turbojet engines, and this aircraft was also given the XF-88A designation. In its new configuration, it made its first flight on May 1, 1950, out of Lambert Field, after which it was ferried to Edwards AFB on May 22 (Muroc AFB had been renamed Edwards AFB on January 27, 1950).

Then, between June 29 and July 7, 1950, the fly-off competition between the McDonnell XF-88A, Lockheed XF-90A, and the North American YF-93A was held. The three airframe contractors then waited for the USAF's decision. (The YF-93A was a late entry into the Penetration Fighter competition because the USAF thought it might be suitable for its requirements.)

In its damaged state, XF-88A number two arrived by truck in St. Louis on July 11, 1950, followed by XF-88A number one on August 3, which had flown home.

A letter from USAF AMC on September 11, 1950, notified McAir that its XF-88A Voodoo had been ranked number one of the three Penetration Fighter contestants. However, the program had been cancelled. It was nice to know of their win, but, without an F-88 production contract forthcoming, the announcement held little value.

Meanwhile, McAir had placed both XF-88As into flyable storage; the number one airplane would still become the XF-88B.

Believing its XF-88 was a sound design, McAir kept designing numerous derivatives. This effort ultimately paid off in the follow-on F-101 Voodoo program.

A full-scale engineering mockup of Lockheed TDN L-188 (YF-94) outside the Skunk Works (Advanced Development Projects) building in Burbank, California. The lightning bolt and triple red stripes were not incorporated in the actual design for the YF-94. *LM via Denny Lombard*

XF-88/-88A SPECIFICATIONS

Propulsive system
Two non-afterburning J34-WE-13 turbojet engines replaced by two MAC Short Afterburner-equipped J34-WE-15 turbojet engines

Length	54 feet, 1.5 inches
Height	17 feet, 3 inches
Wingspan	39 feet, 8 inches
Empty weight	12,140 pounds
Gross weight	18,500 pounds
Maximum speed	1.18 Mn attained
Climb rate	2,413 feet per minute
Maximum Range	1,700 miles

Armament
Six 20mm M39 cannon (XF-88A number two only)

XF-88B

Now designated XF-88B, the number one Voodoo made its first flight out of Lambert Field on March 14, 1953, with McDonnell engineering test pilot Phillip W. "Phil" Houghton at the controls. Testing for NACA lasted until 1957, during which time a number of supersonic propeller types had been mounted to the Allison XT38 engine. The highest speed attained was 1.12 Mn in a power dive, one of the fastest speeds—if not *the* fastest speed—ever recorded on a propeller-driven airplane flight (its two turbojet engines had been shut down for the test).

Unfortunately, neither the XF-88A nor the XF-88B airplanes survived, as they both were eventually scrapped.

SUMMARY

Like its XF-90A and YF-93A contemporaries, the XF-88A fell short of USAF requirements. Although the XF-88A was declared the winner of the Penetration Fighter competition, this didn't matter: the USAF had no further need for such a fighter, and the F-88 program was cancelled. Instead of a Penetration Fighter, the USAF now wanted a "Strategic Fighter." It was this follow-on competition that led to McDonnell's F-101 Voodoo program, which was met by the USAF with open arms.

LOCKHEED YF-94, YF-94B, YF-94C (YF-97A), AND YF-94D STARFIRE

The Lockheed Aircraft Corporation F-94 series of aircraft were built to primarily serve as nighttime (all-weather) fighter interceptors.

With the failure of the F-87 Blackhawk and the delay of the F-89 Scorpion, the USAF was hard-pressed to fill its immediate need for all-weather fighter interceptors.

Since the Lockheed T-33A (TF-80C before June 11, 1948) was already in production—and, as a two-seater, it had adequate room for the required electronic equipment and weapons—this plane became a natural choice to fill in as a stopgap.

Pilot Tony LeVier and Flight Test Engineer Glenn Fulkerson cruise during the YF-94's first flight. A second modified TF-80C (48-373) became YF-94 number two. *LM via Denny Lombard*

YF-94 (ETF-80C, ET-33A)

Two two-place, tandem-seat TF-80C airplanes (USAF serial numbers 48-356 and 48-357) were used as the service test aircraft for the F-94 program. The first of these (48-356) was initially flight-tested on April 16, 1949, by Tony LeVier, with Lockheed flight test engineer Glenn Fulkerson in the back seat.

These TF-80C airplanes, as service test F-94s, were initially designated ETF-80C. After the TF-80C was redesignated T-33A, they became ET-33As. They were never officially designated YF-94, but are recognized as such in the record.

YF-94A AND YF-94B

On July 1, 1949, the first production F-94A made its first flight. The second production F-94A-5-LO (USAF serial number 49-2497) was selected to serve as the service test YF-94B under USAF contract number AF-1847, which was approved on August 26, 1949.

YF-94C (YF-97A)

A single YF-94C (formerly YF-97A) was ordered on USAF contract number AF-11205 approved on July 21, 1950 (USAF serial number 50-955). It was created from the last production F-94B-1-LO.

Tony LeVier made a successful first flight on the YF-97A on January 18, 1950, at Edwards AFB. Its designation was changed to YF-94C on September 12, 1950. The F-94C was the first version of the F-94 to *officially* bear the name Starfire on an official basis.

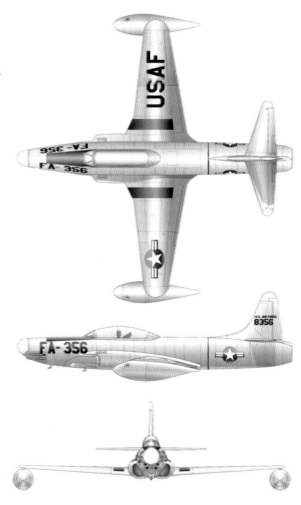

The first of two YF-94 prototypes as it appeared on its first flight on April 16, 1948, in three-view. *G. De Chiara © 2015*

ABOVE: Two YF-94C airplanes were created from modified F-84Bs. *NMUSAF*

RIGHT: Generic model of a B-53 superimposed over a photograph to show how it would appear flying above the clouds. One of the earliest attempts to field an airplane with forward-swept wings, the B-53 offered a configuration with advantages and disadvantages that were not yet fully understood. It was to be powered by three non-afterburning 4,000-lbf General Electric J35 turbojet engines mounted side by side in its aft fuselage section. *NMUSAF*

YF-94D

The F-94D was to be a dedicated fighter-bomber version of the F-94C fighter-interceptor, with tremendous firepower for ground target strafing. Features would have included an eight-gun nose, and Vulcan cannon.

SUMMARY

The F-94 Starfire series of aircraft worked in unison with the F-89 Scorpion series until the advent of dedicated all-weather, all-missile and/or all-rocket-armed interceptor aircraft.

CONVAIR XB-53

XB-53 (XA-44)

The Convair XB-53 was originally designed as an attack aircraft with the designation XA-44. The aircraft was unique in that it was to have a thirty-degree forward sweptwing with an eight-degree dihedral (upward slant). Convair engineers used forward-sweptwing research data that had been captured from Germany in World War II as the basis for the design. The wing was mounted at the far aft portion of the fuselage, also serving as the horizontal stabilizer. The design had a conventional vertical stabilizer and rudder, but changes in pitch and roll were accomplished with wing-mounted control surfaces. The elevators were on the inboard wing, while the ailerons were outboard. The wingtips were also designed as variable incidence control surfaces.

The XB-53 design called for three General Electric J35 turbojet engines of 4,000-lbf each, mounted side by side within the aft fuselage. Estimates indicated a top speed of approximately 580 mph. Classified as a medium bomber, the XB-53 would have carried up to twelve thousand pounds of bombs. It was also designed to carry forty high velocity aerial rockets (HVAR) mounted on underwing pylons.

Convair diverted funding initially allotted to the XB-46 program to work on the XA-44/XB-53 project. Two aircraft were ordered (USAF serial numbers 45-59583 and 45-59584) but neither were completed before the XB-53 program was cancelled.

SUMMARY

The proposed forward-sweptwing Convair XB-53 was originally designed as an attack aircraft, with the designation XA-44. None of the proposed USAF turbojet powered aircraft—the XA-43 (XP/XF-87), XA-44 (XB-53), and XA-45 (XB-51)—were ever built.

The USAF offered no funding for a dedicated jet-powered attack aircraft until much later, with its Ling-Temco-Vought A-7D Corsair II (a development of the US Navy A-7 program) and the Fairchild-Republic A-10A Thunderbolt II. Other than the A-7 and A-10, no other USAF attack types were built.

Three-view of the proposed Convair XB-53 (XA-44), illustrating its dynamic forward-sweptwing construction.
G. De Chiara © 2015

XB-53
SPECIFICATIONS

Propulsive system
Three non-afterburning, axial-flow, 4,000-lbf General Electric J35 turbojet engines

Length	70 feet, 5 inches
Height	23 feet, 9 inches
Wingspan	85 feet, 0 inches
Wing area	1,370 square feet
Wing dihedral	8 degrees (upward angle)
Empty weight	31,760 pounds (estimated)
Gross weight	60,000 pounds (estimated)
Maximum speed	580 mph (estimated)
Maximum Range	2,000 miles (estimated)

Armament
XA-44, twenty .50-caliber machine guns, forty 5-inch-diameter high velocity aerial rockets (HVAR) housed in pods (twenty HVARs in each pod) mounted on underwing pylons; XB-53, two tail-mounted .50-caliber machine guns or 20mm cannon

Payload
2,000 pounds (estimated)

Crew	Four

REPUBLIC XP/XF-91, XF-91A, AND XF-91B THUNDERCEPTOR

Shortly after VJ-Day and the end of World War II, the USAAF Air Materiel Command closed its ledger books on numerous aircraft programs. Yet, due to the overriding importance of some of these programs, several survived the mass cutback.

One of the surviving aircraft programs, Secret Project MX-808, was intended to field a high-speed, high-altitude day fighter-interceptor capable of "meeting and defeating any high-speed, high-altitude bomber aircraft that any potential adversary might produce," for the exclusive purpose of defending US airspace.

To this end, the USAAF sent out an ITB to the industry in December 1945. With the ITB came the following specific operational requirements (SORs):

- 25.5 minutes of combat duration in a number of steps
- 2.5 minutes time-to-climb to 47,500 feet (19,000 feet per minute)
- Fifteen minutes' cruise time duration at 456 knots indicated air speed (kias), or 525 miles per hour
- Three minutes' combat duration at an average airspeed of 688 kias, or 790 miles per hour
- Five minutes' descent time from 47,500 feet to landing

This demanding SOR calls for an all-out performance aircraft. Bidding airframe contractors opted to employ a combination turbojet/rocket motor propulsive system. They decided that, since airborne loiter time was not a parameter, their designs should be propelled in this fashion.

Color view of the second XF-91 (46-681), revealing its strange contours after V-tail installation. *COAM*

XF-91 number two fuels up before takeoff. *NMUSAF*

The preliminary design team at Republic, headed by Georgian-born chief designer/engineer Alexander Kartveli, proposed its Model AP-31, a unique design featuring a dual-propulsive system comprised of a single afterburning, axial-flow, 5,000-to-7,950-lbf (with afterburning) General Electric J47-GE-9 (Model TG-190) turbojet engine and a single four-chamber 7,820-lbf total (two chambers providing 4,000-lbf, 2,000 each; two chambers providing 3,820-lbf, 1,910 each) Curtiss Wright XLR27-CW-1 rocket motor. (Republic had designed an afterburner section that added 33 percent more power in this period.)

The exotic AP-31 offering also featured sweptback flying surfaces with variable incidence wing and horizontal stabilizers, which allowed a high angle of attack for takeoffs and landings, and a low angle of attack for high-speed, high-altitude flight. Its unique wing featured inverse taper, whereby it was wider at the tip than the root. This design was intended to reduce wingtip stall due to the loss of lift that is normal with wings that are tapered conventionally—that is, wider at the root than at the tip. Since its wing was also thicker at the tip than at the root, the main landing gear retracted outward into the wing rather than inward into the wing.

Republic received authorization via teletype on March 22, 1946, to proceed with its Model AP-31 program and to construct a full-scale engineering mockup. Then, just three months after its XP-84 Thunderjet had made its first flight (February 28, 1946), on May 29, 1946, Republic was awarded a contract (USAAF contract number AC-14583) to produce engineering data and drawings, wind tunnel models, a static test article, and two experimental airplanes designated XP-91 for flight testing. The full-scale engineering mockup was inspected in May 1947, and in January 1949 the engineering inspection was held.

Under USAF AMC Project MX-809, the DOD approved USAF contract number AC-14583 on January 27, 1949, for two experimental XF-91-RE airplanes (no longer XP-91). On February 24, 1949, the first example rolled out of the RAC's Farmingdale factory. It was subsequently shipped to Muroc AFB to

Republic test pilot Carl Bellinger sits in the cockpit of XF-91 number one. *NMUSAF*

undergo preparations for its scheduled flight-test program. Carl A. Bellinger was the chief experimental test pilot assigned to the project. Then, with Bellinger at the controls on May 9, 1949, XF-91 number one (46-680) made a successful first flight with an interim non-afterburning J47-GE-7 turbojet engine. The airplane handled so well, Bellinger remained airborne for a full forty minutes instead of the expected ten to fifteen minutes. He was impressed with the Thunderceptor, the official name it had been given by this time.

The first flight of XF-91 number one—using the RAC-designed and -built afterburner section—took place on October 17, 1949. The test pilot enjoyed a reported 33 percent increase in total thrust with the use of the RAC afterburner, and level attitude flight on turbojet power alone (with afterburner) produced a top speed of 0.90 Mn (667 miles per hour).

Three days earlier, on October 14, 1949, the second XF-91 (46-681) made its first flight with Bellinger at the controls.

During a later test flight of the second XF-91 (46-681), on June 30, 1951, an engine malfunction just ninety seconds into the takeoff roll caused the tail section to erupt into flames. Test pilot Bellinger was not aware of the situation until chase pilot Major Yeager informed him of it. This gave Bellinger enough time to abort the takeoff, come to a stop, and escape the burning airplane safely. The tail section was burned off by the time the firemen and firefighting equipment arrived on scene. Bellinger reported afterward that Yeager had saved his life that day.

At the outset, the second of the two XF-91 airplanes was identical to XF-91 number one, but this was to change with the late addition of its "butterfly" or V-type tail assembly installed after the fire that destroyed its original tail assembly.

Difficulties with delivery of the Curtiss Wright 7,820-lbf XLR27-CW-1 rocket motor led RAC to change to a single Reaction Motors 6,000-lbf four-chamber (1,500-lbf from each chamber) XLR11-RM-9 rocket motor.

The XLR27-CW-1 was developed under USAAF AMC Secret Project MX-1512 while the XLR11-RM-1 (the first version of what became the -9) was developed under the MX-1555 designator.

As had been announced by Headquarters Air Force Flight Test Center, on December 9, 1952, the XF-91 became the first US combat-type aircraft to go supersonic in level attitude flight. With Republic test pilot Russell Morgan "Rusty" Roth at the controls, the first of the two XF-91 airplanes climbed to an altitude of thirty-five thousand feet before leveling off for its speed run. Roth first turned on the J47's afterburner to boost its thrust to maximum, then switched on the rocket motor: full thrust. The airplane, already cruising straight and level at more than 600 mph, leapt forward and shot past the speed of sound to register 1.07 Mn (793 mph).

XF-91 number one as it appears today at the National Museum of the United States Air Force. Note the large-capacity external drop tanks—each holds sixty gallons jet petroleum fuel, 218 gallons of liquid oxygen (LOX), and 265 gallons of water/alcohol (WALC). *NMUSAF*

Roth said after the flight, "The plane we fly is a production-type airplane that takes off under its own power, uses a rocket motor to achieve supersonic speed, and lands under its own power."

The president of the Republic Aviation Corporation, Mundy Ingalls Peale, said, "It may be considered as important in the future as was the step from the reciprocating engines to jet power."

Peale added that, although the XF-91 was technically still an experimental machine, it "is a combat-ready plane. It is not a purely research plane."

XP/XF-91A

Republic also planned a XF-91A version with a 5,200-pound thrust J47-GE-21 and a single XLR11-RM-9 rocket motor (totaling 6,000 pounds of thrust) that would reach 1,126 mph (1.7 Mn) at fifty thousand feet, but this configuration was never tested.

The XF-91A (RAC Model AP-31N-1) was a proposed version of the earlier XF-91 Thunderstreak to be considered in the 1954 Interceptor

The pair of Thunderceptor airplanes parked side by side reveal their respective final configurations. On the left, XF-91 number one (46-680) shows its radar nose (now XF-91A) while XF-91 on the right (46-681) shows its V-tail. *AFTC/HO*

XF-91 SPECIFICATIONS

Propulsive system
One afterburning, axial-flow, 7,950-lbf (with afterburner) General Electric J47-GE-9 turbojet engine; one Reaction Motors four-chamber 6,000-lbf (1,500-lbf each chamber) XLR11-RM-9 rocket motor

Length	43.3 feet
Height	18.1 feet
Wingspan	31.3 feet
Empty weight	14,140 pounds
Gross weight	28,300 pounds

Maximum speed
985 miles per hour at 47,500 feet, maximum power with engine afterburning and rocket motor full thrust

Ceiling	48,700 feet

Armament
Four 20mm cannon (200 rounds ammunition each)

program competition under USAF ARDC Project MX-1658. Other proposals, such as the Convair YF-102 and Republic's own XF-103, moved to the head of the pack and further development of the XF-91A was terminated.

SUMMARY

The Republic Aircraft Corporation XF-91 Thunderceptor was unique. It was the first non-research combat-type airplane to exceed the speed of sound in level attitude flight, albeit with the addition of rocket power.

It was also unique for its inverse taper wings (thicker at their tips than their roots), variable incidence wings, combined turbojet and rocket motor propulsive system, and for its investigation of a V-type tail.

Although no production order was forthcoming for any F-91 airplanes, the Thunderceptor proved to be a significant learning tool for future combat aircraft designs.

MARTIN XB-51 (XA-45)

What began life as four-engine turbopropjet- and turbojet-powered attack aircraft, designated XA-45, evolved into a three-engine turbojet-powered medium-class bombardment aircraft, designated XB-51 under US Air Force Air Materiel Command Project MX-838.

With Martin chief test pilot Orville Edwin "Pat" Tibbs at the controls, the premier XB-51 made a successful first flight from Middle River, Maryland, on October 28, 1949. Nineteen days later, on November 15, 1949, USAF contract number AC-14806 was approved for the two experimental XB-51 airplanes, USAF serial numbers 46-685 and 46-686. The second of the two XB-51 airplanes made its first flight on April 17, 1950. Both examples were soon ferried cross-country to Edwards AFB in California.

The experimental XB-51 medium-class, low-level attack bomber was unique in that it was powered by an uneven number of turbojet engines (three), whereas all other jet bombers came with even sets of two, four, six, or eight turbojet engines. Since its three engines were non-afterburning, axial-flow, 5,200-lbf General Electric J47-GE-13 turbojet engines, the XB-51 was capable of reaching a level attitude top speed of more than six hundred miles per hour at sea level.

The specifics are identical for both of the XB-51 aircraft. They measured eight-five feet one inch in length and seventeen feet four inches high, with a wingspan of fifty-three feet one inch. Their maximum speed and altitude was 645 mph and 40,500 feet respectively, and they cruised at 532 mph, with a maximum range of 1,600 miles. Their operational armament was to consist of eight nose-mounted 20mm cannon, and they were to carry either 10,400 pounds of bombs or eight five-inch high-velocity aerial rockets (HVARs) internally on a rotary launcher. Like the Boeing B-47 Stratojet and the Martin XB-48, the

ABOVE: Both XB-51s take off in formation. *NMUSAF*

BELOW: The first of two XB-51 airplanes (46-685)—Martin Model 234—shown in three-view with its original factory nose art. *G. De Chiara © 2015*

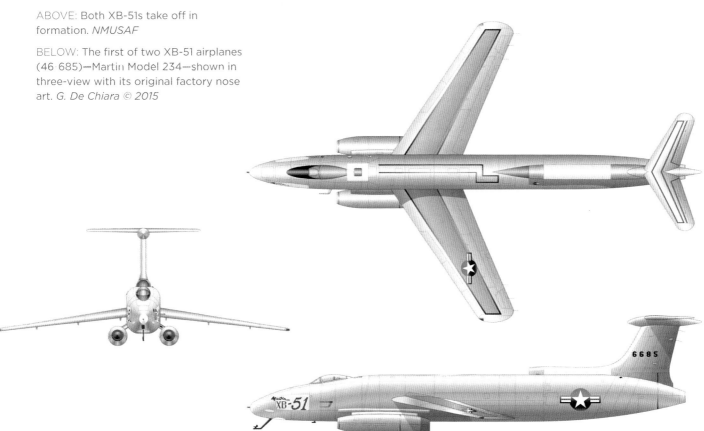

XB-51 number one prepares to land against a cotton-cloud backdrop. *AFTC/HO*

XB-51 number one cuts across the sky. Originally meant to be an experimental attack-type designated XA-45, the first flight of XB-51 number one took place on October 28, 1949. *AFTC/HO*

XB-51 likewise featured a bicycle-type landing gear arrangement with outrigger wheels/tires for taxiing balance on the ground. Their gross takeoff weight was 55,930 pounds. Production B-51s would have a two-man crew, with a pilot and bombardier/navigator. Both its wing and horizontal tail plane featured variable incidence. All flying surfaces were swept back to alleviate compressibility and to increase its speed.

In the end, the Martin Model 234 (first known as the XA-45 and finally the XB-51) became one the most thoroughly tested USAF aircraft that never went into production. Both examples were operated by numerous contractor and government pilots, copilots, bombardier/navigators, and flight test engineers throughout their lengthy tenures at Edwards and elsewhere.

For all that effort, neither example has survived. On May 9, 1952, the second XB-51 (46-686) crashed during high-speed, low-level maneuvers about one quarter mile east of Edwards AFB; its pilot, USAF Maj. Neil Hill Lathrop, was killed. Some four years later, on March 25, 1956, the first XB-51 (46-685) crashed on takeoff at El Paso International Airport, killing both crewmembers in

XF-90 number one (46-687) soars during a flight test near Muroc Lake. *NMUSAF*

the ensuing fire; the pilot was USAF Maj. James O. Rudolph, and the flight test engineer was USAF Staff Sgt. Wilbur R. Savage. The cause was attributed to a premature rotation on takeoff and an unrecoverable stall.

LOCKHEED XP/XF-90 AND XF-90A

The Lockheed Aircraft Corporation XP/XF-90 offered direct competition to the McDonnell XP/XF-88 during the USAAF/USAF Penetration Fighter competition. The North American YF-93A was a late entry, but it also participated.

Lockheed investigated numerous XP-90 configurations under several temporary design numbers (TDN), which included TDN L-153, L-167, and L-169. In the end, the TDN L-153 was built.

Lockheed XP-90 proposal number one called for two afterburning, axial-flow, 4,900-lbf (with afterburning) Westinghouse XJ34-WE-32 US Navy-type turbojet engines. Number two (alternate) featured two afterburning, axial-flow, 6,100-lbf (with afterburning) Westinghouse XJ46-WE-2 USN-type turbojet engines. Neither of these engines was ever used.

Famed test pilot Tony LeVier smiles for the camera after what appears to be just another leisurely test hop on the number one XF-90. *LM*

XP/XF-90

On April 11, 1946, Lockheed received teletype authorization to proceed with its long-range fighter program. On June 5, 1946, it received a letter contract to purchase two XP-90 airplanes under USAAF AMC Secret Project MX-812; USAAF contract number AC-14563.

Lockheed offered its TDN L-169-1 design on December 20, 1946.

A full-scale engineering XP-90 mockup was built, and the USAAF gave a preliminary inspection on June 1, 1947. The full and final USAAF 689 Engineering Board inspection of the XF-90 (as designated after June 11, 1948) mockup was completed in April 1949. Following the engineering inspection, the order for two XF-90 airplanes was finalized under USAAF AMC Secret Project MX-812; USAAF contract number AC-14563. Approved by the DOD on January 25, 1949, the planes had already been issued USAAF serial numbers 46-687 and 46-688. The estimated first flight date was June 1948.

The first XF-90 (46-687) was taken up on its first flight on June 3, 1949. The second XF-90 (46-688) made its first flight at the newly named Edwards AFB in March 1950.

XF-90A

Both XF-90 airplanes were redesignated XF-90A after they were refitted with afterburner-equipped, 4,200-lbf (with afterburning) J34-WE-15 turbojet engines.

The maximum speed attained in level attitude flight by an XF-90A was 0.90 Mn (668 mph) with the afterburning J34 engines installed. XF-90A number one achieved a maximum speed of 1.12 Mn (830.4 mph) on May 17, 1950, during one of the last of fifteen dive tests flown by Tony LeVier with its afterburning J34-WE-15 engines.

On May 24 and June 28, 1950, the first of the two XF-90A airplanes (46-687) was involved in two separate mishaps. The first of these was a landing accident where the nose landing gear collapsed at Edwards AFB with USAF test pilot Capt. John C. Newman at the controls. The second accident, also at Edwards AFB, saw Lockheed test pilot Thomas V. "Tom" Gibson at the helm of XF-90A number one when an engine fire caused him to make a forced belly landing.

According to the USAF in an Air Intelligence Review released on July 5, 1949:

> The new Lockheed XF-90 jet penetration fighter is a heavyweight slugger combining speed, range and maneuverability. Designed to fly and fight deep within enemy territory, seeking out specific military objectives or targets of opportunity in the air and on the ground, the XF-90 is a sweptwing, needle-nosed airplane using two turbo-jet engines relying on speed and maneuverability to avoid trouble.
>
> Although it is a single-seat fighter, the XF-90 is almost as heavy as the familiar DC-3 (C-47) transport plane. Its appearance presents a striking departure from earlier Lockheed aircraft. The 40-foot wings sweep back from the fuselage at a streamlined 35-degree angle, with rudder and elevator surfaces swept back comparably. The single cockpit is located well ahead of the wings for visibility. Into it were built many cockpit installation features developed by the Joint Cockpit Standardization Committee, with the general arrangement coming in for intensive study by pilots and technicians.
>
> Every precaution was taken during the design of the XF-90 to make it as invulnerable as possible to battle damage, since the nature of its mission requires it to operate far behind enemy lines where it would be subject to heavy punishment. Simplified maintenance was highly sought-after and problems of producibility were also taken into account in the design of the XF-90, with the result that in spite of its heavy structure the airplane readily lends itself to assembly-line production.
>
> Extensive flight test data was obtained through use of six steel and plastic models of the XF-90, which were dropped from high-

Three-view of XF-90 number one (46-687) with its distinctive wingtip fuel tanks. *G. De Chiara © 2015*

The first of two XF-90s heads skyward on an early test hop circa July 1949. *LM/Code One*

XF-91 SPECIFICATIONS

Propulsive system
Two non-afterburning axial-flow 3,000-lbf Westinghouse Electric XJ34-WE-11 turbojet engines (Model 24C); later two afterburning 4,100-lbf, maximum power, J34-WE-15 turbojet engines (Model 24C)

Length	56.2 feet
Height	15.8 feet
Wingspan	40.0 feet
Empty weight	
Gross weight	

Maximum speed
665 mph-normal; 1.12 Mn (attained) in a steep dive

Combat Ceiling	39,000 feet
Range	2,300 miles

Armament
Six belly-mounted (three on either side of centerline) 20 millimeter cannon, eight 5-inch HVARs or two 1,000-pound bombs

flying aircraft and studied by radar and radio-recording as they plummeted.

The XF-90 has conventional landing gear and incorporates a pilot ejection seat. It is equipped with cabin pressurization and air-conditioning.

SUMMARY

With the limited power available, the XF-90 proved to be too heavy and too slow, though its sleek design was deceptive. Tony LeVier was forced to point its nose straight down in near ninety-degree full-power dives to make it go fast—this was unacceptable.

In a letter to the author, LeVier states:

> The XF-90's general performance was very poor. It wasn't a hell of a lot better with afterburner. The plane would have had a good chance for success had it had good engines, which it didn't. The aircraft was the only one in the competition that met USAF specifications in regard to structural strength [it was stressed for 12.5g], so it was much heavier than the other two contenders. The XF-90 was the just the third [combat fighter-type] plane in the world to dive supersonic. The plane was also strong as hell, you could not overstress it in flight.

With two afterburning, 4,100-lbf J34-WE-15s (providing 8,200-lbf maximum power in afterburner), its final propulsive system didn't offer enough power to propel this airplane to supersonic speed in level attitude flight; it simply weighed too much, at some twenty-seven thousand pounds during a flight test. In fact, just to get off the ground,

Tony LeVier pulled off a perfect emergency belly landing on Frenchman Flat north of Mercury, Nevada. The aircraft—XF-90 number two (46-688)—was blown apart during several nuclear bomb tests in 1952, and the remnants of the airplane remained there until 2003, when the Air Force museum recovered its remains for future restoration. *NMUSAF*

it often had to use four 1,000-lbf Aerojet General twelve-second-duration assisted takeoff (ATO) units.

During flight testing, LeVier dove the airplane on fifteen different occasions; he ultimately hit a maximum speed of 1.12 Mn on May 17, 1950. To address its inadequate performance, Lockheed offered up several F-90 derivatives: (1) Model 190-33-02, powered by two afterburning Allison J33-A-29 turbojet engines; (2) Model 290-34-03, powered by two afterburning Westinghouse Electric 6,100-lbf J46-WE-2 turbojet engines or two 4,900-lbf Westinghouse J46-WE-32; and (3) Model 390-35-02, powered by a single afterburning General Electric J47-GE-21 turbojet engine. The USAF rejected these proposals and cancelled the F-90 program.

Only one XF-90, the second, survives today. It awaits restoration at the National Museum of the United States Air Force in Dayton, Ohio. It was rescued in 2003 from Frenchman Flat in Nevada, where it had been exposed to fallout from nuclear weapons testing. The first example of this plane was purposely destroyed during static structural strength tests at the NASA facility in Cleveland, Ohio.

NORTH AMERICAN YF-93A (YP-86C)

During fiscal year 1946 (FY46), the USAAF had a trio of turbojet-powered pursuit types in full-scale production: the Lockheed P-80 Shooting Star, the Republic P-84 Thunderjet, and the North American P-86 Sabre. The USAAF's Air Materiel Command had established requirements for a number of new pursuit types. These included an advanced day interceptor, an all-weather (night) pursuit, a parasite pursuit, and a penetration pursuit.

These requirements led to the orders for fourteen experimental pursuit aircraft during FY46 as follows: two McDonnell XP-85 parasite pursuits; one Curtiss XP-87 all-weather pursuit; two McDonnell XP-88 penetration

YF-93 number one (48-317) banks left during its first flight on January 24, 1950. *NMUSAF*

Formerly designated YP-86C, the prototype YF-93A was investigated as both an all-weather (nighttime) fighter and a long-range penetration fighter. This excellent top-down view shows YF-93A after its NACA/NASA engine air inlets were replaced with conventional cheek-type inlets. *NMUSAF*

pursuits; two Northrop XP-89 all-weather pursuits; two Lockheed XP-90 penetration pursuits; two Republic XP-91 interceptor pursuits; and three Convair XP-92 interceptor pursuits.

Moreover, additional P-80, P-84, and P-86 production orders were forthcoming. It was a banner year for airframe and powerplant contractors involved with the development and production of turbojet-powered pursuit-type aircraft. Powerplant contractors were happier still with the successes enjoyed by airframe contractors, who continued to develop their production aircraft to address USAAF AMC requirements as they were published.

North American Aviation Inc. in particular had proceeded with the development of its P-86 Sabre to meet two of the FY46 requirements set forth by the USAAF. Without major redesign, NAA felt, the P-86s airframe could be modified to match two specification proposals. For this reason, NAA's preliminary design organization was tasked with producing an all-weather interceptor (AWI) and a penetration pursuit, both of which would be based on the P-86 airframe, especially its propulsive system. The penetration pursuit is the subject of this section.

Although the specifications established by the AMC for the penetration pursuit program called for two turbojet engines, NAA opted to design its proposal with just one, as the P-86 airframe didn't offer enough room for two engines. The single turbojet engine they selected was the upcoming centrifugal-flow Pratt & Whitney Model JT7, which had been designated J48 under USAAF AMC Secret Project MX-1447, a British design that would be produced in the United States under license. Actually an Americanized Rolls-Royce R.Ta.1 Tay, the J48 was to have an afterburner section and produce as much as 8,000-lbf with afterburning. This was at least 2,000-lbf more power than two Westinghouse J34s would produce for the XP-88 and XP-90 penetration pursuits being developed by McDonnell and Lockheed, respectively. What's more, one J48 would weigh considerably less than two J34s, and, of course, use less fuel—so precious to the penetration pursuit mission. Since centrifugal-flow

Some exterior details of the NACA (now NASA) flush engine air inlet. *NMUSAF*

turbojet engines have larger diameters than axial-flow engines, a deeper and wider fuselage was called for than was used in the P-86A. With the required fuselage redesign to accommodate the promising J48 Tay engine in mind, the redesign began.

To kill two birds with one stone, NAA engineered fuselage cheek-type engine air inlets so that the nose of the aircraft could be free to house all-weather radar and electronics. In this way, the aircraft could be developed for penetration pursuit and all-weather interceptor roles at the same time. Suddenly, NAA's new P-86 variant would have a new airframe *and* a new powerplant. It was therefore given a new NAA Charge Number, NA-157, and was designated YP-86C in-house, as it still had the wings and tail planes of the P-86A. (The proposed P-86B, NA-152, of which 190 examples were to be built, had been cancelled and transferred onto the P-86A contract.)

With its project 8,000-lbf (in afterburner) power, lighter weight, all-weather adaption, six nose-mounted 20mm cannon armament, and advertised speed of 700-plus mph in level flight, NAA proposed its NA-157 (a C version of the P-86) to the newly established US Air Force on September 20, 1947. It was offered as either a penetration pursuit or an all-weather interceptor. After a relatively short evaluation period, the USAF ordered 120 examples—two service test airplanes and 118 production aircraft—on December 17, 1947 (USAF contract number AC-21672).

A three-view drawing depicting the first of the two YF-93 examples built (48-317). The plane appears in its original configuration, with the NACA flush-type engine air inlets. *G. De Chiara © 2015*

Since the type differed from the basic P-86 line, it received a new designation. The two service test airplanes were designated YP-93A, and the production aircraft were designated P-93A. At this time, McDonnell and Lockheed had been developing their XP-88 and XP-90 penetration pursuit prototypes for some eighteen months. It was a late start for NAA's entry, although it was being developed as a dual-role aircraft—even though the P-93 was specifically ordered as a penetration pursuit. The author can only assume, therefore, that the P-93 was only ordered in case of the unlikely failure of all all-weather interceptor and penetration pursuit prototypes. In fact, this worst-case scenario nearly came to pass as, among all of the prototypes ordered during FY46, only the XP-89 went on to become a production airplane. Coincidentally, another version of the Sabre, the YP-86D (formerly YP-95A), went on to serve as the P-89's all-weather interceptor teammate.

At the same time, NACA (now NASA) had engineered a flush-type engine air inlet system to provide improved airflow for turbojet engines, specifically centrifugal-flow types. Since NAA wanted to have cheek-type engine air inlets for the solid nose configuration on its YP-93A, it decided to incorporate the flush-type recessed engine air inlets for two reasons: better airflow to the J48 engine and reduced parasite drag. But this decision proved a mistake.

Lockheed employed the flush-type engine air inlets on its modified XP-80B, redesignated P-80R and named *Racey*. This was the airplane that USAAF Col. Albert G. "Al" Boyd flew in October 1946 in an attempt to establish a world speed record. After the flush-type inlets were replaced with more conventional P-80 style air inlets (albeit enlarged), Boyd recaptured the word speed record for the United States on June 19, 1947. By this time, however, the YP-93A design was frozen and NAA proceeded to develop the airplane with NACA flush-type inlets. The reason for this turn of events remains unclear, because NAA was well aware of Lockheed's flush inlet experience.

On June 11, 1948, based on an earlier USAF request, an order from the US Department of Defense specified that the "P for pursuit" prefix be changed to "F for fighter" and the YP-93A became the YF-93A.

In late November 1949, the first of two YF-93A airplanes was completed in Inglewood, California, and it was subsequently trucked to Muroc AFB for preflight ground tests before its maiden flight. With NAA test pilot George S. Welch at the controls, the airplane made a successful first flight on January 24, 1950. By this time, both of McDonnell's XF-88s were flying, as were both of Lockheed's XF-90s. North American's YF-93A, therefore, it was flying into a rather strong "head wind."

(There is a minor discrepancy in the date of this airplane's first flight: January 24 or 25, 1950. The USAF claims it was on January 24, while the Air Force Flight Center History Office at Edwards AFB claims January 25.)

Subsequent test hops, especially where high angles of attack were flown, showed that the flush-type engine air inlets starved the J48 engine and caused compressor stalls and flameouts.

It was then decided to complete YF-93A number two with larger cheek-type air inlets, with the goal of eliminating the problem of engine air starvation. The new inlet system worked and, for NACA evaluation, YF-93A number one retained its flush inlets. It was later determined by NACA that, while the flush inlets reduced drag, the reduction of air flow to the engine was unacceptable.

While the two YF-93As were being developed and flight-tested exclusively for the penetration fighter role, the USAF planned to procure one example as an all-weather fighter. It therefore ordered one such airplane under NAA Charge Number NA-166, designated F-93A on May 5, 1949 (USAF contract number AC-21672), amended, now excluding the previous order for 118 production F-93As under the penetration fighter criterion. Shortly thereafter, though, due to the success of NAA's own F-86D version of the Sabre, Lockheed's F-94 Starfire, and Northrop's F-89 Scorpion, the contract for the one-of-a-kind two-seat F-93A all-weather fighter was abruptly cancelled.

The USAF contract for the pair of YF-93A airplanes (USAF serial numbers 48-317 and 48-318) was approved by the DOD on February 25, 1949.

In the end, both YF-93A airplanes were turned over to NACA and received NACA aircraft numbers 139 and 151, respectively, for its flight test programs with the type. YF-93A number one was scrapped out in 1956, and YF-93A number two was last seen at the NACA aircraft storage yard on Naval Air Station Moffett Field, California, in June 1956, where it had been partially stripped.

YF-93A SPECIFICATIONS

Propulsive system
One afterburning 8,120-lbf (with afterburning) Pratt & Whitney J48-P-1 centrifugal flow turbojet engine (Pratt & Whitney Model JT-7)

Length	44 feet, 1 inch
Height	15 feet, 8 inches
Wingspan	38 feet, 9 inches
Empty weight	14,035 pounds
Gross takeoff weight	26,516 pounds
Maximum speed	708 mph (attained)
Combat Ceiling	45,500 feet

Climb rate
11,960 feet per minute (maximum power)

Maximum Range	2,000 miles

Armament
Six 20mm cannon (1,380 rounds of 20mm ammunition, 230 rounds each; provision for two thousand-pound bombs, or eight five-inch T-200 HVARs or sixteen five-inch T-38 HVARs)

YF-93A/F-93A
PRODUCTION BREAKDOWN

YF-93A:	2-48-317 and 48-318	118 additional cancelled
F-93A:	48-49-001 to 49-048	cancelled
F-93A:	9-49-059 to 49-068	cancelled

SUMMARY

Both the North American YF-93A and the Lockheed XF-90/-90A entries had come up short in the USAF penetration fighter competition when the USAF announced on August 15, 1950, that the XF-88/-88A entry from McDonnell was the best of the lot. Thus the two YF-93A airplanes became surplus to the USAF. They were, however, still of value, and NACA took them under its wing during early fall 1950 for further flight test evaluations at its facility on Edwards AFB.

BOEING XB-52, YB-52, AND XRB-52 STRATOFORTRESS

The Boeing Airplane Company B-52 Stratofortress is one of the most important aircraft ever produced. First flown in April 1952, the B-52 entered service on June 29, 1955, and remains in operation today.

The B-52 was put into service with the US Air Force Strategic Air Command (SAC) as a heavy-class strategic bombardment aircraft under WS-101A and WS-101L, intended for pure bombardment and reconnaissance bombardment, respectively.

The Stratofortress was designed and built under USAF AMC Secret Project MX-839 to replace the piston-powered, turbojet-augmented Convair B-36 series of heavy-class strategic bombardment aircraft.

XB-52 AND YB-52

One experimental XB-52 and one prototype YB-52 were ordered under USAAF letter of intent to purchase contract number W33 (038) AC-15065, which was approved by the DOD on November 17, 1948. The planes were issued USAAF serial numbers 49-230 and 49-231.

The XB-52 climbs during a test flight out of Boeing's Plant 2 facility adjacent to Boeing Field in Seattle, Washington. *NMUSAF*

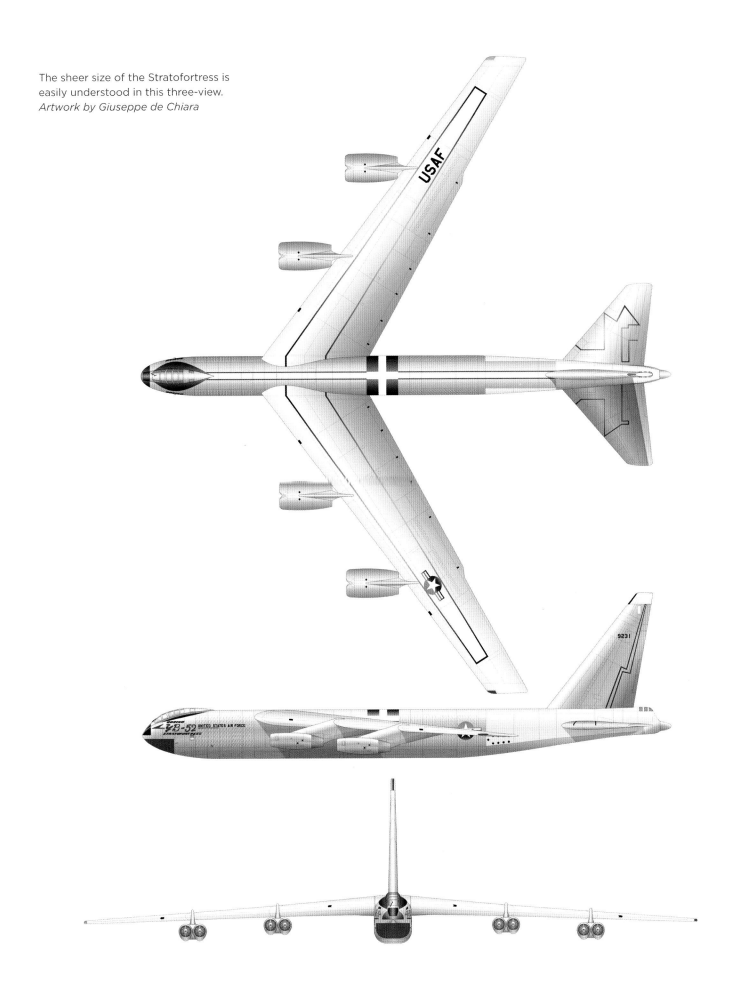

The sheer size of the Stratofortress is easily understood in this three-view.
Artwork by Giuseppe de Chiara

The XB-52 makes its maiden voyage on October 2, 1952. *NMUSAF*

The XB-52 passed through numerous preliminary design stages before the final configuration was decided.

The model 464-67 with eight Pratt & Whitney YJ57-P-3 engines and 330,000 pounds proved to be the final choice for the designs of the XB-52 and YB-52 airplanes.

SAC Commander-in-Chief Gen. Curtis Emerson LeMay profoundly disliked the B-47 style in-tandem pilot/copilot seating arrangement on these aircraft, leading him to demand side-by-side seating on production aircraft. Beginning with the B-52A variant, all future B-52s had side-by-side pilot/copilot seating.

The experimental XB-52 was secretly rolled out of the Seattle Plant 2 factory on November 29, 1951, and was moved into the company-owned Boeing Flight Center hanger across State Highway 99 on Boeing Field.

During preparations for its first flight, the XB-52's pneumatic system failed during a full-pressure test outside the Boeing Developmental Center hangar. The resulting blowout severely damaged the wing trailing edge, which required that the aircraft be moved back into the Plant 2 production facility for extensive and lengthy repairs.

At the same time, the prototype YB-52 had been finished and moved over to Boeing Field on March 15, 1952. Since the XB-52 was still being repaired, the YB-52 would fly first: on April 15, 1952, with Boeing chief engineering test pilot Alvin Melvin "Tex" Johnston and USAF Air Research and Development Command Lt. Col. (later Brig. Gen.) Guy Mannering Townsend III at the controls, the YB-52 made a two-hour, fifty-one-minute shakedown flight and landed at Boeing's flight test facility at Moses Lake, in eastern Washington, where it would undergo further test flights.

The first of two prototype B-52H aircraft rolled out with four dummy Douglas GAM-87 Skybolt air-launched ballistic missiles (ALBM) attached. Boeing eventually produced another 100 production B-52Hs. *NMUSAF*

Finally, on October 2, 1952, after nearly a year's delay due to extensive modifications and repairs, the experimental XB-52 took flight. Again with Johnston at the helm, it was flown for two-plus hours before it too landed at Moses Lake.

Following contractor test flights at Moses Lake, both aircraft were ferried to Edwards AFB for general USAF evaluations. The experimental XB-52 was

XB-52 AND YB-52 DESIGN STAGES

June 28, 1946:	462-six Wright T35-W-1	360,000 pounds
January 7, 1947:	464-22-four Wright T35-W-3	480,000 pounds
December 8, 1947:	464-35-four Wright T35-W-5	280,000 pounds
May 1948:	464-eight Westinghouse Electric J40-WE-6	USAAF asked for a study on the use of J40 turbojet engines as interim propulsion until the availability of Pratt & Whitney J57 engines
July 1948:	464-40-Boeing proposal to USAAF using eight interim J40-WE-6 engines	280,000 pounds
October 21, 1948:	464-49-eight Pratt & Whitney YJ57-P-1	330,000 pounds
October 24, 1948:	464-49-eight Pratt & Whitney YJ57-P-3	330,000 pounds
March 29, 1950:	464-67-eight Pratt & Whitney XJ57-P-1	390,000 pounds
October 6, 1950:	464-67-eight Pratt & Whitney J57-P-1/-3	390,000 pounds without fuel for assisted takeoff (ATO) rocket motor unit(s)

XB/YB-52 SPECIFICATIONS

Propulsive system	Eight axial-flow, non-afterburning, 8,700-lbf Pratt & Whitney J57-P-3 turbojet engines
Length	152 feet, 7 inches
Height	48 feet, 2.5 inches
Wingspan	185 feet, 0 inches
Empty weight	155,200 pounds
Gross takeoff weight	390,000 pounds
Maximum speed	611.5 mph
Combat range	7,014 miles with ten-thousand-pound payload
Combat Ceiling	46,500 feet
Armament	Two .50-caliber M2 heavy machine guns in tail-mounted turret with six hundred rounds ammunition each (not installed)
Payload	25,000 pounds

brought to the Wright Air Development Center (WADC) on Wright-Patterson AFB in Dayton, Ohio, in 1957, where it was used as a Pratt & Whitney J75 turbojet engine testbed aircraft. On January 28, 1958, the prototype YB-52 was donated to the US Air Force Museum on Wright-Patterson AFB.

On October 22, 1965, President Lyndon B. Johnson signed the National Highway Beautification Act into law. A pet project of his wife, "Lady Bird" Johnson, the goal of this act was to remove unsightly signs and metallic derelicts from motorists' view on US highways, to replacing such eyesores, she said, "with flower beds." The parked XB-52 and YB-52 aircraft, potentially seen as "junk" by drivers passing near Wright-Patterson AFB on the adjacent freeway, were ordered scrapped.

XRB-52

No experimental XRB-52 airplanes were ordered or built, though the type was studied under USAAF AMC Secret Project MX-1608. Numerous B-52B and B-52C airplanes were modified to serve as RB-52s, designated RB-52B and RB-52C, respectively, under WS-101L. This precluded the need to acquire any experimental XRB-52 aircraft.

SUMMARY

In addition to the XB-52 and YB-52 airplanes, seven production versions of the Stratofortress were built: the B-52A through the B-52H, culminating in 744 airplanes built. The last Stratofortress built, a B-52H (USAF serial number 61-0040), rolled off the assembly line on October 26, 1962.

No X or Y prefixes were applied to the B-52A through B-52H airplanes, but several examples of each variant were used for service test activities.

Apart from the B-52H, all versions of the Stratofortress have been retired. A total of 102 B-52Hs were built, with the first example making its first flight on July 10, 1960. The B-52H first entered service with USAF SAC on May 9, 1961, and it remains on duty with the Air Force Global Strike Command.

Several newer bombers were built to replace this airplane—but none of them ever could. Moreover, since most of its current pilots are in the twenty- to thirty-year-old age range, the B-52H is much older than they are. In addition, when fitted for underwing stores, the B-52H can carry payloads equal to or greater than any other bomber in the Air Force Global Strike Command.

Nicknamed "BUFF" (for "Big Ugly Fat Fella"), the B-52H Stratofortress continues to receive modifications and is expected to soldier on until 2044. If this holds true, the B-52H will be eighty-two years old at its retirement, at which point it will be the longest-lived combat aircraft in history.

The first of three preproduction B-52A aircraft rolls out of the Plant 2 factory in Seattle, Washington. These three B-52A aircraft were used extensively as service test airplanes; they were not considered production aircraft. *PMB Collection*

BOEING XB-55

The Boeing Airplane Company's XB-55 program was short lived to say the least: it came and went in a mere six months. Officially initiated on July 1, 1948, the program was cancelled on January 27, 1949, by order of USAF AMC while it was still in its design phase.

Appearing more as a throwback, the XB-55 program had begun in October 1947, two months before the first flight of the prototype B-47—the very plane it was supposed to replace.

Boeing beat out twelve other airframe contractors and was awarded a letter of contract on July 1, 1948, to proceed with its Model 474 offering, designated XB-55 under USAAF AMC Project MX-1022.

Boeing offered its Model 474 to the USAF as a follow-on to the B-47, with proposed features such as the capability of achieving long-distance flight at a speed near 500 mph and the use of more economical turbopropjet engines instead of the much thirstier turbojet engines.

The Boeing XB-55 was essentially a new bomber design with sweptback flying surfaces loosely related to its B-47 and upcoming B-52 designs but with four turbopropjet engines rather than six or eight. The XB-55 program

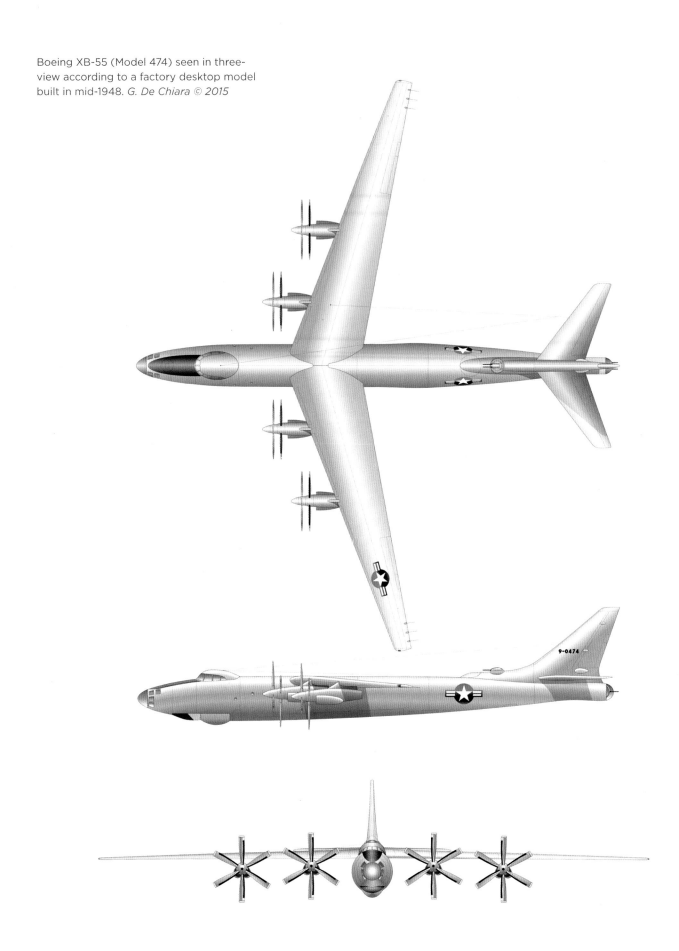

Boeing XB-55 (Model 474) seen in three-view according to a factory desktop model built in mid-1948. *G. De Chiara © 2015*

proceeded even though these better all-turbojet-powered bombers were on the near horizon. Nevertheless, the XB-55 program was soon cancelled while it was still in its design phase, prior to any orders for the type; no USAF serial numbers were issued.

Oddly, one version of the XB-55 (Boeing Model 479) featured a six-jet propulsive system whereby the six axial-flow Westinghouse J40 turbojet engines were arranged identically to the B-47, with two inboard and one outboard, under either wing. This did not serve USAF's purposes, and the turbopropjet-powered version went forward.

SUMMARY

The USAF never did field a turbopropjet-powered bombardment aircraft in any weight class, whether light, medium, or heavy. When aerial refueling became commonplace in the late 1940s and early 1950s, the need for slower, longer ranging turbopropjet-powered bombers became unnecessary, and subsonic turbojet-powered bombers whose speed was 100 to 150 mph faster became status quo.

Interestingly, the Russian Air Force fielded its own large turbopropjet powered strategic bomber, the Tupolev Tu-95 Bear. The venerable Tu-95 remains in service today and is expected to continue flying until the year 2040. Powered by four fourteen-thousand-shaft horsepower engines spinning dual contra-rotating propellers, its top speed is 575 mph with an unrefueled range of 9,400 miles. The Bear (its NATO code name) is an amazing aircraft, and it has been produced in many variants. It is almost famous because of its testing of US air defenses on many occasions while flying off the coast of Alaska, with countless intercepts since the mid-1950s.

XB-55 SPECIFICATIONS

Propulsive system
Four axial-flow, 5,600-shaft-horsepower Allison T40-A-2 turbopropjet engines (Model 500), each spinning two contra-rotating, fourteen-foot-diameter propellers

Length	118 feet, 11 inches
Height	33 feet, 8 inches
Wingspan	135 feet, 0 inches
Gross weight	168,000 pounds
Maximum speed	490 mph
Maximum range	5,000 miles
Payload	24,000 pounds of bombs
Armament	12 20mm cannon
Crew	Ten

The first of two prototype YB-60 airplanes cruises during its first flight on April 18, 1952. *NMUSAF*

The YB-60 prepares to touch down during sunset in this rare color photograph at Edwards AFB. *AFTC/HO*

YB-60 SPECIFICATIONS

Propulsive system
Eight axial-flow, non-afterburning, 8,700-lbf (maximum power) Pratt & Whitney YJ57-P-3 turbojet engines

Length	175 feet, 2 inches
Height	50 feet, 4 inches
Wingspan	206 feet, 4 inches
Gross takeoff weight	153,016 pounds

Maximum speed
508 mph at 39,250 feet and maximum power

Maximum range (unrefueled)	6,191miles
Ceiling	44,650 feet
Payload	72,000 pounds

Armament
Ten 20mm cannon with 3,600 total rounds of ammunition

Crew	Nine

Had it been built and put into service, the Boeing B-55 would most likely have evolved into a similar aircraft. We will never know, for the B-55 program was obsolescent before it had a chance to become passé.

CONVAIR YB-60

The Convair B-60 was the Boeing B-52's only direct competition (albeit unofficial) and, like the B-52, this aircraft came with eight podded, axial-flow, non-afterburning 8,700-lbf (maximum power) Pratt & Whitney YJ57-P-3 turbojet engines. Its design was initiated on March 5, 1951.

Originally designated YB-36G-1-CF and YB-36G-5-CF, the two airplanes bore Convair Model 6. No known MX- or WS-number was issued by Wright Air Development Center (WADC).

As originally proposed by Convair to the USAF, the aircraft were to be modified production Block 1 and Block 5 B-36F airplanes with sweptback flying surfaces and facilities to mount either turbojet engines or turbopropjet engines. The remainder of the aircraft would be all B-36F, including the flight deck, complete with its unsightly cockpit canopy.

On November 2, 1949, the DOD approved USAF contract number AF-2182 for the manufacture and flight test of two service test YB-60 airplanes. The first example (USAF serial number 49-2676) was designated YB-60-1-CF (formerly B-36F-1-CF), while number two (USAF serial number 49-2684) was designated YB-60-5-CF (formerly B-36F-5-CF).

YB-60

The first YB-60 was rolled out of the Fort Worth, Texas, factory on April 6, 1952. It then underwent numerous systems checks and engine run-ups prior to taxi tests and first flight. Convair chief engineering test pilot Beryl A. Erickson (Fort Worth Division) was in the left seat when YB-60 number one finally made its first flight on April 18, 1952, three days after the YB-52 had flown for the first time. According to the USAF, its first flight was expected to occur in November 1951, but the actual first flight had slipped some five months due to the late delivery of YJ57-P-3 engines; these didn't arrive until April 6, 1952.

Subsequent flight tests demonstrated engine surge, control system buffeting, rudder flutter, and electrical engine-control system problems. Moreover, at its demonstrated 508 mph maximum speed, the YB-60 was some 142 mph slower than the 650 mph speeds demonstrated by both the YB-52 and XB-52.

SUMMARY

The YB-60 wasn't the first turbojet-powered version of the B-36 series. From the B-36D on, all production B-36s were built with four axial-flow 5,200-lbf (maximum thrust) General Electric J47-GE-19 turbojet engines, with two mounted side by side in nacelles under either wingtip, to help speed the Peacemaker along.

Both YB-60 airplanes were completed, but only number one entered flight-testing. It was flown twenty times (totaling sixty-six hours) by Convair and another four flights (totaling fifteen hours) by the USAF. The YB-60 program was cancelled on August 14, 1952, but flight-testing was allowed to continue until this program was also cancelled on January 20, 1953. The USAF didn't accept the YB-60s until June 24, 1954, at which time it ordered them to a scrap yard to partially recoup the total program cost of $14,366,022 million. Both of the YB-60 airplanes were stripped of their engines and other usable equipment and then scrapped-out during the summer of 1954.

Convair had offered the USAF serious plans to evaluate four turbopropjet engines on one of the YB-60s, in the unlikely case the J57 engine proved unworthy of production, but these never moved forward.

As it turned out, the Boeing B-52 Stratofortress was a much better choice for the USAF SAC. A more aerodynamic, considerably faster aircraft, it was more capable of carrying out its primary mission: strategic bombardment.

Three-view revealing the YB-60 as it appeared on its first flight.
G. De Chiara © 2015

The handsome B-57A number one flies toward the camera in this top-down shot. *NMUSAF*

MARTIN B-57A CANBERRA

The airplane produced for the Glenn L. Martin Company's B-57 tactical bombardment airplane program for the USAF Tactical Air Command was unique, beginning its life not as an American design but as the B.2 Canberra, which was built by English Electric Company Ltd. in Great Britain and operated by the Royal Air Force (RAF).

By order of the British Ministry of Defense (MOD), the RAF loaned one B.2 Canberra (RAF serial number WD932) to serve as pattern and/or prototype aircraft for the B-57 program, on loan for six months. The plane arrived at Andrews AFB, Maryland, on February 24, 1951. This occasion marked the first time that a turbojet-powered airplane had crossed the Atlantic Ocean nonstop without refueling. The airplane took off from Aldergrove, Northern Ireland, and landed at Gander, Newfoundland, piloted by Roland Beamont. He then flew it from Gander to Andrews AFB after refueling.

Martin requested a second B.2 Canberra (RAF serial number WD940) in June 1951. The airplane arrived on August 31, 1951, after setting a transatlantic speed record from Aldergrove to Gander in four hours, eighteen minutes; its average speed during this flight was 481.12 mph.

B-57A

On September 3, 1953, USAF contract number AF-22617 was approved by the DOD for two B-57A Canberra airplanes (USAF serial numbers 51-17352 and 51-17387). No experimental (X) or service test (Y) B-57s were generated for this series, thanks to the two B.2 Canberra pattern/prototype aircraft used by the USAF, Martin, and their subcontractors. The first and subsequent seven Martin-built Canberras were therefore designated B-57A. All eight of these B-57As were used as service test aircraft, however, and none of them was ever placed into USAF service.

The "B for Bomber" Missile and Rocket Fiasco

Beginning in the mid-1950s and from 1960 to September 18, 1962, the US Department of Defense had designated USAF unmanned missiles and rockets as "bombardment-type air vehicles." These erroneous "B for bomber" designations included the following craft:

- Northrop B-62 Snark
- Bell B-63 Rascal
- North American B-64 Navajo
- Convair B-65 Atlas
- Martin B-68 Titan (not to be confused with the Martin XB-68 Tactical Bomber)
- McDonnell B-72 Quail
- Fairchild B-73 Bull Moose
- Douglas B-75 Thor
- North American B-77 Hound Dog
- Boeing B-80 Minuteman
- Martin B-83 Bullpup
- Douglas B-87 Skybolt

(The designations B-88 through B-99 and higher were not used.)

After this designation debacle, all new bomber designations were to begin with Designation Number 1A.

The only two "real bomber" invaders into this missile/rocket mess were the Douglas B-66 Destroyer and the Lockheed B-69. The designation B-69 was applied to seven specially built, composite-powered (using both piston and turbojet engines) planes: the Lockheed RB-69A special reconnaissance aircraft, used in clandestine ELINT (*EL*ectronic *INT*elligence) gathering operations. These seven airplanes were former US Navy P2V-7U Neptune maritime aircraft on loan to the USAF.

The designation B-71 was reserved for the Lockheed RB-12/RS-12, which represented two other proposed versions of the Lockheed A-12 Mach 3-plus reconnaissance aircraft. Lockheed had proposed its RB-12/RS-12 as an alternative to the proposed (but never built) North American XRB-70/XRS-70 Valkyrie. The RB-12/RS-12 designs were offered as both the RB-71 and the RS-71, although these designations were never officially adopted.

The designation B-72 was to be applied to the Convair NX-2 (Model 54) nuclear-powered bomber, but this never materialized. The number 72 was ultimately assigned to the McDonnell GAM-72 Quail.

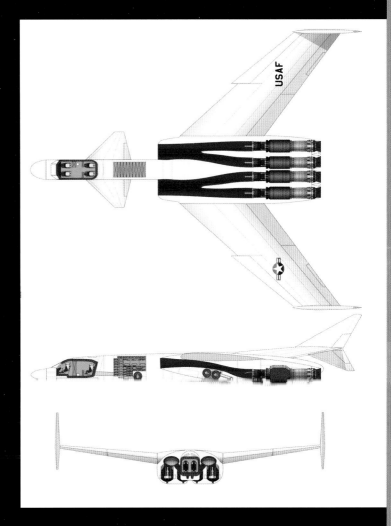

Inboard profile showing the full nuclear propulsion system on the NX-2 in great detail. Four GE J87 engines were designed to be mounted in a pair of side-by-side two-packs. *G. De Chiara © 2015*

The Northrop SM-62 Snark was originally designated B-62. *USAF*

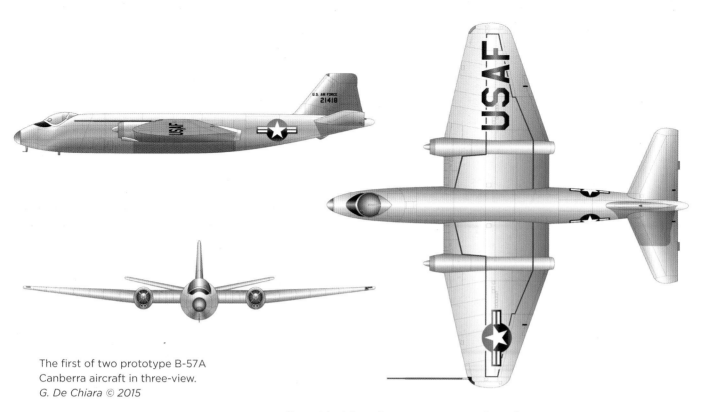

The first of two prototype B-57A
Canberra aircraft in three-view.
G. De Chiara © 2015

Strangely, Martin's own XB-51 was the only competition ever placed against the RAF-provided B.2 Canberra aircraft—and the XB-51 proved to be least capable of the two.

SUMMARY

The Martin B-57 Canberra program was a success, with no fewer than 403 aircraft built in sixteen variants. The rare occurrence of the US Armed Forces procuring foreign weaponry (especially aircraft) was a gamble undertaken by both Headquarters USAF and by the Glenn L. Martin Company. It paid off, opening the door for other foreign aircraft orders from US airframe contractors—witness the success of the US Marine Corps AV-8 Harrier (a.k.a. the "Jump Jet") program.

Although the B-66 looked a lot like its US Navy A3D (later A-3) Skywarrior counterpart, it was a different airplane packed with USAF amenities. *NMUSAF*

In the final analysis, the USAF Tactical Air Command blessed off on the Glenn L. Martin Company's procurement of a strong English Electric Company Ltd. airplane design and actually made it better.

A T-33A flies chase during the first flight of the premier RB-66A on June 28, 1954. *NMUSAF*

DOUGLAS B-66A AND RB-66A DESTROYER

The Douglas Aircraft Company B/RB-66 Destroyer program began when the USAF issued general operational requirements (GOR) for a medium-class tactical bomber to double as a reconnaissance platform.

On February 12, 1952, letter contract AF33 (600)-9646 initiated the procurement of five test airplanes designated RB-66A. Since the USAF B-66 was to be a close copy of the A3D Skywarrior (also built by Douglas, for the US Navy), it was decided that no experimental XB-66 or service test YB-66 airplanes were required: only an initial batch of five full-scale development (FSD) RB-66A airplanes would be ordered to serve as the test and evaluation aircraft.

These aircraft were developed under USAF ARDC Secret Project MX-1934, Douglas Model 1326, Weapon System 308A/L. Douglas suggested the name Destroyer (originally applied to its earlier XSB2D-1 of 1943), which was readily approved by the USAF.

On January 26, 1953, the DOD approved the USAF contract for the five Douglas RB-66A FSD airplanes (USAF serial numbers 52-2828 to 52-2832), to be built in Long Beach, California.

Douglas test pilot George R. Jansen piloted RB-66A number one on its June 24, 1954, maiden flight, from Long Beach, California, to Edwards AFB.

SUMMARY

The success of the five FSD RB-66A test aircraft led to the production of B-66B and RB-66B aircraft for the USAF Tactical Air Command.

RB-66A SPECIFICATIONS

Propulsion system
Two axial-flow, non-afterburning, 10,200-lbf Allison J71-A-11 turbojet engines

Length	75 feet, 2 inches
Height	23 feet, 7 inches
Wingspan	72 feet, 6 inches
Gross weight	79,000 pounds
Maximum speed	585 mph
Ceiling	43,000 feet
Crew	Three

Payload
Fifteen thousand pounds of conventional/nuclear bombs (tactical mission); eight thousand pounds of photo-flash bombs for night photography (photographic reconnaissance mission)

Armament
Two tail-mounted M24 20mm cannon in radar-directed barbette

FOUR

CENTURY SERIES

There is a peculiar gratification on receiving congratulations from one's squadron for a victory in the air. It is worth more to a pilot than the applause of the whole outside world. It means that one has won the confidence of men who share the misgivings, the aspirations, the trials, and the dangers of aeroplane fighting.

— Capt. Edward V. "Eddie" Rickenbacker, USAAS

The Century series was an interesting mix of popular US Air Force fighters, including daytime (fair-weather) fighters, fighter-bombers, and nighttime (all-weather) interceptors. It was this significant group of fighters that spawned many new innovations: the first level attitude flight supersonic fighter (North American F-100 Super Sabre), the first 1,000-mph fighter (McDonnell F-101 Voodoo), the first fighter to employ area rule in its design (Convair F-102 Delta Dagger), and the first doublesonic fighter (Lockheed F-104 Starfighter). Two of these aircraft—the F-102 and F-106—were armed with all missiles and an air-to-air rocket equipped with a nuclear warhead (the Douglas AIR-2 Genie).

Four of the Century Series fighters flying in formation, illustrating their vastly different configurations. Clockwise from bottom: North American F-100, Lockheed F-104, McDonnell F-101, and Convair F-102. *AFTC/HO*

Five of the Century series fighters. Clockwise from bottom: Lockheed F-104 Starfighter, North American F-100 Super Sabre, Convair F-102A Delta Dart, McDonnell F-101 Voodoo, and Republic F-105 Thunderchief. The Convair F-106 Delta Dart and the North American F-107 are absent from this group. *AFTC/HO*

The Lockheed F-104 Starfighter, Republic F-105 Thunderchief, Convair F-106 Delta Dart, and the North American F-107 were all capable of doublesonic speeds.

Most aviation historians don't include the F-107 in this group of aircraft because it didn't go into production. It did, however, undergo evaluations under Weapon System 306B as an alternative to the F-105 (WS-306A), for which reason it is included in this chapter. Thus the USAF Century series of fighter aircraft begins with the F-100 and, in this case, ends with the F-107.

The cancelled and never-produced Republic XF-103, North American F-108A Rapier, and the Bell "YF-109" (D-188A) would most likely have been considered part of this group, as would the McDonnell F-110A "Spectre" (which became the F-4C Phantom II), the General Dynamics F-111, and, especially, the F-103. Nevertheless, these planes are not considered part of the Century series, though they are significant aircraft and are discussed elsewhere in this work.

What Happened to the F-95, F-96, F-97, F-98, and F-99 Fighter Aircraft?

YF-94C number one in color, with Day-Glo red livery. *LM via Denny Lombard*

YF-96A was the first designation authorized for what became the Republic YF-84F; designation took place on September 8, 1950. Still designated YF-96A, the prototype YF-84F takes off for the first time on June 3, 1950. It was named Thunderstreak on December 2, 1952. *AFTC/HO*

The F-95, F-96, and F-97 planes became the F-86D, F-84F, and F-94C fighter aircraft, while the F-98 and F-99 designations were assigned to air-to-air and ground-to-air guided missiles.

- The F-95A, to be built by North American Aviation, Inc., was to be an all-weather (nighttime), all-missile- and/or rocket-armed version of the F-86 Sabre without either machine guns or cannon. The USAF decided it wasn't different enough from the F-86 to warrant the new designation, however, and it was instead ordered into production as the F-86D Sabre.

- The F-96A, to be built by the Republic Aircraft Corporation, was to be a daytime air superiority type fighter with sweptback flying surfaces. Also declared too similar to its F-84 brethren, it was redesignated the F-84F. To distinguish it from the straightwing Thunderjet variants, Republic named it Thunderstreak, a name approved by the USAF.

- The F-97A, to be built by the Lockheed Aircraft Corporation, was to be an all-rocket-armed, all-weather (nighttime) fighter interceptor. Again, according to the USAF, it wasn't different enough from its F-94 cousins and it became the F-94C. It was, however, the first of the F-94 variants to carry the official name Starfire.

- The F-98, to be built by the Hughes Aircraft Company, was to possess both heat- and radar-guided air-to-air guided missiles. Common sense eventually prevailed, and the US Department of Defense changed the F (for "fighter") prefix of the F-98 to GAR (for "guided aerial rocket"). This prefix was likewise incorrect, and GAR was eventually changed to AIM (for "air-intercept missile").

- The F-99, to be built by the Boeing Airplane Company, was to be a ground-to-air guided missile. Its F-99 designation was also deemed incorrect, and the F-99 was redesignated IM-99 (finally CIM-10) to better portray its mission as a ground-launched interceptor missile intended to down enemy aircraft.

ABOVE: The designation XF-98 was the 1951 classification applied to the XAAM-A-2 Falcon air-to-air guided missile in 1951 developed under USAF ARDC Secret Project MX-904. In 1955, the XF-98 designation was changed to XGAR-1 under MX-904-A; XAIM-4 in 1962. Several Falcon AAMs are attached to the wingtip missile/rocket pods on this F-89. *NMUSAF*

BELOW: Four CIM-10 (last designation change) BOMARC missiles ready for launch. The XF-99A and YF-99A designation for Boeing Model 624 was the nomenclature given to the Boeing/Michigan Aeronautical Research Center (BOMARC)—developed ground-launched experimental and prototype aircraft intercept missiles. In 1955, these designations were changed to XIM-99A and YIM-99A under USAF ARDC Secret Project MX-1599. *NMUSAF*

ABOVE: The first of two YF-100s is shown here on an early test hop, with its forty-five-degree sweptback flying surfaces evident. The secret North Base area of Edwards AFB lies directly below. Before the F-100 received its official name, Super Sabre, the USAF referred to it as Sabre-45, the name North American had used during the selling process. In fact, Sabre-45 was still being used on May 22, 1953, the day before the first YF-100 first flew. It was only after that date that the name Super Sabre was adopted. *NMUSAF*

RIGHT: Before its first flight on May 25, 1953, the premier YF-100 Super Sabre had to perform a number of low-, medium-, and high-speed taxi tests near the North Base area on Rogers Lake to check its engine performance, flying surfaces (ailerons, wing flaps and slats, rudder, and stabilator), instrumentation, nose wheel steering, brakes, and more. *NMUSAF*

NORTH AMERICAN YF-100 (SABRE 45) AND YF-100A SUPER SABRE

With the success of its turbojet-powered Sabre series of fighters (all of them subsonic in level attitude flight), North American Aviation, Inc. set out in early 1949 to create a version of its Sabre that would be capable of routine supersonic speed in level attitude flight. First, though, it would have to find a suitable Mach 1–rated powerplant.

In January 1949, then, NAA became aware of the new axial-flow turbojet engine that had been under development by Pratt & Whitney since fall 1947: the Model JT3 Turbowasp. It was the first original turbojet design from Pratt & Whitney, which had been producing the J42 (JT6) and J48 (JT7), both employing license-built Rolls-Royce centrifugal-flow designs. Pratt & Whitney's new goal at first was to catch up with the turbojet engine competition in the

United States, then surpass it. For its JT3 program, Pratt & Whitney chose to develop a dual-compressor, axial-flow design that would develop 7,500-lbf. It also planned to develop an afterburner section for its first original design. Unknown to Pratt & Whitney at the time, it had begun development of what was to become a hallmark turbojet engine.

On February 3, 1949, having selected Pratt & Whitney's JT3 engine for propulsion, NAA initiated preliminary design studies to increase the level-flight speed capability of its Sabre design to Mach 1-plus, purely as a company-sponsored program.

Since supersonic performance was the primary goal, NAA designers settled on forty-five-degree aftward swept flying surfaces. For this reason, the company dubbed the design effort project Sabre 45.

The project Sabre 45 design team was headed by Raymond H. "Ray" Rice, vice president in charge of engineering, and Edgar O. "Ed" Schmued, assistant chief engineer for design. Schmued was famed for his earlier designs of the Mustang, Twin Mustang, and Sabre. Rice was also instrumental in the design and development of these classic fighters.

Early Sabre 45 designs were closely related to the F-86D and F-86E versions of the Sabre; the first of these was an Advanced F-86D (NAA Design D-169), proposed to the USAF unsolicited on August, 25 1950. It was proposed as a transonic, single-seat, all-weather interceptor with forty-five-degree sweptback flying surfaces. (In aeronautics, "transonic" refers to the condition of flight in which a range of velocities of airflow exist surrounding and flowing past an air vehicle or an airfoil that are concurrently below, at, and above the speed of sound in the range of Mn 0.8 to 1.0; that is, 600–768 mph at sea level.)

NAA's proposed Advanced F-86D was rejected by the USAF immediately, instead suggesting that NAA design an advanced day fighter. The second offering, again unsolicited, was the Advanced F-86E (NAA Design D-153),

YF-100A SPECIFICATIONS

Propulsive system
One axial-flow, afterburning, 13,200-lbf (with afterburning) Pratt & Whitney XJ57-P-7 turbojet engine

Length	46.2 feet
Height	14.4 feet
Wingspan	36.8 feet
Empty weight	18,279 pounds
Gross takeoff weight	28,965 pounds
Maximum speed	729.6 mph or Mach 1.10
Ceiling	52,600 feet
Maximum range	1,622.6 miles

Armament
Four M39 revolver-type 20mm cannon with 1,100 rounds of 20mm ammunition (275 rounds each), YF-100A number two only

which the firm offered as a supersonic, single-seat, day interceptor with forty-five-degree sweptback flying surfaces. This proposal was likewise rejected, but the USAF suggested that it be further developed as an air superiority day fighter.

Shortly after NAA proposed its Advanced F-86E, under the project Sabre 45 banner, in August 1950, it became apparent that an all-new fuselage would be required to attain the desired speed. By simply attaching flying surfaces with ten-degree additional sweepback onto an existing airframe and adding a more powerful engine, NAA found through wind tunnel tests that a fuselage with a higher fineness ratio than the F-86E was needed to get the additional 200 mph required. In this way NAA settled on a brand-new design for project Sabre 45: its Charge Number NA-180.

By this time (late 1950), Pratt & Whitney was boasting that its Model JT3 turbojet engine—now designated J57—would produce as much as 10,000-lbf without afterburning, and 15,000-lbf with afterburning. North American's revised Sabre 45 design, it felt, could now easily attain supersonic speed in level flight. And with this in mind, NAA went back to the USAF with yet another unsolicited proposal in early December 1950. This time, NAA's proposal for a supersonic day air superiority fighter was not rejected.

Instead, NAA's new offering was well received, and, on January 19, 1951, the USAF ordered two service test NA-180 airplanes under contract number AF-6545. They received USAF serial numbers 52-5754 and 52-5755 and would be developed under USAF ARDC Secret Project MX-1894. On December 7, 1951, the Sabre 45 was officially designated YF-100, and the first of the Century series was conceived.

During the manufacture of NAA's two YF-100s, there were enough design changes to warrant a revised designation. Therefore, the Super Sabre (as it had been officially named) became the YF-100A on March 22, 1952.

When the first service test YF-100A appeared in April 1953, it no longer looked anything like its F-86 predecessor. Featuring an oval-shaped engine air inlet orifice (resembling a fish mouth), design could feed the required amount of air to the face of its new afterburner-equipped Pratt & Whitney XJ57-P-7 turbojet engine housed within its aft fuselage section. It also featured low-set wings and stabilators (combined horizontal stabilizers and elevators), and a high vertical tail (its vertical stabilizer and rudder). In addition, it sported a clamshell-type cockpit canopy that raised up and down, rather than sliding aftward and forward, which offered excellent visibility and good cockpit ingress and egress.

ABOVE: The short tail used by early production F-100As proved to be deadly. On the right, the short tail is easily seen, while another F-100A (52-5778) (on the left) features a longer tail. The taller tail eliminated the F-100's inertia coupling problem and made the airplane safe for USAF to fly pilots on a daily basis. *NASA*

LEFT: One TF-100C (NA-230) was built as a dedicated supersonic pilot training and transition test aircraft, created from a modified F-100C (54-1966). The success of this program earned NAA production orders for its F-100F that doubled as a pilot trainer/transition airplane and a combat aircraft. *AFTC/HO*

During development, the USAF decided that YF-100A number one would be used for aerodynamic and performance tests, while the number two would perform armament evaluations. For this reason, the second Super Sabre came equipped with the AN/APG-30 ranging radar and fire control system, and it was armed with four nose-mounted M-39 20mm cannon (two on either side of its fuselage).

The full-scale engineering mockup was inspected and approved by the USAF ARDC 689 Engineering Board on November 14, 1951, with only a few requests for alterations (RFA).

After its unofficial rollout at NAA's Inglewood, California, facility, the premier YF-100A was trucked to North Base at Edwards AFB on an eighteen-wheeler for manufacturer's phase I flight test evaluations. NAA's chief engineering test pilot George Swartz "Wheaties" Welch would demonstrate its prowess and reveal whether it truly deserved the name <u>Super</u> Sabre.

On May 25, 1953, Welch did just that during his fifty-seven-minute test hop on the YF-100A (which, curiously, had "YF-100" painted on its nose). In straight and level flight at thirty-seven thousand feet, Welch said "Bingo!" over his hot mike to ground control: he had just hit Mach 1.10, exceeding the speed of sound. This was verified by USAF test pilot Maj. Frank Kendall "Pete" Everest Jr.,

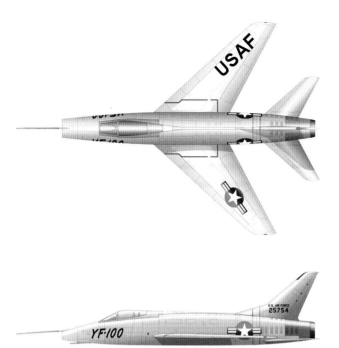

Three-view of the YF-100 number one
(52-5754). *G. De Chiara © 2015*

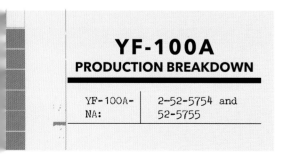

YF-100A
PRODUCTION BREAKDOWN

YF-100A-NA:	2-52-5754 and 52-5755

who flew chase. Supersonic speed wasn't unprecedented in 1953, but in straight and level flight it was—especially by a combat-type airplane.

Welch flight-tested the number two YF-100A (the armament testbed) for the first time on October 14, 1953. After a few more contractor test hops it was ferried to Eglin AFB, Florida, where it would undergo armament tests.

The F-100 was hot. On October 29, 1953, piloted by now Lt. Col. Pete Everest, the first YF-100A established a new world speed record of 755.149 mph (0.97 Mn) on a fifteen kilometer (9.3 miles) course over the Salton Sea in California. On the same day, Welch conducted the first flight on the number one F-100A, a thirty-minute test hop during which he easily exceeded 1.0 Mach.

North American Aviation's aerodynamicists were headed by Edward J. "Ed" Horkey, dubbed the "Prince of Smooth Air" by Joseph V. Mizrahi, founder and editor-in-chief of *Airpower* and *Wings* magazines. Horkey thought the F-100's tall vertical tail was unnecessary, and, in a weight-saving venture, the group engineered a shorter tail, a trait of all early production F-100As. Almost immediately, both NAA and USAF pilots reported stability and control problems with the short vertical tail. And this situation would only get worse.

While flying the ninth production F-100A (52-5764) on October 12, 1954, in an all-out effort to check the short tail problem during a maximum g and maximum speed demonstration, Welch had split-S dived from forty-five thousand feet toward Rosamond Lake, alongside Rogers Dry Lake, which was his reference point. On his way down, the airplane became unstable and broke apart near its cockpit area. Welch ejected but died of his injuries.

The ensuing crash investigation blamed the tragedy on the aerodynamic phenomenon called "inertia coupling." Further investigation found that more vertical tail area—equal to the tail area on the two service test YF-100As—would have fixed the problem. After a grounding period, during which the taller vertical tails were ordered back onto the production F-100As, the short-tailed Super Sabre airplanes were banned.

SUMMARY

The F-100 Super Sabre, first of the Century series, was the first USAF fighter capable of flying at supersonic speed in level attitude flight.

While it was fraught with aerodynamic difficulties early on (with deadly results), it was ultimately developed into a reliable combat aircraft that was used heavily in the Vietnam War.

In the end, North American Aviation, Inc. produced 2,294 F-100 Super Sabre airplanes in five main variants for the USAF, at both its Los Angeles, California, and Columbus, Ohio, facilities. These included 2 service test YF-100As, 203 F-100As, 476 F-100Cs, 1,274 F-100Ds, and 339 F-100Fs. From these came two other USAF sub-variants, including six RF-100As for the "Slick Chick" program, and one TF-100C.

Finally, several proposed but unbuilt F-100B variants eventually metamorphosed into the F-107 program, discussed below.

McDONNELL F-101A AND F-101B VOODOO

On September 29, 1954, the USAF Strategic Air Command terminated its requirement for a long-range bomber escort fighter. On the same day, the fighter that SAC had developed for the mission made its maiden flight at Edwards AFB, California. Flight-tested by McDonnell chief engineering test pilot Robert C. "Bob" Little, the F-101A hit Mach .90 (9/10 the speed of sound) in level attitude flight. This was only the beginning of the story for McDonnell's second version of the Voodoo.

The McDonnell Aircraft Corporation F-101 owes its existence to a USAF/SAC requirement for a strategic fighter. The strategic fighter program was a follow-on to the penetration fighter program, was won by the firm's earlier XF-88 Voodoo. This plane served as the direct predecessor to what became the F-101, also named Voodoo.

Following the penetration fighter competition (which ended in mid-1950 without a production contract for any of the three contenders forthcoming), the DOD instructed the USAF to use existing Republic F-84s as long-range bomber escort fighters. Strategic Air Command, as the air arm of the USAF affected by the order, was not pleased with this order. Instead, SAC needed a

ABOVE: The premier F-101A during its first flight on September 29, 1954. with McDonnell chief engineering test pilot Bob Little under glass. The F-88 number one made its first flight on October 20, 1948, almost six full years earlier. Note the W-shaped wing planform and large wing flaps. Its two afterburning P&W J57-P-13 turbojet engines propelled early production F-101As to 996.5 mph (1.5 Mn) in level flight at thirty-five thousand feet. *GSLASM via Mark Nankivil*

RIGHT: Wing flaps down and speed brake doors open, the ninth of twenty-nine preproduction F-101A-17-MC Voodoos (53-2426) lands at Edwards AFB. It has just come from its world speed record flight of 1,207.6 mph on December 12, 1957, during Operation Fire Wall. USAF Maj. Adrian E. Drew of the 27th Fighter Bomber Wing (seated) sits in the cockpit. *AFTC/HO*

very long range fighter to escort its growing fleet of B-36s, and it felt that the F-84 was not satisfactory.

To get the fighter it really needed, SAC outlined a set of requirements for what it called a strategic fighter, in essence an advanced penetration fighter, in January 1951. General Operational Requirement 101 (GOR 101) was issued to the industry the following month, under weapon system 105A or WS-105A. Interested parties had until May 1, 1951, to complete their respective proposals.

Among the entries to meet SAC's proposal was one from Northrop, based on its F-89; one from North American, based on its F-93; two from Lockheed, based on its F-90 and F-94; three from Republic, two based on its F-84 and one on its F-91; and one from McDonnell, based on its F-88. McDonnell's F-88 design had been declared the winner of the penetration fighter competition on September 11, 1950. Upon review, USAF/SAC deemed the revised F-88 design (designated in-house F-88K) acceptable, and it was therefore advanced.

In its effort to create a suitable strategic fighter airplane, McAir increased fuselage volume in order to accommodate a pair of Pratt & Whitney Model JT3 (J57) afterburning turbojet engines and to provide the fuel required to maintain the thousand-mile combat radius. To do this, both the frontal area and the length of the fuselage were increased over that of the F-88. The length of the fuselage was increased 10.5 feet and the frontal area was increased by 11.5 square feet. It also added a ten-square-foot area to the vertical tail to maintain the same degree of static directional stability. In addition, the horizontal tail plane was moved from the base of the vertical tail to near its top, forming a T-tail configuration.

Since McAir elected to propel its revised F-88 with a pair of J57 turbojet engines, more airflow was needed than the J34s in the F-88. To meet this requirement, the duct inlet area of the advanced Voodoo was increased about 2.5 square feet.

The increased power of two J57 turbojet engines would allow the new Voodoo to operate at Mach numbers up to 1.7 in level flight, even more in dives. Since the F-88 had excellent aerodynamic characteristics, up to its maximum 1.18 Mn speed, McAir wisely chose to keep its basic configuration for its proposed strategic fighter (known in-house as Model 36W, or the twelfth variation of the basic F-88 design: the F-88K).

The USAF was sold on McAir's Model 36W, and, without a Y prefix, it ordered twenty-nine service test F-101A-17-MC airplanes in October 1951 on USAF contract number AF-8743, approved by the DOD on May 28, 1953; USAF serial numbers 53-2418 to 53-2446.

F-101A SPECIFICATIONS

Propulsive system
Two axial-flow, afterburning, 15,000-lbf (with afterburning) Pratt & Whitney J57-P-13 turbojet engines

Length	67 feet, 4 inches
Height	18 feet, 0 inches
Wingspan	39 feet, 7 inches
Empty weight	25,374 pounds
Gross weight	51,000 pounds
Maximum speed	Mach 1.5
Combat Ceiling	40,000 feet
Maximum range	1,600 miles

Armament
Four 20mm M39 cannon (two hundred rounds each) and three Falcon AAMs

An extremely rare photograph showing the number one F-101A (53-2418, background) flying in formation with F-101A number two (53-2419, foreground). *AFTC/HO*

The first full-scale development (FSD) F-101B-30-MC Voodoo (56-0232) shown at the McDonnell facility in St. Louis, Missouri, adjacent to Lambert Field. Here, the F-101B follows a high-speed taxi test with its drag parachute deployed. Temporarily known as the F-109, the F-101B *Voodoo* was a dedicated fighter-interceptor. *GSLASM via Mark Nankivil*

F-101B SPECIFICATIONS

Propulsive system
Two axial-flow, afterburning, 16,900-lbf (with afterburning) Pratt & Whitney J57-P-55 turbojet engines

Length	67 feet, 4 inches
Height	18 feet, 0 inches
Wingspan	39 feet, 7 inches
Empty weight	28,778 pounds
Gross weight	52,400 pounds
Maximum speed	1.7 Mach number
Combat Ceiling	40,300 feet
Maximum range	2,000 miles

Armament
Three Falcon AAMs, two Genie rockets

Two service test photographic reconnaissance examples were added onto this contract soon after. Due to its many differences, the designation was officially changed from F-88K to F-101A, but the name Voodoo was retained. As stated above, the production start included the manufacture of twenty-nine F-101A-17-MCs, as well as two YRF-101As for research, development, test, and evaluation (RDT&E).

The DOD approved amended USAF contract number AF-8743 on June 4, 1954, for these two YRF-101A airplanes (WS-105L), McAir Model 36X; USAF serial numbers 54-149 and 54-150.

Production of the RDT&E F-101As and RF-101As moved ahead. Before long, the USAF's Tactical Air Command (TAC) found that the Voodoo was a good candidate to fill its upcoming fighter-bomber role. Moreover, the USAF's Air Defense Command (ADC) thought it might make a good supplement interceptor until the ultimate interceptor appeared. The latter would lead to the creation of Weapon System 217A, that is, the B version of the Voodoo.

Although the Voodoo was an instant success, it had its share of developmental problems. The first of these was engine compressor stalling, which called for a redesign of the airplane's engine air inlet and duct system. The second problem was related to aerodynamics, whereby the F-101's T-tail did not receive enough airflow at high angle of attack (AOA), causing loss of lift and severe nose pitch-ups, especially at high speed. The airflow problem, though serious, was never fully corrected. Instead, performance limitations were placed on the airplane.

The continued success of WS-105A/L (F-101A and RF-101A) prompted TAC to order the C version. Both the F-101A and F-101C versions were armed with four nose-mounted M39 20mm cannon, aimed through the K19 gun sight. Each version incorporated the

Hughes Aircraft MA-7 fire control system, M-1 toss-bomb computer, and low-altitude bombing system (LABS) equipment. The C version differed in its 7.33g stress capability, whereas the A version had a 6.33g stress rating.

The second F-101B Voodoo (56-0233), a Block 40, flies near St. Louis circa July 1957. The F-101B used two afterburning P&W J57-P-55 turbojet engines to attain its maximum speed of 1.7 Mn (1,127.7 mph) at thirty-five thousand feet in level attitude flight. *GSLASM via Mark Nankivil*

F-101B

In its continuing attempt to field an interim all-weather interceptor, the ADC was authorized to procure the B version of the Voodoo, in response to developmental delays, which allowed it to supplant the F-102A until the arrival of the F-106A, the ultimate all-weather interceptor.

On June 30, 1956, the DOD approved USAF contract number AF-29841 for full-scale development (FSD) F-101B-30-MC (USAF serial number 56-232) under WS-217A.

Since the two-seat F-101B (McAir Model 36AT) was vastly different than its F-101A predecessor, McAir requested its designation as F-109 instead of F-101B. This request was turned down, and the plane retained its F-101B designation.

First flown on March 27, 1957, by McAir test pilot Robert C. "Bob" Little, the F-101B Voodoo incorporated the higher powered J57-P-55 turbojet engine, Hughes MG-13 radar, and fire control system. Its armament was comprised of three AIM-4 Falcon air-to-air and, ultimately, two MB-1 Genie air-to-air rockets.

F-101A, RF-101A, AND F-101B
PRODUCTION BREAKDOWN

F-101A-1-MC:	5-53-2418 to 53-2422
F-101A-5-MC:	8-53-2423 to 53-2430
F-101A-10-MC:	6-53-2431 to 53-2436
F-101A-15-MC:	10-53-2437 to 53-2446
YRF-101A-MC:	2-54-149 and 54-150
F-101B-30-MC:	1-56-232

SUMMARY

Affectionately known as the "One-Oh-Wonder" to those associated with it, the F-101 Voodoo was McAir's first fighter plane to enter service with the USAF. In total, McAir produced 807 Voodoos (all versions) for the USAF Air/Aerospace Defense Command and Tactical Air Command before its production ended.

Never classified as a penetration fighter like the F-101, the McAir F-88 variant was considered a tactical fighter, fighter-bomber, fighter-interceptor, and reconnaissance fighter.

CONVAIR YF-102, YF-102A, YF-102B (F-106A), YF-102C, AND F-102X DELTA DAGGER

For the first years after World War II, the probability of an aerial attack on North America seemed low: only the United States had atomic bombs and the means to deliver them. US Army Air Force planners knew better, reasoning that they only had a temporary monopoly on nuclear weaponry. Though it might be years away, any future aggression against the United States would most likely begin with an aerial bombardment of its strategic targets. Billy Mitchell's prophecy had come true: air power was indeed the key to successful warfare. This had been proven in the war just concluded.

Four years after the close of World War II, in 1949, the threat of aerial bombardment from the USSR loomed just over the horizon. During that year, the Soviet Union revealed that it was developing its own atomic bombs and the means to deliver them. The so-called Cold War had turned hotter, and the security blanket over North America had gotten thinner. In this year, then, the USAF initiated a plan to field an all-weather, all-missile, rocket-armed supersonic high-altitude interceptor airplane, to be fully operational in the year 1954. The goal was to create an aircraft that could operate above fifty thousand feet at supersonic speeds in level attitude flight, day or night, in good or bad weather. With this goal in mind, the 1954 interceptor program was born under USAF Air Materiel Command Secret Project MX-1554, Weapon System 201A.

Earlier in 1947, in what ultimately turned out to be a closely related development under AMC Secret Project MX-1179 (code-named Project Dragonfly), the Hughes Aircraft Company was given the go-ahead to develop its air-to-air Falcon guided-missile family, both heat- and radar-guided types. Since the unmanned Falcon missiles were at first considered aerial interceptor air vehicles, they came with the fighter-like designation F-98. But this erroneous designation quickly gave way to the more appropriate GAR designation (for guided-aircraft rocket), with subsequent dash numbers and suffixes, and, finally, the acronym AIM was applied (for air intercept missile), again with subsequent dash numbers and suffixes.

In May 1950, Hughes was selected to fully develop this family of Falcon missiles, and its proposed MA-1 all-weather search radar and missile fire control system—all to be built into WS-201A. It was time to find a suitable airframe to use the advanced radar, fire control, missiles, and rockets all provided by Hughes; and, to find a suitable powerplant to propel the entire package.

On June 18, 1950, the USAF issued an RFP to the industry coinciding with the 1954 interceptor program under WS-201A. A number of responses were received, and on July 2, 1951, those from Convair, Lockheed, and Republic were accepted as viable.

- Convair had proposed its Model 8, essentially a larger version of its XF-92A but with cheek-mounted engine air inlets and a solid nose to house the radar and fire control system. It retained a delta wing and tail planform similar to the XF-92A, but it sported a longer and higher volume fuselage to house all of its missiles and rockets internally.

- Lockheed proposed its Temporary Design Number L-205 (Model 99), which appeared to be an early version of what became the F-104 but with a dorsal engine air inlet on centerline and a solid, cone-shaped nose section to house its radar and fire control systems. Powered by a single afterburning General Electric J53 turbojet engine, it would feature thin, straight wings like the F-104's and carry Hughes Falcon missiles internally within a missile bay.

- Republic offered its Model AP-57, which appeared more like a rocket ship than a fighter airplane. It featured a ventral engine air inlet, sharply raked delta wing and tails, a solid cone-shaped nose to house its radar and fire control systems, and was to be powered by a dual-cycle Wright Aeronautical turboramjet propulsive system.

While Lockheed's Model 99 was cancelled in favor of its Model 83 (F-104), the Convair Model 8 entry showed immediate promise, and Republic's Model AP-57 (F-103) went forward into development—not for WS-201A, but for WS-304A on a longer-term contract. Convair's entry got the nod, and it won the appellation of "1954 interceptor."

The "1954 interceptor" concept appeared shortly after the creation of the US Air Force from the former US Army Air Forces in the years after World War II. The rearrangement of command led to the creation of several new organizations within the Air Force, including Tactical Air Command (TAC) and Air Defense Command (ADC), groups that shared an interest in fighter aircraft. Offensive duties constituted TAC's primary task, but it required capable fighters in order to protect their offensive capability. ADC was tasked solely with the defense of the United States from enemy attack, employing long-distance bombers in particular. Soon after their formation, TAC and ADC started moving away from each other in terms of requirements.

In early 1949, ADC drew up an RFP for an advanced interceptor capable of attacking Soviet bombers that were expected to enter service in the early- to mid-1950s. At that point, several turbojet-powered interceptors were already in development, typically adaptations of existing fighters like the F-86 Sabre (the F-86D) and the F-94 Starfire (adapted from the TF-80C); the exception was the all-new F-89 Scorpion. While these were all subsonic aircraft, it was felt they would not offer sufficient performance to effectively intercept high-speed jet bombers such as the Bison that was known to be under development in the Soviet Union.

To counter such threats, the ADC prepared a request for an entirely new aircraft expected to perform "hot war" interceptions at supersonic speeds. For reasons that remain unclear, ADC also demanded that the entire system be flown and operated by a single pilot. Given the complexity of the radar systems—and the aircraft themselves—this would require lowering pilot workload via an advanced autopilot system that was integrated with the fire control system. Given the perceived threat, the new designs were expected to enter service in 1954.

On August 31, 1951, Convair received a contract to produce engineering data, wind tunnel models, and a full-scale engineering mockup of its Model 8. Since it was a derivative of the XF-92, the airplane was the service test designation YF-102 instead of the experimental designation XF-102.

The several Model 8 YF-102 offerings included:

- Model 8, YF-102, with a single afterburning Westinghouse Electric XJ40-WE-1 turbojet engine
- Model 8-80, YF-102, with a single afterburning Wright Aeronautical XJ67-W-1 turbojet engine
- Model 8-82, YF-102, with a single afterburning Pratt & Whitney YJ57-P-11 turbojet engine

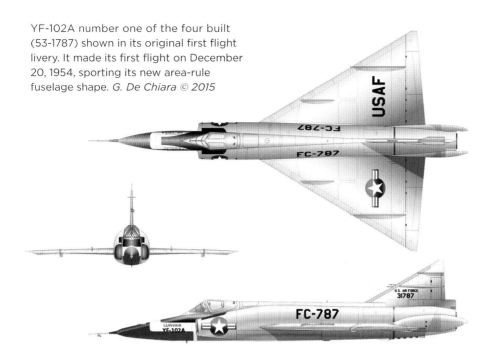

YF-102A number one of the four built (53-1787) shown in its original first flight livery. It made its first flight on December 20, 1954, sporting its new area-rule fuselage shape. *G. De Chiara © 2015*

The last, Model 8-82, received the nod.

The mockup was approved with a number of request for alterations (RFAs). Once these were addressed, the USAF ordered two flyable YF-102s and a static test article under USAF contract number AF-5942 on December 19, 1951. This initial order was upped to ten airplanes shortly thereafter on the same contract. The airplane received the unofficial nickname "Machete," though it was later given the official name Delta Dagger.

Meanwhile, development had slowed on the proposed Hughes MA-1 radar and missile/rocket fire control system. To get the YF-102 into the air, it was decided to substitute the operational Hughes MG-3 radar/fire control system as an interim measure.

Additionally, the production F-102A was now known as the Interim 1954 Interceptor, an appellation it held until the MA-1 system came on line to be incorporated in the proposed F-102B (WS-102B).

By August 1952, the F-102 program hit a new snag. While its development had largely proceeded on time, wind tunnel evaluations led to the discovery of a serious problem: touted as a supersonic interceptor, the F-102A's wind tunnel tests showed that it would not be able to surpass 0.90 Mach number in level flight; it would be *subsonic*, no better than current interceptor aircraft. Worse, there were indications that the USAF was ready to cancel the contract. Fortunately, NACA had an aerodynamic trick up its sleeve.

Earlier in 1951, the brilliant NACA aerodynamicist Richard T. "Dick" Whitcomb and his colleagues ran a series of tests in the new eight-foot-diameter wind tunnel at the NACA facility in Langley, Virginia.

YF-102 (MODEL 8-82) SPECIFICATIONS

Propulsive system
One afterburning, axial-flow, 14,800-lbf (with afterburning) Pratt & Whiney J57-P-11 turbojet engine

Length	52.5 feet
Height	18.0 feet
Wingspan	37.0 feet
Empty weight	17,974 pounds
Gross takeoff weight	28,583 pounds

Maximum speed
1.2 Mach number (projected; never attained; 0.90 Mn attained)

Combat Ceiling	54,400 feet
Maximum range	1,350 miles

Armament
Six XGAR-1A (later XAIM-4A) Falcon missiles in bay; ninety-six 2.75-inch diameter FFARs in bay and bay doors, twenty-four 2.75 inch diameter FFARs in bay doors

The premier TF-102A lifts off for another service test flight out of Edwards AFB. Developed under WS-201L, the instructor and student sat side by side in the TF-102A cockpit. *NMUSAF*

Their initial tests in the transonic wind tunnel (the only one of its kind at the time) were designed to eliminate unwanted drag in the transonic-to-supersonic speed regimes, that is, 0.75 to 1.2 Mn. They discovered that, by pinching in the waist of a fuselage (fore-to-aft on a wing-root chord plane)—and thereby taking away area—zero-lift drag-rise numbers near the speed of sound would be greatly reduced, if not eliminated. Since their discovery dealt with the removal of fuselage area, they called it the "area rule theory." After they had proven their theory, it was given (on a top-secret basis) to Convair aerodynamicists working at NACA Langley at the time. They immediately returned to San Diego, California, taking their non-area ruled wind tunnel model with them for modification.

The discovery of this aerodynamic gold mine was too late for application on the first batch of F-102s (the ten YF-102s on order). Area rule would have to be applied in stages until the first A-variant airframe (number 11) was built with new tooling. Since that airframe and three subsequent airframes would be set aside for complete area rule treatment, test, and evaluation, they were redesignated YF-102A.

The first non-area ruled YF-102 airplane (USAF serial number 52-7994) was completed in September 1953 and secretly trucked to the North Base facility at Edwards AFB for its preflight ground tests and inaugural flight. Then, on October 24, 1953, with Convair chief engineering test pilot Richard Lowe "Dick" Johnson at the controls, the Delta Dagger made its first flight. As feared, though, it couldn't slice through the air supersonically in level flight, only in a dive. A November 2 attempt to make it fly supersonic in level flight led to the engine flaming out during takeoff and crash of the airplane. Johnson survived, but YF-102 number one (52-7994) was lost. YF-102 number two (52-7995), an identical plane to number one, made its first flight on January 11, 1954.

Two other mishaps were recorded during flight test activities, both at Edwards AFB. The first, with Johnson at the controls, was a takeoff accident involving a YF-102 (52-7995) on March 13, 1954. The second incident, on September 10, 1954, involved another YF-102 (53-1781) in a landing accident, caused this time by a malfunction; Alonzo B. "Brew" a.k.a. "Tref" Treffer was in the cockpit.

In addition to area rule, Convair made other changes to their F-102 design. These included a sharply raked, V-shaped windscreen to help reduce drag, larger area engine air inlets to improve airflow, larger area vertical tails for improved high-speed stability, and a longer nose section to improve its fineness ratio and make more room for its radar and fire control system. The number six YF-102 (53-1782) was the first to incorporate all these changes; still, it could not reach level flight supersonic speed.

After new jigs and tooling were added to produce airframes with area rule, Convair produced YF-102A number one (53-1787). It featured complete area ruling, which was created by its pinched-in waistline and the addition of aft fuselage bulges (one on either side). These bulges, immediately nicknamed "love handles," were necessary as the Delta Dagger's fuselage could not be pinched in as much as required due to a lack of internal volume, already filled with devices such as the fuel tanks, engine, and so on. Thus, the aft fuselage had to be fattened with these bulges to create the illusion of a narrowed waistline, much as when a belt is pulled tight. The "new" Delta Dagger also received the uprated 16,000-lbf (with afterburning) J57-P-23 turbojet engine. In its new guise, the first YF-102A rolled out in San Diego on November 15, 1954.

Five days later, on December 20, 1954, YF-102A number one made its first flight, after which it was declared airworthy. It was scheduled to make an all-out assault on supersonic speed the next day: Convair was looking for an early Christmas present, and it got one.

After rotation and immediate landing gear and wing flap retractions, Johnson simultaneously pulled back on the control stick and pushed forward on the engine throttle. Then, in full afterburner in a thirty-degree climb, the YF-102A went supersonic—effortlessly. With this, fourteen long months of frustration were over.

Continued flight-testing on YF-102A numbers one through four (53-1787 to 53-1790) were mostly flown by Johnson, Ellis D. "Sam" Shannon, and William J. "Bill" Martin. They demonstrated consistent 1.2 Mn speed in level attitude flight. Now it was clear that NACA's area rule had saved the once-faltering F-102 program. Production of these aircraft continued.

Meanwhile, development of the F-102B moved forward. Because its numerous changes had become too vast for its inclusion in the F-102 program, it was redesignated F-106 (see below).

In an effort to evaluate the upcoming Hughes MA-1 radar and fire control system, the sixteenth production F-102A (53-1806) was modified to test the armament package, including the nuclear-warhead-equipped Douglas AIR-2A Genie air-to-air rocket, destined for the F-102B/F-106A; for this purpose, the airplane was designated YF-102C. Also used to first evaluate the MG-10 system that eventually replaced the MG-3 system, it was a one-of-a-kind airplane—not a service test airplane for another version of the F-102, even though an F-102 variant was proposed.

And it would remain only a proposal: the F-102X (F-102C) was to be powered by an uprated afterburning, axial-flow Pratt & Whitney

YF-102A SPECIFICATIONS

Propulsive system
One afterburning, axial-flow, 16,000-lbf (with afterburning) Pratt and Whiney J57-P-23A turbojet engine

Length	68.3 feet
Height	21.2 feet
Wingspan	38.1 feet
Empty weight	19,283 pounds
Gross takeoff weight	27,950 pounds
Maximum speed	1.2-plus Mach number
Combat Ceiling	55,500 feet
Maximum range	1,140 miles

Armament
Twenty-four 2.75-inch diameter FFARs in bay doors; six GAR-1D, six GAR-2, or three GAR-1D and three GAR-2 Falcon missiles

YF-102, YF-102A, F-102B, AND YF-102C
PRODUCTION BREAKDOWN

YF-102:	10-52-7994 and 52-7995, 53-1779 to 53-1786
YF-102A:	4-53-1787 to 53-1790
F-102B:	0-redesignated F-106A
YF-102C:	1-53-1806, modified F-102A

J57-P-47 turbojet engine. Since the F-106A had been ordered into production around this time, the program for F-102Cs was abandoned.

The final version of the Delta Dagger was the Model 8-12 two-seat TF-102A for pilot training and transition duties. No service test YTF-102As were ordered, but several early production TF-102As were used as service test aircraft.

SUMMARY

The success of area rule, with the four YF-102As in particular, led to a total procurement of one thousand Delta Dagger aircraft. The USAF eventually procured 889 single-seat YF-102s (ten), YF-102As (four) and F-102As (875), and 111 two-seat TF-102As.

More important, though: their success led to the creation of the F-106 Delta Dart, widely considered the best interceptor ever put into USAF service.

LOCKHEED XF-104 AND YF-104A STARFIGHTER

In mid-1950, after its XF-90 was eliminated from the USAF penetration fighter competition, Lockheed went to work on an all-weather fighter-interceptor, which it studied under Temporary Design Number 265 (TDN 265). This fighter was to be powered by a single afterburning, 15,000-lbf (with afterburning) General Electric J53 turbojet engine, but it was not continued.

In May 1952, Lockheed was offered a contract to build two fighter-interceptor prototypes with a gross takeoff weight of thirty-two thousand pounds, powered by two Wright Aeronautical J65 turbojet engines. Lockheed declined to participate because the USAF had insisted on a contract stipulation forfeiting all patent rights, which would have permitted the government to give production rights to any airframe contractor.

The XF-104 number one Starfighter's wings were so thin they weren't thick enough to make room for fuel tanks. For this reason, wingtip fuel tanks were adopted to make up for this shortcoming. The design and performance of the F-104 was ahead of its time. Lockheed test pilots Tony LeVier and Herm Salmon loved to fly it, and USAF pilots said it was a real "hot rod." *NMUSAF*

The MX-1554 and MX-1554-A
The "1954 Interceptor" Competition

Several airframe contractors answered the MX-1554/-1554-A "1954 interceptor" RFP with offerings.

- Chance Vought Aircraft offered its Model V-371, featuring a long, sleek fuselage with delta-shaped horizontal and vertical stabilizers. It had a trapezoidal-shaped wing planform from its fuselage to its wingtip-mounted engine nacelles, then sharply raked-back delta-shaped wingtips outboard of its engine nacelles. Its two turbojet engines, housed in outboard wing nacelles (one on either wingtip), were to be afterburning Pratt & Whitney J57 turbojets (other engines were also considered). It had tricycle landing gear. Its length was eighty-four feet and its height was twenty-four feet, two inches. Little else is known about this design.
- Convair offered its Model 8 series in answer to MX-1554, which became the WS-201A (see F-102).
- Lockheed Aircraft Corporation's TDN L-205 (Model 99) was to be powered by a single afterburning, 15,000-lbf General Electric XJ53-GE-X10 (later J79) turbojet engine, two flying prototypes were ordered (USAF serial numbers unknown), but it was not continued.
- McDonnell Aircraft Corporation's example was a pure interceptor version of its F-101A *Voodoo* (Model 36AG), powered by two afterburning, 21,500-lbf (with afterburning) Wright J67-W-1 turbojet engines.
- North American Aviation offered two designs, a single-engine type and a twin-engine type. The single-engine design had three weapon bays, a wingspan of forty-one feet, seven inches, length seventy feet, four inches, and a gross takeoff weight of 35,400 pounds, with a top speed of 1,000 mph. The twin-engine design was similar in size, with a gross takeoff weight of 46,666 pounds and a slightly higher top speed of 1,040 mph. Both designs were offered in 1951.
- Northrop Aircraft Corporation's example has left little trace in the record. The Northrop Project Number N-65 was proposed to the USAF for its "1954 interceptor" program.
- Republic Aircraft Corporation, initially offered as XF-91A (MX-1658), was to be powered by a single afterburning turbojet engine—an advanced version of the XF-91 Thunderceptor, Republic Model AP-31N-1. Its second competitive offering was the Model AP-54, which was based again on its earlier XF-91. A third offering was its much more radical Model AP-57, a rocket-ship-like design to be propelled by

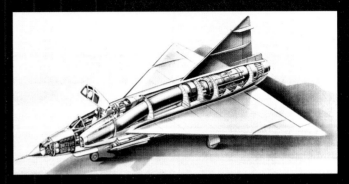

XF-102, designated on September 11, 1951, was the short-lived experimental fighter designation of what became the prototype Convair YF-102 (Model 8). Since the design was an outgrowth of the cancelled XP-92, the *X* for experimental was replaced with *Y* for prototype. *USAF*

Since its new design was not based on any previous aircraft, XF-103 was the designation applied to Republic's Model AP-57 on September 11, 1951. The XF-103, along with the XF-cum-YF-102, were co-developed under Weapon System 201A and WS-204A (formerly MX-1554 and MX-1554-A), respectively. *USAF*

a combined turbojet and ramjet engine, called a turboramjet. This became the XF-103 under MX-1554-A, which in turn was offered as the WS-304A.

Three of these contractors were selected on July 2, 1951, to move forward to mockup stage: Convair, Lockheed, and Republic. All were authorized to build full-scale engineering mockups and two flyable airplanes for USAF ARDC evaluation. For reasons that remain unclear—most likely funding and/or the advent of the F-104—the Lockheed Model 99 design was later eliminated from the competition and its two prototypes were cancelled.

XF-104 SPECIFICATIONS

Propulsive system

One Buick-built axial-flow, afterburning 7,330-lbf Wright Aeronautical J65-B-3 turbojet engine (XF-104 number one only); one 7,500-lbf, Wright Aeronautical-built J65-W-7 turbojet engine (both XF-104s after XF-104 number one was refit with the -7 engine)

Length	49.17 feet
Height	13.49 feet
Wingspan	21.94 feet (without wingtip fuel tanks)
Empty weight	11,500 pounds
Gross takeoff weight	16,700 pounds

Maximum speed

1.79 Mach number (attained)

Armament

One 20mm T-171E-3 (later M61A1) Vulcan cannon (XF-104 number two only)

In addition, Lockheed's Advanced Design Group (ADG) (later Lockheed Advanced Developments Company (LADC)) was then under contract to develop test data and design aids on the Douglas X-3 Stiletto, which it had received from the USAF. Moreover, it was developing TDN CL-246, an uncomplicated single-seat, single-engine fighter with a thin, straight wing. As an unsolicited proposal, this design was submitted to the USAF on an official basis by none other than Kelly Johnson in November 1952.

Although the USAF did not have a requirement for the enticing CL-246 design, its seductiveness was enough for the USAF to invent a GOR calling for a new lightweight air superiority type of day fighter to ultimately replace the F-100, beginning in 1956. To be fair, though, competitive bids for this GOR were invited from all airframe contractors.

Lockheed had already met all USAF requirements for a high-speed, lightweight fighter by designing an aircraft with an empty weight of just 11,500 pounds. It was a lean machine, every ounce of extra weight being stripped from its airframe and equipment. In fact, it actually proved to be half the weight of the aircraft proposed by the participating airframe contractors. Moreover, the Lockheed Skunk Works had concentrated its efforts on a fighter plane with a high thrust-to-engine-weight ratio because of the supersonic speed regime in which it would fly and fight. It was to be, in the truest sense, a "star fighter," which is the origin of its official name.

Under Weapon System 303A then, USAF ARDC Secret Project MX-1853, the USAF ordered two experimental XF-104 prototypes on March 12, 1953 (USAF contract number AF-23362); USAF serial numbers 53-7786 and 53-7787. The full-scale engineering mockup was inspected and approved on April 30, 1953, which cleared the way for the production of the two prototypes.

In order to obtain the best possible speed from its airframe, Lockheed chose to propel its XF-104 aircraft (Model 083-92-01) with a single Buick Motor Company-built axial-flow, afterburning Wright Aeronautical J65 turbojet engines rated at 15,000-lbf (with afterburning). The J65 was an Armstrong Siddeley Sapphire of British design produced in the United States under license (initially, only the non-afterburning 10,200-lbf version of the J65 was available to the XF-104 program).

After being secretly trucked overnight from Burbank to Edwards AFB in late January 1954, the number one XF-104 was prepared for its initial flight test activities. A high-speed taxi run followed by a short test

XF-104 number two (53-7787) soars on an early test flight without its wingtip-mounted fuel tanks attached. *LM-COM*

hop was scheduled for February 28, 1954, with Lockheed ace test pilot Anthony William "Tony" LeVier at the controls. As scheduled, he reached rotation speed and pulled back on the stick, raising the aircraft some five feet off the dry lakebed—the Starfighter had flown, if unofficially. Its first official flight didn't take place until March 4, 1954. During this first flight of some twenty minutes at low speed, the landing gear refused to retract, at which point LeVier landed. After some adjustments were made, LeVier took off again and, once again, the landing gear would not retract.

Once it was realized that this was due to low pressure in the hydraulic system, the problem was soon corrected. LeVier made flights three and four on March 26, 1954, without landing gear difficulties. Further flight tests found that the premier XF-104, powered by the non-afterburning J65-B-3 turbojet engine, could not exceed Mach 1.0 in level flight. In a shallow dive and without aerodynamic discrepancies, though, the XF-104 easily exceeded the speed of sound.

The afterburning, 10,200-lbf (with afterburning) Wright aeronautical J65-W-7 turbojet engine was installed in XF-104 number one in July 1954, at which point its performance increased dramatically. In level flight it could attain 1.49 Mn at 41,000 feet, Mn 1.6 in a dive, and it could zoom-climb to fifty-five thousand feet.

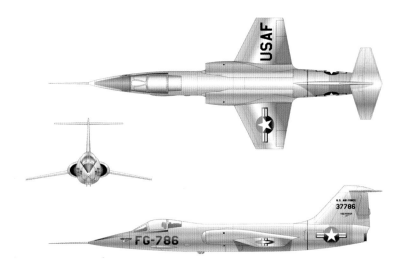

Three-view illustration showing what XF-104 number one looked like on the day of its first flight in early March 1954. G. De Chiara © 2015

Fitted with the J65-W-7 from the start, the second XF-104 made its first flight on October 5, 1954, with Lockheed test pilot Herman Richard "Fish" Salmon under its cockpit canopy. On March 25, 1955, powered by the J65-W-7, the XF-104 number two reached a top speed of 1.79 Mn (1,324 mph) at sixty thousand feet, with Lockheed test pilot Ray Goudey flying.

On April 18, 1955, a near-fatal accident darkened the XF-104 program during a gun-firing test on XF-104 number two, with Salmon at its controls. When he pressed the trigger, severe vibration began, caused by a gun malfunction. The resulting vibrations were enough to shake loose the emergency ejection hatch on the belly of the airplane, just below Salmon's seat. Since he was at high altitude (fifty thousand feet) and in a high-altitude pressure suit, his suit pumped up and over his face, obscuring his vison. Remembering LeVier's earlier experience when a 20mm round exploded prematurely, Salmon thought the same thing had happened. Not realizing his ejection seat hatch was gone, he elected to eject and parachuted to safety. XF-104 number two was lost.

XF-104 number one was also lost due to severe tail flutter during a flight on July 11, 1957, while it was being used to fly chase on a service test Y F-104A. At the controls of the number one XF-104, Lockheed test pilot William C. "Bill" Park survived his emergency bailout.

Earlier, seventeen service test YF-104As were ordered to be built to fully investigate the assets of the Starfighter. They were assigned USAF serial numbers 55-2955 to 55-2971 under USAF contract number AF-27378, approved March 30, 1955.

Seven preproduction F-104A-1-LOs (USAF serial numbers 56-730 to 56-736), along with the seventeen YF-104As and two XF-104s, were variously revised during testing to solve different problems revealed during the long running test program. These changes addressed issues in armament, pitch-up, low-speed handling, and engine problems. Following these revisions, twenty six aircraft participated in the Starfighter growth program to mature the F-104 enough for full-scale production of Block 5 F-104A-5-LO Starfighters. The seven additional test F-104A-1-LOs were ordered on USAF contract number AF-30756, approved March 2, 1956.

When the wraps were finally taken off the Starfighter, some two years after it had flown, Lockheed rolled out the second service test YF-104A built for public viewing on February 16, 1956. By the time of its first public appearance, people were listening to a new sound called rock 'n' roll: it was the rocking fifties—and when the Starfighter rolled out, it literally "rocked" those in attendance.

Air enthusiasts at the demonstration were astounded by what they saw: the craft didn't seem to have wings, let alone the capacity to fly. Moreover, the covers faired over its engine air intakes on either side of the fuselage hid its half-cones; these covers were immediately dubbed "flight falsies"; today these remind some of the bullet-shaped brassiere once worn by the singer Madonna. It featured downward-angled wings, a high T-tail, and a rocket-like fuselage that was much longer than its wingspan was wide. In truth, it looked more like a missile with a cockpit than an airplane. It appeared to be going fast indeed, even though it was sitting static on the ramp.

SUMMARY

The Lockheed Aircraft Corporation F-104 Starfighter was a bona fide hot rod in its day, earning its nickname: "The Missile with a Man in It." It was light, ultra-sleek, and extremely powerful. When some other fighters were struggling to exceed Mach 1.0 in level attitude flight, the F-104 easily blazed past Mach 2.0 on the level. Additionally, it could zoom-climb to altitudes never attained by its contemporaries. Its engine power-to-airframe weight ratio was outstanding, which allowed it to establish numerous speed, altitude, and time-to-climb records.

YF-104A SPECIFICATIONS

Propulsive system
One axial-flow, afterburning 14,800-lbf (with afterburning) General Electric J79-GE-3/-3A turbojet engine

Length	54.8 feet
Height	13.5 feet
Wingspan	29.9 feet
Empty weight	13,400 pounds
Gross takeoff weight	26,000 pounds
Maximum speed	2.2-plus Mach number
Combat Ceiling	55,000 feet
Maximum range	1,400 miles

Armament
One 20mm M61A1 Vulcan cannon, two AIM-9 Sidewinders

XF-104 AND YF-104A
PRODUCTION BREAKDOWN

XF-104:	2-53-7786 and 53-7787	55-7787 crashed to destruction on April 18, 1955, test pilot Salmon survived his bailout; 53-7786 crashed to destruction on July 11, 1957, test pilot Ray Goudey survived his bailout
YF-104A:	17 -55-2955 to 55-2971	
F-104A-1-LO:	7 -56-0730 to 56-0736	these preproduction F-104As were used as additional service test aircraft

Lockheed/NASA NF-104A

Three F-104A Starfighters were modified to serve NASA as high-altitude trainers (USAF serial numbers 56-0756, 56-0760, and 56-0762), with the Aerospace Research Pilots School (ARPS) created at Edwards AFB to provide this training. These were designated NF-104A and known as aerospace trainers (ASTs), optimized for astronaut training.

The three NF-104A aircraft were powered by both a turbojet engine and a liquid-fueled rocket motor. While the former was an afterburning, axial-flow, 14,800-lbf (with afterburning) General Electric J79-GE-3B turbojet engine, the latter was a 6,000-lbf (maximum power) Aerojet Rocketdyne AR2-3 (LR121) rocket motor. The rocket motor could be throttled from 3,000-lbf to 6,000-lbf, with a maximum burn time of two minutes (120 seconds). A reaction control system (RCS) for flight in the upper atmosphere was also provided.

USAF Col. Chuck Yeager waves from the cockpit of an NF-104A after a test flight on December 4, 1963. Oddly, cockpit ingress and egress were from the starboard (right-hand) side of the airplane rather than the port (left-hand) side, as is more common. *NMUSAF*

The trio of NF-104A AST aircraft were all highly modified F-104A-10-LO airplanes. They measured fifty-four feet, nine inches long, thirteen feet, six inches high, with a wingspan of twenty-five feet, nine inches and a wing area of 212.8 square feet. Their empty weight was 13,500 pounds, and their maximum takeoff weight was 21,400 pounds. Their top speed is listed at 2.2 to 2.4 Mn.

Numerous zoom-climb records were established by these aircraft, starting with its first flight, which reached 118,860 feet on October 23, 1963, with USAF Lieutenant Colonel Robert W. "Bob" Smith at the controls. Lt. Col. Smith, the primary pilot assigned to the NF-104A program, was flying again on December 6, 1963, when the plane reached an altitude of 120,800 feet.

On December 10, 1963, the second NF-104A (56-0762), with ARPS Commandant Charles E. "Chuck" Yeager (then a USAF Colonel, later Brigadier General) at on its controls, the AST went out of control at an altitude of 104,000 feet and entered a flat spin. Col. Yeager managed to eject successfully but was badly burned on his face by his ejector seat's rocket motor; the aircraft was destroyed in the crash. An investigation later showed that the cause of the crash was a spin that resulted from excessive angle of attack and lack of aircraft response.

NF-104A (56-0756) rockets up and away from Edwards AFB. *LM via Denny Lombard*

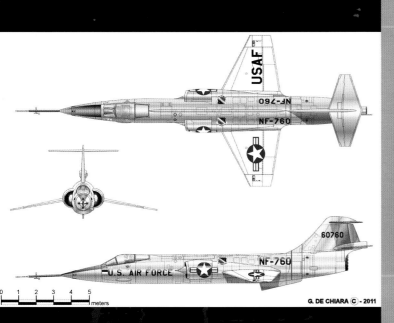

NF-104A (56-0760)—Lockheed Model CL-586—in three-view.
G. De Chiara © 2015

NF-104 (G-762, formerly 56-0762) in three-view.
G. De Chiara © 2015

The excessive angle of attack was not due to pilot error, but caused by a gyroscopic condition set up by the J79 engine spooling after shutdown for the rocket-powered zoom-climb phase.

The highly successful NF-104A AST program ended in 1971. One survivor, NF-104A number one (56-0760), is mounted on a pylon outside the USAF Test Pilot School on Edwards AFB.

CL-1200-2/X-27 SPECIFICATIONS

Propulsive system	One afterburning, axial-flow, 25,000-lbf (with afterburning) Pratt & Whitney TF30-P-100 turbofanjet engine
Length	57 feet, 3 inches/53 feet, 2 inches
Height	17 feet, 2 inches/16 feet, 2 inches
Wingspan	29 feet, 2 inches/28 feet, 7 inches
Empty weight	17,885 pounds/17,250 pounds
Gross weight	35,000 pounds
Maximum speed	2.57 Mn/2.19 Mn
Ceiling	60,000-plus feet

REPUBLIC YF-105A AND YF-105B (JF-105B) THUNDERCHIEF

The turbojet-powered airframe production business was brisk for the Long Island—based Republic Aircraft Corporation in 1950. At that time, the first production blocks of a total run of 3,025 F-84G Thunderjets were rolling off its Farmingdale assembly line. Further, Republic had recently received a contract to produce its F-84F Thunderstreak, a new sweptback-winged version of the Thunderjet. Even with all of this activity, the USAF came knocking again.

In June 1950, the USAF asked Republic Chief Designer and Engineer Alexander Kartveli to initiate a design program under Weapon System 306A for an advanced all-weather supersonic (1.5 Mn maximum) fighter-bomber for its Tactical Air Command to help replace subsonic fighter bombers in the period from 1955 to 1960. The aircraft was

Lockheed CL-1200 Lancer and the X-27

The Lockheed CL-1200 Lancer was a late-1960s company-funded proposal for a new, advanced version of the F-104. Engineered in the Skunk Works (Lockheed Advanced Development Projects), the Lancer was intended for export and was in direct competition with the McDonnell Douglas F-4 Phantom II, Northrop F-5E Tiger II, General Dynamics F-16, Northrop F-17, Northrop F-20A Tigershark, and the Dassault Mirage F1, among others.

Lockheed hoped to capitalize on its F-104 production experience through commonality of parts and systems. The company also sought to minimize expenses by reusing jigs, tooling, and its existing factory facilities. Lockheed was also experienced in consortium production and further hoped to continue this arrangement with the CL-1200. It was projected that CL-1200 deliveries could begin in 1972.

Borrowing heavily from the F-104 design, the new type featured a new high-mounted, increased span wing and low-mounted, enlarged tail planes. Both features were incorporated to improve flight handling characteristics and short-field performance. The CL-1200-1 was to have an uprated version of the F-104 engine, the General Electric J79, with a later variant known as the CL-1200-2, which was to be powered by a Pratt & Whitney TF-30 turbofan.

The CL-1200-1 was entered in the International Fighter Aircraft competition, but, since the Northrop F-5-21 was named as the winner in November 1970, the primary market for the Lancer was lost and the project terminated with no examples completed.

The X-27 was an experimental designation assigned by the USAF to a proposed high-performance research aircraft derived from the CL-1200 Lancer project. The X-27 was to have tested advanced technology, high-performance engines and equipment, though the X-27 project did not proceed beyond the mockup stage.

The CL-1200-2 was a proposed development of the X-27 for entry into the lightweight fighter (LWF) competition in 1972. The CL-1200-2 was discontinued when General Dynamics and Northrop designs were given contracts for the YF-16 and YF-17. The design was similar to the X-27, but it had round intakes with shock cones, arched bi-convex wings, and a different vertical tail.

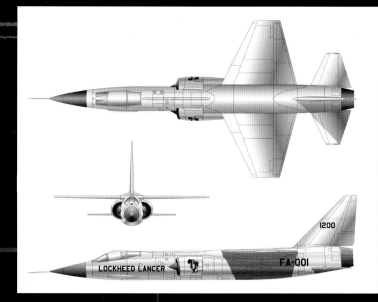

X-27 *Lancer* in three-view. *G. De Chiara © 2015*

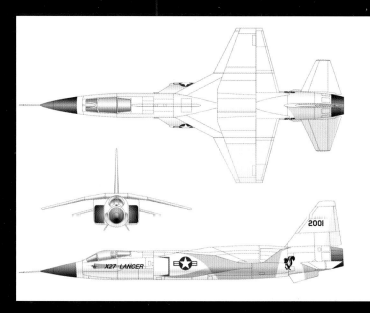

Advanced mockup of the Lockheed X-27 Lancer in its stylish white livery—a noteworthy configuration study of its high shoulder-mounted wings and downward-angled stabilators. *G. De Chiara © 2015*

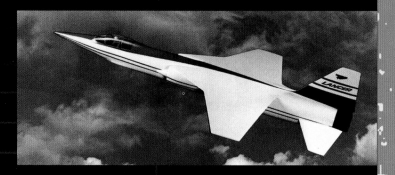

Lockheed's illustration of its proposed Lancer shows yet another slight change in its configuration. *LM via Denny Lombard*

Color view of YF-105A number two (54-0099) on the ramp at Edwards AFB. Like YF-105A number one (54-0098), this aircraft was powered by a single P&W J57-P-25 turbojet engine that produced 16,000-lbf with afterburning. *AFTC/HO*

also to double as an all-weather air superiority fighter, capable of carrying a single nuclear weapon. It was to fly in early 1955, powered by a single afterburning, axial-flow, 14,000-lbf (with afterburning) Allison J71 turbojet engine.

Following two years of design and engineering work on what Republic called its Model AP-31, (during which time it examined over 100 configurations based on its Thunderjet, Thunderstreak, and Thunderflash series of aircraft), the company settled on the thirty first version of AP 63 for its initial submission to the USAF as its AP-63-31. Republic submitted this proposal in April 1952.

The design submitted by Republic was loosely based upon its RF-84F Thunderflash design: it featured large area wing root engine air inlets and a solid nose. It had provision for several different propulsive systems, including the Allison J71, Wright Aeronautical J67, Pratt & Whitney J57, and others. It also featured an internal weapon bay, five external attachment points for ordnance, and provision for a single, nose-mounted General Electric T-171E-3 (now General Dynamics M61A1) Vulcan 20mm six-barrel, rotary-action, Gatling-type cannon. The design was eagerly approved and found immediate backing within USAF ranks.

The USAF awarded Republic a contract on September 10, 1952, to produce engineering data, wind tunnel models, weapon system and cockpit mockups, a full-scale engineering mockup of the airframe, a static test article, and production cost and time estimates (USAF contract number AF-22512). Since the United States was embroiled in the Korean War at the time, the USAF placed an initial order for 37 XF-105As, 9 YRF-105As, and 153 production F-105s. The type was to be operational by late 1955 and, although Republic favored the Pratt & Whitney J57, the USAF specified the Allison J71.

As development on the F-105 program proceeded at Republic, several important changes came about. First, the USAF changed its mind and gave Republic the go-ahead to use the J57 instead of the J71 (a good decision, since

ABOVE: The first YF-105A at Edwards AFB on May 5, 1956, having made its first flight more than six months earlier. There were no production F-105As, and production began instead with the F-105B variant. The most-produced version, the F-105D, was given the official name, Thunderchief, on June 19, 1956. Several propulsion systems were considered for the F/RF-105A, beginning with the P&W J57, then Allison J71, Wright J67, and finally the P&W J75. The P&W J57, which powered the two prototype YF-105As, was an interim engine, standing in until the J75 became available for the F-105B. *NMUSAF*

RIGHT: An aerial refueling method appropriately called the "Buddy System," under evaluation at Edwards AFB. Some of these tests were flown with J57-powered prototype YF-105A number two and J75-powered service test F-105B number three (54-0102). *AFTC/HO*

the J71 never lived up to expectations). Second, the Korean War ended, which led to program renegotiation. Third, the initial order for 199 aircraft was reduced to forty-six: thirty-seven XF-105As and nine YRF-105As.

The full-scale F-105 engineering mockup for WS-306A was inspected in October 1953, passing with just a few RFAs. At this point, the F-105 no longer resembled the F-84 family and had taken a look all its own. In January 1954, the USAF ordered that Republic plan on using the upcoming Pratt & Whitney J75 (Model JT4), which was a much-improved J57 (Model JT3), to be rated for 2 Mn.

In February 1954, the USAF changed its F-105 order yet again: two service test YF-105As, to be powered by the J57; four F-105Bs and four RF-105Bs, to be powered by the J75. This was a reduction of thirty-six aircraft. If the type went on to prove itself worthy, Republic was assured production orders would follow. Before that could happen, however, the F-105 would have to meet and defeat a friendly foe: the upcoming North American F-107A.

As production of its two interim J57-powered YF-105As progressed, Republic continued to refine the design. To meet the higher airflow demand of the J75, Republic redesigned the wing root engine air inlets, using Dr. Antonio Ferri's forward-swept "sugar scoop" design. It also redesigned the fuselage to incorporate NACA's area rule and devised a unique ram air inlet housed within the base of the vertical tail's leading edge to cool the afterburner section and actually boost the aircraft's top speed by some 15 percent. Other design features included a ventral fin to improve high-speed, high-altitude stability; a four-segment, cloverleaf speed-brake assembly around the variable-afterburner exhaust nozzle; and a ten-segment lateral (roll) control spoiler system (five atop either wing).

With the Cold War growing hotter during the mid-1950s, early development and production news on the F-105 was treated covertly. The first good photographs of the aircraft were not published until the May 20, 1957, issue of *Aviation Week* hit the stands. The first cleared photograph had appeared earlier, in the March 25, 1957, issue albeit heavily retouched. The F-105, however, had been flying some nineteen months already, about twelve months with J75 power.

Earlier, in September 1955, the first of two YF-105A aircraft was completed in Farmingdale. It was transported to Edwards AFB, housed inside a Douglas C-124D Globemaster II. After its arrival, the disassembled airplane was reassembled and thoroughly ground tested before being cleared for flight-testing. The number two YF-105A arrived shortly thereafter.

Covertly, then, on October 22, 1955, the first YF-105A completed its first flight, with Republic chief engineering test pilot Russell M. "Rusty" Roth behind the flight control stick. Even with a J57 for power, it flew supersonic in level attitude flight—hitting Mach 1.05 at thirty thousand feet. Additional flights proved its basic airworthiness, and pilots raved about its maneuverability and agility—even with its interim J57 engine. They couldn't wait to see what the J75 powered F-105B would do.

On December 16, 1955, though, the flight-test program encountered its first significant problem: the hydraulic system on YF-105A number one failed and the pilot bellied-in on the dry lake.

YF-105A SPECIFICATIONS

Propulsive system

One afterburning, axial-flow, 16,000-lbf (with afterburning) Pratt & Whitney J57-P-25 turbojet engine

Length	???
Height	17.5 feet
Wingspan	34.9 feet
Empty weight	20,454 pounds
Maximum weight	41,500 pounds
Maximum speed	1.2-plus Mn at 35,000 feet
Maximum range	2,320 miles
Combat Ceiling	31,950 feet

Armament

One rotary-action, six-barrel GE T171E2 Vulcan 20mm cannon with 1,190 rounds of ammunition; four pods filled with 2.75-inch-diameter folding-fin aerial rockets: two pods on inboard pylons under either wing, two pods on outboard pylons under either wing (seventy-six rockets total (nineteen in each pod); eight thousand pounds of bombs in various configurations, one 1,700-lb. nuclear weapon in weapon bay

Preproduction F-105B number three prepares for another takeoff at Edwards AFB. An F-105B set a world speed record of 1,216.48 mph over a sixty-two-mile course on December 11, 1959. *GSLASM via Mark Nankivil*

YF-105B/JF-105B SPECIFICATIONS

Propulsive system
One afterburning, axial-flow, 23,500-lbf (with afterburning) Pratt & Whitney YJ75-P-3 turbojet engine

Length	63.1 feet
Height	17.5 feet
Wingspan	34.9 feet
Empty weight	23,126 pounds
Maximum weight	42,162 pounds

Maximum speed
1.96 Mn (1,300 mph) at 35,000 feet

Maximum range
921 miles, 2,425 pounds of payload

Combat Ceiling	53,390 feet

Armament
One rotary-action, six-barrel General Electric T171D Vulcan 20mm cannon with 1,130 rounds of ammunition; five pods with nineteen 2.75-inch FFARs in each: four under the wings, one under the fuselage on centerline pylon; eight thousand pounds of bombs in various configurations, one 1,700-lb. nuclear bomb in weapon bay

The airplane broke near the cockpit and was not repaired. The number two YF-105A, again flown by Roth, made its first flight on January 28, 1956, and carried on until the first service test J75 powered F-105B arrived in April 1956. It made its first flight on May 26, 1956, flown by Republic test pilot Lindell Eugene "Lin" Hendrix. It was every bit the performer the pilots had anticipated. Near-sonic climb speeds and level flight doublesonic speeds became common, and the plane's overall performance was spectacular. With its heavy ordnance loads as well, the Thunderchief (as it had been officially named) was quickly becoming a genuine fighter-bomber: just what the USAF wanted. But it still had to get past an aggressive challenger from North American, the F-107A.

First flown in September 1956, North American's F-107 had been developed from the firm's F-100 Super Sabre. It was also powered by a single afterburning J75 turbojet engine, featured NASA's area rule, and could carry a nuclear bomb while maintaining the role of air superiority. By early 1957, though, the hard-fought competition between these two marvelous aircraft was won by the Republic F-105. When Republic got the nod, its Thunderchief went into production.

Republic built two service test YF-105A airplanes under USAF contract number AF-22512, approved on June 27, 1955: Republic Model AP-63. These were issued USAF serial numbers 54-098 and 54-099. Oddly, it appears that USAF serial numbers 54-001 to 54-097 had been assigned to ninety-eight cancelled F/RF-105As (this detail remains unverified).

Four service test YF-105Bs were built under the same USAF contract (USAF serial numbers 54-100 to 54-103). These featured modified engine air inlets, area ruling, and single afterburning, axial-flow, 23,500-lbf (with afterburning) Pratt & Whitney YJ75-P-3 turbojet engines.

On May 26, 1956, Republic test pilot Hendrix first took to the air on the number one YF-105B (54-100), Model AP-63. After about an hour in the air, Hendrix was forced to make a crash landing because the

nose landing gear refused to extend. The airplane was damaged beyond repair, but Hendrix was not injured.

The second YF-105B (54-101) made its first flight on January 30, 1957, but it too crash-landed because of another landing gear problem: this time, the main landing gear. The problem was attributed to interference between the auxiliary engine air inlets and the main landing gear. The problem was addressed and soon resolved.

It wasn't until May 14, 1957, that the first production F-105B entered into flight test with Republic test pilot Henry G. "Hank" Beaird Jr. at the controls.

Republic test pilot Lin Hendrix boards F-105B number three, which has been fitted with inert bombs and an auxiliary fuel tank for separation tests. In addition to the F-105B, Republic produced F-105Ds and F-105Fs. The F-105G (Wild Weasel) versions were modified F-105Fs. According to the USAF, the principle mission of the F-105 was that of a fighter-bomber. It carried a variety of payloads internally and externally, including a 1,700-pound nuclear store internally and up to 12,000 pounds externally. Later, in the Vietnam War, the internal weapons bay was sealed and used to carry an auxiliary 350-gallon fuel tank, and all stores (conventional, high-explosive bombs) were carried externally on five droppable pylons—one central on fuselage centerline, two under either wing. *NMUSAF*

YF-105A, YF-105B, AND JF-105B
PRODUCTION BREAKDOWN

YF-105A-1-RE:	2-54-098 and 54-099	
F-105B-1-RE:	4-54-100 to 54-103	
F-105B-5-RE:	1-54-104	
JF-105B-1-RE:	2-54-105	originally ordered as RF-105B
F-105B-5-RE:	2-54-106 and 54-107	
JF-105B-1-RE:	1-54-108	originally ordered as RF-105B
F-105B-5-RE:	2-54-109 and 54-110	
F-105B-6-RE:	1-54-111	
JF-105B-2-RE:	1-54-112	originally ordered as RF-105B

The USAF ordered three prototype RF-105B airplanes with USAF serial numbers 54-105, 54-108, and 54-112 on the same contract as the YF-105As and YF-105Bs. An additional twenty-two production RF-105B aircraft were subsequently ordered as well (believed to have USAF serial numbers 54-113 to 54-134 assigned to them). Before any of the production RF-105Bs (Model AP-71) could be built, however, the USAF cancelled the program. The three prototypes had been completed, minus their nose section camera installations; they were instead used for various test purposes. Under the circumstances, they were redesignated JF-105Bs, with the prefix J referring to its status as "Special Test, Temporary."

Republic proposed a two-seat version of its F-105B for pilot training and transition, while retaining combat capability. In April 1956, the USAF ordered five full-scale development examples, designated F-105C. Its first flight was scheduled for September 1958. A full-scale engineering mockup was also ordered, built, and inspected. In October 1957, though, the F-105C (Model AP-53-5) fell from favor and the program was cancelled before any F-105C FSD airplane could be built.

A further development of the cancelled two-place F-105C—the F-105E (Model AP-63-33), scheduled to fly in November 1959—was proposed, but it, too, was eventually rejected.

In addition to its F-105B variant (with seventy-one built), Republic also manufactured its F-105D (610 built) and F-105F versions (143 built) of the Thunderchief tribe, to bring its total production run to 833 examples.

SUMMARY

What became the F-105, then, had begun its life as a fighter-bomber experimental (FBX) study under USAF ARDC Weapon System 300A and USAF ARDC Project MX-1764, which evolved into WS-306A.

The Thunderchief, especially the F-105D variant (Model AP-63-31), proved to be a bona fide workhorse in the Vietnam War. Nicknamed "Thud," the F-105 became an awesome fighter bomber that could carry and deliver huge payloads of any air-to-ground weapon in the US arsenal.

Other versions of the Thunderchief, such as the F-105G (Model AP-63-31), modified F-105Fs), served in the Wild Weasel III program, whereby it ferreted out radar installations and ground-to-air missile batteries and then destroyed them.

CONVAIR F-106A AND F-106B DELTA DART

The Convair F-106 Delta Dart was a vast improvement on the F-102 Delta Dagger. Appearing quite similar to its counterpart to the untrained eye, the redesigned, redesignated, and renamed F-106 Delta Dart proved to be the essence of what an all-weather, high-speed, high-altitude interceptor should be. In fact, before its retirement from front line service, it was the best single-engine fighter interceptor ever produced.

ABOVE: F-106A number one (56-451) made its first flight at Edwards AFB on December 26, 1956, with Convair test pilot Dick Johnson at the controls. Seventeen prototype F-106As were built without the prefix Y to serve as preproduction test aircraft. *NMUSAF*

LEFT: F-106A number two (56-452) soars during an early test flight near Edwards AFB. Unlike the prototype F-102s, the prototype F-106s were manufactured from the outset with area rule. *NMUSAF*

BELOW: The Delta Dart in three-view. While it had retained the delta wing of the F-102A, its vertical tail did not. Since it had become a doublesonic airplane, its designers adopted a new tail (with more height and area) to help its high-speed stability. *G. De Chiara © 2015*

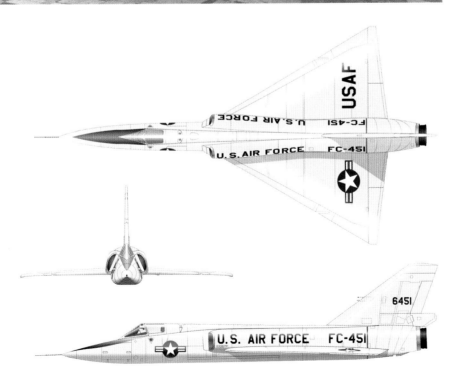

ABOVE: F-106A number three (56-0453) screams skyward in a near ninety-degree vertical climb during performance tests. As a dedicated all-weather interceptor, the F-106A had a rate of climb of 9,340 feet per minute on an area intercept mission. Its time to climb to fifty thousand feet (area intercept) was 15.5 minutes.

ABOVE, RIGHT: With Day-Glo florescent paint applied for better visibility, F-106A number one flies by during a later test flight. AFTO/110

Since the F-102 Delta Dagger had become the "1954 all-weather interceptor," then dubbed the "Interim Interceptor," its F-106 Delta Dart follow-on earned title the "Ultimate Interceptor." Indeed, the F-106 became what the F-102 was to be and, since the F-106 did not enter service until 1959, the whole 1954 interceptor/interim interceptor/ultimate interceptor program was a full five years late.

The Convair Division of the General Dynamics Corporation in San Diego, California, manufactured 277 single-seat F-106A and 63 two-seat F-106B Delta Dart airplanes between 1957 and 1961. The Delta Dart had a long tenure with the USAF's Air Defense Command (ADC), over twenty nine years. (The Air Defense Command was renamed Aerospace Defense Command on January 15, 1968.)

The Delta Dart was a fast airplane, so fast that, on December 15, 1959, an F-106A (56-0467), the seventeenth and last preproduction airplane built, established a new single-engine world absolute speed record. Piloted by USAF Maj. Joseph William "Whistlin' Joe" Rogers, the plan reached 1,525.96 mph (2.31 Mn) at an altitude of forty thousand feet, during what was called Operation Firewall. This record for a single-engine airplane has never been broken. Major Rogers, who was promoted to colonel before his retirement, was serving as project officer for F-106 integration at Air Defense Command headquarters at the time of his record flight.

F-102B

The designation F-102B (Model 8) was the temporary designation applied to what became the F-106A. The F-102B was originally optimized to use a single afterburning, axial-flow Wright Aeronautical YJ67-W-1 rated at 21,500-lbf with afterburning, 13,300-lbf military power, and 11,700-lbf normal. With a design initiated in September 1950 as the Convair "1954 AWI"

ABOVE: The main landing gear on F-106A number one, just about to lift off the main runway at Edwards AFB. *NMUSAF*

LEFT: The ninth F-106A (56-0459) lifts off for another test hop at Edwards AFB in summer 1957. This particular Delta Dart is on permanent display at the McChord Air Museum at Joint Base Lewis-McChord in Tacoma, Washington. It formerly flew with the 318th Fighter Interceptor Squadron at McChord AFB. *AFTC/HO*

BOTTOM: The prototype two-seat F-106B (57-2507) didn't receive a Y prefix, but it was used extensively for service test activities. It made its first flight on April 9, 1958, with Convair's Dick Johnson in control. The first twelve F-106Bs off the production line served as test aircraft. *NMUSAF*

(1954 all-weather interceptor), the plane's original contract date was August 1951, with the first example estimated to fly in August 1955.

As of November 2, 1953, according to the USAF Standard Aircraft Characteristics (SAC) F-102B of that date, the J67-powered F-102B was to measure 52.5 feet long, 18.0 feet high, with a wingspan of 37.0 feet. Its projected empty weight was 18,654 pounds, while its gross takeoff weight was 28,583 pounds. It was to employ the Hughes MX-1179 electronic and control system. It was to carry six Hughes XGAR-1A Falcon guided AAMs in its weapon bay, whose doors were to house twenty-four (twelve per door) unguided 2.75-inch-diameter FFARs; as many as ninety-six FFARs could have been carried in the doors (twenty-four) and weapon bay (seventy-two) in place of the six Falcon AAMs. Maximum speed was projected to be 2.4 Mn, with a combat ceiling of fifty thousand feet.

The F-102B designation stuck until at least April 23, 1956, when USAF SAC F-102B (Model 8-23) was released. By this time, though, the airplane was different from its initial designs; by its first flight, a little more than eight months later in December 1956, it had been redesignated F-106A.

F-106A AND F-106B:
MODELS 8-24 AND 8-27, 8-31 AND 8-32

As the F-102A was just starting to enter service, beginning on May 1, 1956, the first of seventeen preproduction F-102B test airplanes was slowly nearing completion at the Convair aircraft production facility in San Diego, California. No longer just a mere follow-on to the Delta Dagger, the F-102B was altogether a different airplane, which led to its redesignation F-106A on June 17, 1956. It also received a new official name at that time, Delta Dart, to further distinguish it from its predecessor.

The first F-106A (56-0451), Model 8-24, was completed and rolled out in late November 1956. It was subsequently trucked to Edwards AFB from San Diego for systems evaluations, followed by a series of engine runs and low-, medium-, and high-speed taxi tests in preparation for its first flight. After about a month of preparations, on the day after Christmas 1956, it was time to flight-test the first Delta Dart. Convair chief engineering test pilot Richard Lowe "Dick" Johnson completed his preflight checks, strapped in, and prepared to fly her. Once airborne, however, its maiden flight was cut short due to a malfunction of its speed brake doors: they would not close once extended. In addition, its ram air turbine (RAT motor) experienced frequency fluctuations. F-106A number two (56-0452) made its first flight exactly two months later, on February 26, 1957, again with Johnson at the controls.

On August 3, 1956, the USAF authorized production of an in-tandem, two-seat combat trainer version of the Delta Dart, which it designated F-106B

The proposed F-106X featured highly modified rectangular engine air inlets and canard foreplanes. It was designed to be powered by a single afterburning, axial-flow, 30,000-lbf (estimated with afterburning) Pratt & Whitney Model JT4B-22 turbojet engine, an advanced version of the P&W J75 (Model JT4A). *LM/Code One*

(it had been temporarily designated TF-106B, Model 8-27). Although it was to be a pilot training and transition version, it retained the tactical function of the single-seat F-106A, with a few differences. In part, these included a second cockpit with a complete set of flight controls; a longer, one-piece cockpit canopy to cover and seal the two cockpits; rearranged fuselage fuel tanks; and a different fire control system. The first of sixty-three tandem-seat F-106B airplanes (57-2507), to be used for service test duties, this example made its first flight at Edwards AFB on April 9, 1958, with Convair test pilot John M. "Fitz" Fitzpatrick at the controls.

F-106X: F-106C AND F-106D PROPOSALS

The F-106X was proposed as an alternative to the Lockheed F-12B improved manned interceptor (IMI). As such, two versions of the Delta Dart were proposed: the single-seat F-106C (Model 8-28) and tandem-seat F-106D (Model 8-29) variants. As proposed, these two aircraft featured a large, triangular canard foreplane planform, to be powered by a special version of the Pratt & Whitney J75 known as Model JT4B-22, which was itself fed by large-volume, cheek-mounted, rectangular air inlets.

On September 23, 1958, the F-106C/D (F-106X) program was terminated. Nevertheless, two single-seat production F-106A airplanes (57-0239 and 57-0240) were set aside for modification to incorporate in part the forty-inch-diameter radar dish within their extended length nose sections. Even though the F-106C/D (F-106X) program had been cancelled, these two airplanes were redesignated YF-106C.

Convair test pilot John Fitzpatrick made the first test-flight—as well as nine subsequent flights—on the first of the two YF-106C (57-0239) airplanes before it was permanently parked. YF-106C number two never flew.

NASA pilots who didn't fly on a daily basis used an F-106B (59-0158) for practice. These seven pilots—photographed at Edwards AFB on January 20, 1961—are America's original seven Mercury astronauts, having flown this particular tandem-seat F-106B to keep their flying skills sharp. From left to right: Malcomb Scott Carpenter, Leroy Gordon "Gordo" Cooper Jr., John Herschel Glenn Jr., Virgil Ivan "Gus" Grissom, Walter Marty "Wally" Schirra Jr., Alan Bartlett Sheppard Jr., and Donald Kent "Deke" Slaton. *NASA*

F-106A SPECIFICATIONS

Propulsive system
One afterburning, 24,500-lbf thrust Pratt & Whitney J75-P-17 turbojet engine, Model JT4A-28; early F-106A airplanes used J75-P-9 engines

Length	70 feet, 7 inches
Height	20 feet, 3 inches
Wingspan	38 feet, 1 inch
Empty weight	23,648 pounds
Maximum takeoff weight	38,729 pounds
Maximum speed	2.3 Mn (1,525 mph)
Climb rate	38,500 fpm
Ceiling	54,100 feet

Combat radius
477 miles without external fuel tanks; 705 miles with two 230 US gallon external fuel tanks

Armament
One Douglas AIR-2A Genie and four Hughes AIM-4F or AIM-4G Falcons

F-106E AND F-106F PROPOSALS

With the demise of the triplesonic Lockheed F-12B program by order of the USAF on January 5, 1968, the door was opened for Convair to propose two additional advanced versions of the Delta Dart post-F-106C/D (F-106X). On September 3, 1968, Convair proposed its single-seat F-106E and tandem-seat F-106F versions of its Model 8 series.

(The USAF had ordered ninety-three production F-12B airplanes for its ADC on May 14, 1965, as well as the funding for them—$90 million, approved by Congress. These aircraft were never built.)

The projected single-seat F-106E was to incorporate an improved look-down/shoot-down radar and fire control system, which could carry and launch the Hughes AIM-47A Falcon.

The proposed tandem-seat F-106F would be capable of combat and also serve as a pilot trainer and transition aircraft.

Neither of these Delta Dart offerings was needed by the USAF, and the F-106E/F proposals never went beyond the drawing boards.

SUMMARY

It was not until March 18, 1958, that an F-106A actually used the MA-1 automatic weapon control system (AWCS). The MA-1 AWCS weighed about 2,500 pounds and was comprised of some 170 black boxes tightly packed within the nose section of the F-106A (now Model 8-31). About 200 major components were contained in the nine subsystems of the MA-1 AWCS. Nearly all of the electronic equipment—including the communication receiver and transmitter, the gyro compass, automatic direction finding, and some of the electronic counter-counter measure (ECCM) elements—were part of the MA-1 AWCS complex.

Modifications were approved on February 12, 1969, to include the installation and testing of a clear-top cockpit canopy and a single belly-mounted 20mm M61A1 Vulcan cannon. The modifications were successful and approved for employment four days ahead of schedule on September 25, 1969. For these tests, an F-106A-100-CO (58-0795) from Tyndall AFB, Florida, was used.

The first combat-ready F-106A was delivered earlier in July 1959 to the 498th FIS, and the F-106 became operational with this unit in October 1959. The last F-106 was retired in August 1988 from the 119th FIS, a New Jersey Air National Guard unit. Adding up all of its time in service, the F-106 was operational for twenty-nine years and one month.

Ten Delta Darts were accepted by the USAF in December 1960, which ended production at 277 F-106As and 63 F-106Bs, respectively. The 340th and last Delta Dart built, an F-106A (59-0148), was delivered to the USAF's ADC on July 20 1961. All sixty-three of the tandem-seat F-106Bs (now Model 8-32) had been delivered prior to this date.

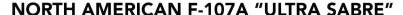

Three additional contracts—one for 100 F-106As, one for 38 F-106As, and one for 35 F-106Bs—were cancelled. The respective USAF serial numbers for these were 58-0799 to 58-0899, 59-0166 to 59-0204, and 59-0205 to 59-0240.

F-106 pilots affectionately called their Delta Darts the "Six," and, after M61 cannon were installed, the nickname changed to the "Six-Shooter." Whatever they were called, the F-106s played a central role in defending the United States from possible penetration by enemy aircraft for nearly thirty years.

Prototype F-107A number one (55-5118) in three-view. *G. De Chiara © 2015*

NORTH AMERICAN F-107A "ULTRA SABRE"

Most aviation historians do not recognize the North American F-107A as a bona fide part of the USAF Century series. Since the type was built and flown in direct competition with one of the "accepted" members of this family, discussion of this type remains appropriate here.

What began life as the B version of the North American F-100 Super Sabre evolved into the A version of the North American F-107, similar to its immediate predecessor in some ways but quite different overall. Intended for service with the USAF Tactical Air Command, the somewhat odd-looking F-107A was to be an all-weather, doublesonic fighter-bomber and/or fighter-interceptor, fully capable of performing the air superiority role. It was designed for maximum climb and roll rates, altitude and speed, and maneuverability and agility.

A number of factors contribute to the unusual appearance of the F-107A. Its bifurcated, dorsal-mounted engine air inlet and ducting system is the most

The second of three F-107A airplanes (55-5119) with an inert special store in place. In operation, these special (nuclear) stores were to be attached in their ventral semi-recessed cavities on centerline. *NMUSAF*

distinct, complete with variable geometry air inlet ramps (located just aft of its cockpit canopy), where the required amount of air was fed to the face of its single, afterburning Pratt & Whitney J75 turbojet engine for propulsion to a maximum level flight speed of 2.3 Mn, and nearly 1.0 Mn straight up—quite a performance for 1956–57. This was the primary reason for the rather unorthodox engine air inlet system built into the F-107A, engineered to obtain the best possible engine performance from the 24,500-lbf (with afterburning) J75 engine it incorporated.

Although the configuration of the F-107A borrowed much from its F-100 predecessor (particularly its wing and tail planform and its similar all-movable slab-type stabilator), the novelty of the F-107A was most apparent in its more advanced aerodynamic flying surface, with an all-movable vertical tail that doubled as a vertical stabilizer and rudder. Another novel feature was its specially hinged twelve-segment wing-mounted spoiler system (six above the wing, three on either side of the fuselage, six below the wing, three on either side of the fuselage), which replaced conventional ailerons to enhance maneuverability. Its fuselage was area-ruled, and both speed brakes and a parachute retard landing aid were provided.

For its fighter-bomber role, production F-107s were to have either four nose-mounted 20mm cannon (two on either side of the nose) or a single fuselage-mounted rotary-action six-barrel 20mm M61A1 Vulcan cannon (the latter being preferred), in addition to five store stations: two under either wing, and one semi-recessed ordnance station under the fuselage on centerline. For its fighter-interceptor and/or air superiority role, production F-107s were to carry a varied assortment of guided (infrared-guided and/or radar-guided) air-to-air missiles and rockets. It was to be capable of in-flight refueling.

Prior to the first flight of the North American YF-100 Super Sabre on May 25, 1953, North American initiated an in-house program to investigate its growth potential. That action was due to USAF TAC's hint that it would need two different, doublesonic tactical fighter types, an all-weather fighter-bomber with nuclear weapon capability, and an all-weather fighter-interceptor capable of both point and area defense.

In anticipation of these requirements, NAA came up with two company-funded projects: its Charge Number NA-211 on October 7, 1953, designated F-100B (I); and its Charge Number NA-212 on October 20, 1953, designated F-100B (the uppercase letter I in parentheses after F-100B stood for "interceptor," indicating F-100B as an interceptor). NAA's preliminary design and engineering ended in a combination all-weather, doublesonic fighter-bomber and fighter-interceptor aircraft. Several versions of each type were investigated, including cheek-type, dorsal-type, nose-type, and ventral-type engine air inlet systems. For examples, the cheek- or side-type inlets F-100B (I) interceptor was preliminary design D238, while the F-100B fighter-bomber was preliminary design D221.

These design modifications ended in a follow-on to the orthodox A-version of the F-100 Super Sabre, but, by the time it was offered to the USAF, it had become a radical departure from the F-100A, which was just about to enter production.

As proposed, the final design (the fighter-bomber NA-212 version with dorsal engine air inlet system) was the offering accepted by the USAF. The NA-212 full-scale engineering mockup passed its inspection in September 1953. Impressed, it ordered three service test prototypes on October 20, 1953, under USAF contract number AF-27787; USAF serial numbers 55-5118 to 55-5120. Due to the numerous changes differences between it and the F-100A, however, the proposed F-100B was redesignated F-107A. The popular and official name Super Sabre was not passed on to the F-107A, but those closely associated with it dubbed it the "Ultra Sabre"; no name was ever officially adopted for this plane. On August 10, 1954, the USAF upped its F-107A order to nine airplanes (same contract, amended). The last six of these would be preproduction types for use in further service tests duties, including armament, performance, and mission-related evaluations. It was to directly challenge the Republic F-105 Thunderchief in the fighter-bomber competition under USAF Weapon System 306B (the F-105 under WS-306A).

F-107A SPECIFICATIONS

Propulsive system
One axial-flow, afterburning 24,500-lbf (with afterburning) Pratt & Whitney J75-P-9 turbojet engine

Length	61 feet, 10 inches
Height	19 feet, 8 inches
Wingspan	36 feet, 7 inches
Empty weight	22,700 pounds
Maximum weight	39,800 pounds
Maximum speed	2.3 Mn

Armament
Four M39 20mm cannon or one M61A1 cannon; wide variety of tactical ordnance (including nuclear)

F-107A
PRODUCTION BREAKDOWN

F-107A-NA:	3-55-5118 to 55-5120	
F-107A-1-NA:	6-55-5121 to 55-5126	cancelled

North American commenced NA-212 production design on May 1, 1955, and the first service test F-107A (55-5118) was completed in early August 1956 without any fanfare. It was subsequently transported to Edwards AFB and, after its arrival to the secretive North Base facility there, the airplane went through a series of preflight tests.

On September 10, 1956, with NAA chief engineering test pilot Joel Robert "Bob" Baker at the controls, F-107A number one made a successful first flight. Even though its maiden flight occurred eight-plus months later than the premier YF-105A, this didn't matter: on May 26, 1956, the J75-powered F-105B had made its first flight, four months earlier.

NAA test pilot James O. "J. O." Roberts took F-107A number two (55-5119) into the air for its first time on November 28, 1956. F-107A number three (55-5120) was first flown on February 18, 1957, by NAA test pilot Alvin Swauger "Al" White.

The "paper" fly-off (the F-105-versus-F-107 competition) began in earnest in December 1956. The F-105 was judged best in March 1957, and the order for six additional F-107As was cancelled; the entire F-107 program was axed on March 22, 1957. Nevertheless, the trio of F-107As continued flight-test activities, piloted by both USAF and NAA test pilots through November 1957.

North American proposed a couple of F-107 derivatives, including a two-seat fighter-bomber and a single-seat, area- and point-interceptor with an M61 Vulcan cannon in place of the M39 20mm cannon.

During these subsequent flight tests, the F-107A trio had demonstrated outstanding performance. NACA put in a request for them to evaluate certain technologies at its facility on Edwards AFB, including a new right-hand, sidearm flight control stick. The USAF gave NACA F-107A numbers one and three on December 1, 1957, and, five days later, it gave YF-107A number two to the Air Force Museum in Dayton, Ohio, where it can be seen today.

BELOW: F-107A number two looks lean and mean on the ramp with three external fuel tanks in place. The number two F-107A made its first flight on November 28, 1956, with NAA test pilot J. O. Roberts under glass. *NMUSAF*

Despite its advanced configuration, the F-107A was not a major leap forward in the state of the art. Both the USAF and NACA got their money's worth out of these prototypes, though, as they provided priceless flight-test data to each entity.

Ignoring a proposed NAA fighter—the F-108 Rapier—which was cancelled before going into production, the F-107A was the last *new* fighter-type airplane to be produced by NAA.

SUMMARY

The F-107A competed against the J75-powered Republic F-105B Thunderchief in the late-1956/early-1957 fighter-bomber competition. Although the F-107A performed well, it was outclassed by the YF-105B: the Thunderchief could carry almost twice the payload of the F-107A.

The F-107A's dorsal variable geometry engine air inlet system that fed air to its J75 engine was unique in that no earlier or later USAF jet-powered aircraft employed such an arrangement. Instead of an enclosed weapons bay, such as that found on the F-105, the F-107 featured a ventral, semi-recessed cavity to hold conventional or nuclear ordnance.

As good as the F-107A was, it couldn't match the YF-105B's load carrying capability. For this reason, as a fighter-bomber it could not compete. Moreover, since the F-106 was proving itself to be an excellent interceptor, the F-107A couldn't match its capabilities in this regard, either. Unfortunately, there was simply no need for it, and the F-107A faded into history.

ABOVE: NAA test pilot Bob Baker controls F-107A number one during its first flight on September 10, 1956. The first of three F-107A airplanes (55-5118), it was used in part to demonstrate zoom climbs. During one of these tests, beginning at a speed of Mach 2.1 at thirty-nine thousand feet, test pilot Al White zoomed to sixty-nine thousand feet—thirty thousand feet higher than expected—before his climb rate dropped to just 100 feet per minute. *NMUSAF*

FIVE

PROPULSIVE SYSTEMS

An airframe can only be as good as its powerplant.

— Anonymous

The propulsive systems used (or intended for use) by the vast assortment of USAF bomber and fighter aircraft range widely. Besides the few rocket motors, ramjet engines, and turbopropjet engines discussed here, these aircraft primarily employed turbojet and turbofanjet engines.

While the early gas turbine (turbojet) engine designs were mostly of the centrifugal-flow variety, these types quickly gave way to the more efficient and powerful axial-flow types.

At the beginning of the period of US turbojet engine development, American airframe contractors heavily relied on propulsive systems produced by powerplant contractors that were for the most part license-built designs from engine manufacturers in Great Britain. With just a few exceptions, this was the status quo in designs produced in the first ten years.

The non-afterburning, centrifugal-flow, 1,250-lbf Type I (Model I-A) General Electric turbojet engine was exclusively used by the Bell XP-59A Airacomet. *GEAE*

The success of an aircraft is 100 percent reliant on its propulsive system. The old adage quoted above remains true to this day. While many other systems are important to jet aircraft, their lifeblood resides in their respective propulsive systems.

The aircraft discussed in this work were propelled by a wide array of propulsive systems, including ramjet-type engines and rocket motors. While it proved impractical to propel any of these production aircraft with ramjet engines and/or rocket motors alone, a number of the prototypes discussed here were used to investigate the feasibility of using such power.

In any event, comparing the minuscule 1,250 pounds force (lbf) generated by the General Electric I-A of 1942 with the awesome 43,000-lbf produced by a Pratt & Whiney F135 today reveals an astounding increase in power. The aircraft considered here owe their very existence to a vast array of propulsive systems.

AEROJET XCAL-200A AND XCAL-2000A-1 "ROTOJET"

Six science-minded "futurists" from the California Institute of Technology (Caltech)—John W. Parsons, Edward S. "Ed" Forman, Dr. Martin Somerfield, Dr. Frank J. Malina, Andrew Haley, and renowned aerodynamicist Dr. Theodore von Kármán—established the Aerojet Engineering Corporation on March 19, 1942 in Azusa, California. With their shared goal of sending rockets into space, these six colleagues had considered forming a corporation since the early 1940s.

Caltech would become the home of NASA's Jet Propulsion Laboratory (JPL), which was in part working on jet-assisted takeoff (JATO) devices for USAAF AMC Aircraft Laboratory under Secret Project MX-121. The Caltech six were part of that program, as well as serving on the USAAF AMC Secret Project MX-806 for ATO liquid-fueled rocket research. These solid- and liquid-fueled ATO and JATO units proved to be invaluable for getting heavy aircraft off the ground.

In January 1943, the Northrop Aircraft Corporation tasked the fledgling Aerojet Engineering Corporation (AEC)

Primary Differences between Centrifugal-Flow and Axial-Flow Engine Types

Centrifugal flow: Imagine a simple case where flow passes through a straight pipe to enter a centrifugal compressor. The simple flow is straight, uniform, and has no vorticity. As the flow continues to pass into and through the centrifugal impeller, the impeller forces the flow to spin faster and faster. According to a form of Euler's fluid dynamics equation (the "pump and turbine equation"), the energy input to the fluid is proportional to the flow's local spinning velocity multiplied by the local impeller tangential velocity.

In many cases, the flow leaving the centrifugal impeller is near the speed of sound (750.58 miles per hour or 1,115.48 feet per second). This typically flows through a stationary compressor, causing it to decelerate. These stationary compressors are actually static guide vanes where energy transformation takes place. As described in Bernoulli's principle, this reduction in velocity causes the pressure to rise, leading to a compressed fluid.

Axial flow: Axial compressors consist of rotating and stationary components. A shaft drives a central drum, retained by bearings, which has a number of annular airfoil rows attached (usually in pairs), one rotating and one stationary, attached to a stationary tubular casing. A pair of rotating and stationary airfoils is called a stage. The rotating airfoils, also known as blades or rotors, accelerate the fluid. The stationary airfoils, also known as stators or vanes, convert the increased rotational kinetic energy into static pressure through diffusion, redirecting the flow direction of the fluid and preparing it for the rotor blades in the next stage. The cross-sectional area between rotor drum and casing is reduced in the flow direction to maintain an optimum Mach number, using variable geometry as the fluid is compressed.

to design and develop a 200-lbf rocket motor, the XCAL-200A, for its powered MX-324 glider, then the XCAL-2000A-1 (a 2,000-lbf rocket motor) for use in its proposed XP-79 Rocket Wing. The three letters CAL in the designations XCAL-200 and XCAL-2000 meant "Caltech," a clever little plug "for our Alma mater" gambit from the Caltech team that went on to found Aerojet.

XCAL-200A

The XCAL-200A was the interim small-scale 200-lbf test engine for the full-scale 2,000-lbf XCAL-2000A-1 rocket motor, which would weigh 427 pounds. Installed in the second of three Northrop MX-334 gliders for flight-test activities, the motor was housed inside the wing trailing edge with its four pressure tanks, two propellant tanks, and controls. Its fuel was monoethylaniline and used red fuming nitric acid as an oxidizer.

Once the XCAL-200A rocket motor was fitted to the M-334 unpowered glider, it became the powered MX-324 aircraft, America's first rocket-powered airplane. This craft made a successful flight on July 5, 1944.

XCAL-2000A-1 "ROTOJET"

The XCAL-2000A-1 Rotojet was to be the experimental 2,000-lbf rocket motor that would propel the experimental rocket-powered Northrop XP-79 Rocket Wing. Neither the XCAL-2000A-1 nor the XP-79 were built.

Nevertheless, the throttleable Rotojet was to be a liquid-fueled rocket motor with four chambers: two offering 750-lbf and two 250-lbf. It was to burn liquid fuel in the ratio of 3.5:1 by weight of oxidizer (red, fuming nitric acid) and fuel (aniline): 8,400 pounds of onboard fuel.

Complete de Havilland Goblin turbojet engine on display at the South African Air Force (SAAF) Museum in Port Elizabeth, South Africa. It is representative of the H.1B Goblin (designated J36) used to propel the Lockheed XP-80. *SAAF Museum*

ALLIS-CHALMERS J36

The Engine Division of the Allis-Chalmers Corporation was to manufacture the de Havilland Halford H.1B Goblin centrifugal-flow turbojet engine in the United States under license as the J36-AC-1 for the Lockheed P-80 Shooting Star. The non-afterburning Goblin engine tested on the XP-80 produced about 2,300-lbf and proved to be inadequate for the P-80 production program. It was, however, just the second turbojet engine to propel a jet-powered type of airplane in the United States.

J36

The J36 (H.1B Goblin) turbojet engine was a part of USAAF AMC Secret Project MX-409, which led to the one-off Lockheed XP-80 airplane. Production J36s were to produce 2,700 to 3,000-lbf—a mere 400 to 700-lbf increase, which was not adequate for production P-80s.

The H.1B measures 107 inches in length, 50 inches in diameter, with a dry weight of 1,550 pounds. It has a single-sided, centrifugal-flow compressor and sixteen combustion chambers ("flame cans").

ALLISON J31, J33, J35, J71, J89, XT31, XT38, XT40

At the start of the Jet Age, the Allison Engine Company (a subsidiary of General Motors) took over the production of a few turbojet engines that had been designed by the General Electric, namely the J33 and J35 engines. It later modified the J35, which became the J71.

Allison also entered the turbopropjet engine business, producing a number of different types, including a few that went into production. These included the T56, which propels the fleet of Lockheed C-130 Hercules aircraft.

Before the T56, though, Allison offered the XT31, XT38, and XT40 engines to propel a number of the aircraft described here.

J31

Allison was to manufacture the General Electric–designed J31 centrifugal-flow turbojet engine for the P-59 under USAAF AMC Secret Project MX-172. This never came to be, however, since the P-59 production program was cut short and cancelled outright after only fifty production airplanes had been built. For this reason, no J31s were built by Allison.

J33

The centrifugal-flow J33 (GE Model I-40) produced a maximum 4,000-lbf without afterburning. Allison produced over seven thousand of these engines before production ended in 1955.

J35

As with the centrifugal-flow J33, Allison also built a large number of axial-flow J35s. These were produced both under General Electric's TG-180 and Allison's 450 model number series. With afterburning, Allison-built J35 engines produced as much as 7,500-lbf; by the time production ended in 1955 (the same year as the Allison J33), Allison had manufactured over fourteen thousand J35 engines.

The non-afterburning 3,750-lbf (maximum) Allison J35-A-5 (Model TG-180-A7) propelled the Douglas XB-43, Convair XB-46, Northrop YB-49, and the North American XP-86.

Six axial-flow non-afterburning 5,000-lbf (maximum power) J35-A-19 (Model 450-D) turbojet engines powered the one-of-a-kind Northrop YRB-49A, which made its first flight on May 4, 1950. The J35-A-21 (Model 450-D10) was an afterburning version of the -19.

The XJ35-A-23 (Model 450-E1) evolved to become the Allison J71.

J71

Originally designated XJ35-A-23 (Model 450-E1), the Allison J71 series of axial-flow J71 turbojet engines—rated at 10,000 plus lbf (with afterburning)—were used in the twin-jet Douglas B/RB-66 Destroyer production program.

J33 PRODUCTION LINES AND USE

J33-A-9	(Model 7E-I-40-A9)	XFP-80A
J33-A-13	(Model 7E-I-40-A17)	cancelled
J33-A-17/-17A		for F-80s
J33-A-19	(Model J33-C1)	for F-80s
J33-A-21	(Model 400-C4)	for F-80s
J33-A-23	(Model 400-C5)	for F-80s and T-33s
XJ33-A-25	(Model 400-C6)	P-80R, water-alcohol injection
J33-A-27	(Model 400-D4)	
J33-A-29	(Model 400-D2)	XF-92A, equipped with afterburner
J33-A-33	(Model 400-D9)	YF-94A, equipped with afterburner
J33-A-35	(Model 400-C13)	F-80s and T-33s

J35 PRODUCTION LINES AND USE

J35-A-9	(Model TG-180-D1; also 450-D1)	cancelled
J35-A-11		same as J35-GE-7 F-84s
J35-A-13/-13A	(Models TG-180-D and TG-180-D1)	F-84s
J35-A-15	(Model TG-180-A9)	YB-49, F-84s
J35-A-17	(Model 450-D4)	F-84s

J35-A PRODUCTION LINES AND USE

J35-A-25	(Model 450-D12)	F-84s

The diameter of the J71 is 39.5 inches, its length (without afterburner section) is 191.4 inches, and its dry weight (without afterburner) is 4,100 pounds.

Two 9,570-lbf YJ71-A-9 turbojet engines powered the five preproduction RB-66A airplanes.

Two 10,200-lbf J71-A-11 or J71-A-13 (slightly uprated version) turbojet engines propelled the B-66B, RB-66B, and RB-66C fleets of Destroyer tactical bomber and photographic reconnaissance aircraft.

J89

The afterburning low-bypass Allison J89 was an alternative propulsive system offered to the XB-70 program. The favored propulsive system for the XB-70 became the General Electric J93, however, which meant that the J89, a turbofanjet design, died in infancy.

XT31

The axial-flow 2,300-estimated-shaft-horsepower (eshp) Allison XT31-GE-1/-3 (Model TG-100) was the first turbopropjet engine built and flown in the United States. It was developed by General Electric under USAAF AMC Secret Project MX-464 for the Convair XP/YP-81 program.

A few years before the XT31 engine was first flown on an XP-81 in December 1945, its further development and production was turned over to Allison.

XT38

The axial-flow XT38 (Allison Model 501) turbopropjet engine produced a maximum 2,250 eshp. It was first flown on the nose of a modified Boeing B-17 engine testbed airplane on April 19, 1949. It was later employed by the XF-88B supersonic propeller testbed aircraft.

The T38 developed into the T40 and, ultimately, became the successful T56.

XT40

The Allison XT40-A-1 turbopropjet engine (Mode 500-B2-1) was comprised of two Allison T38 turbopropjet engines, with 5,850-estimated-shaft-horsepower (eshp) takeoff power. Two XT40-A-1 engines were delivered to the Republic XF-84H Thunderscreech program under USAF contract number AF33 (038)-22106. The two XF-84H airplanes were used in an in-flight USAF supersonic propeller test program.

Two T38-A-1 power sections made up the T40 (Allison Model 500) axial-flow turbopropjet engine. The 5,600-shaft-horsepower T40-A-2 was designed to propel the proposed Boeing XB-55, and it also served on the Republic XF-84H as the XT40-A-1 with 5,850 shaft horsepower.

Another mission for the XT40 was the proposed (never built) fighter-bomber version of the Curtiss F-87 Blackhawk, powered by two XT40 engines, with one on either wing.

The XT40-A-1 is 339.25 inches (28.25 feet) long, with a dry weight of 2,860 pounds. It has four turbine stages, nineteen compressor stages, eight combustor cans, and four turbine stages.

CHEVROLET J35

The production version of the General Electric Aircraft Engines XJ35-GE-1 (Model 7E-TG-180-A1), this engine was built by the Chevrolet Division of the General Motors Corporation to GEAE production drawings. It is 168 inches long with a 40-inch diameter; its dry weight is 2,400 pounds.

Designated J35-C-3 (Model 7E-TG-180-A5), this axial-flow non-afterburning eleven-stage turbojet engine produced 3,750-lbf (maximum). It was utilized by the Douglas XB-43 Jetmaster, Convair XB-46, three versions of the P-84 Thunderjet—the XP-84, XP-84A, and YP-84A, and the North American XP-86 Sabre.

CURTISS-WRIGHT/WRIGHT AERONAUTICAL J65, YJ67/XRJ55, T35, T47, T49, XLR27

The Wright Aeronautical Division of the Curtiss-Wright Corporation produced a number of propulsive systems employed by the aircraft discussed elsewhere in this book.

J65 SAPPHIRE

The axial-flow J65 was produced by Curtiss-Wright under license from Armstrong Siddeley. It was a development of the Armstrong Siddeley Sapphire turbojet engine, which powered a number of US designs, including the Martin B-57 Canberra, Republic F-84F Thunderstreak, Lockheed XF-104 Starfighter, and RF-84F Thunderflash.

YJ67/XRJ55

The Wright YJ67-W-3/XRJ55-W-1 engine combination was developed under USAF ARDC Secret Project MX-1787 (Wright Model TJ32C5) for use by the Republic XF-103.

The XRJ55-W-1 was to be a supersonic, variable-geometry ramjet engine for use with the YJ67-W-3 turbojet engine to form a double-cycle propulsion system in the XF-103. The turbojet engine was to use the ramjet engine as an afterburner to power the aircraft during takeoffs and up to a speed of Mach 2.24. After this speed was reached, the turbojet engine was to be bypassed and the ramjet engine would be used to propel the aircraft to speeds of 3.0 Mn.

The overall length of this combined propulsive system was 44.5 feet, maximum diameter was 4.6 feet, and its dry weight was 7,600 pounds. Total power was to be 32,500-lbf at 3.0 Mn, up to forty-five thousand feet.

T35 "TYPHOON"

Wright Aeronautical received a contract on November 22, 1944, to develop its XT35-W-1 "Typhoon" under USAAF AMC Secret Project MX-580. An axial-flow turbopropjet engine, it was expected to develop anywhere from 5,000 to 6,600 shaft horsepower. The XT35-W-1 was first run on June 24, 1946, and, by March 1948, it had become a candidate engine for the Boeing XB-52 program (Boeing Model 464-17 with four, then six T35-W-3 or -5 engines).

Two additional contracts (dated December 15, 1946, and May 19, 1947) were issued to Wright Aeronautical for the development of two additional versions of the T35: the XT35-W-3 and the XT35-W-5 under Secret Project MX-881. The -3 engine was to produce 7,300 to 8,900 shaft horsepower, while the -5 was to develop 8,900 shaft horsepower.

The T35 did make it to flying stage before it was superseded by the T47/T49 engines. Only a single test article was ever flown, mounted on the nose of a special test JB-17G (44-85813) flying out of Wright Field.

The Wright T35 program was terminated in March 1949 and its technology was transferred to the T47 and T49 programs.

T47

The axial-flow Wright Aeronautical YT47-W-1 (Model TP-51CA1) was an attempt to produce a high-powered turbopropjet engine based on the YJ67-W-1 turbojet engine. It was developed as a possible alternative engine on a B-47 type aircraft.

USAF letter contact AF-18488 covered the initial engineering and development of the YT47-W-1 and YT49-W-1 engines, with the ultimate aim of providing a production-version turbopropjet that incorporated the desirable features of the YJ65-W-1.

The YT47-W-1 had 11,400 shaft horsepower (maximum) and it produced 3,050-lbf of thrust. It was 166.8 inches long and 42.0 inches in diameter, with a dry weight of 4,859 pounds.

The YT47-W-1 program was abandoned, at which point it metamorphosed into the more successful YT49-W-1 program.

Head-on view revealing the XT35-W-1 "Typhoon" turbopropjet engine mounted to the nose of the special test JB-17G, being serviced at Wright Field for ongoing flying evaluations. *NMUSAF*

T49

The axial-flow Wright Aeronautical YT49-W-1 turbopropjet engine (Model TP-51A) was rated at 8,500 shaft horsepower (military power); two of them helped power the two Boeing XB-47D composite-powered Stratojet airplanes. In addition to 8,500 shaft horsepower, it also produced 3,025 lbf exhaust thrust (military power). The YT49-W-1 measures 146.5 inches long and 41 inches in diameter, with a dry weight of 4,466 pounds.

The YT49-W-1 was a direct turbopropjet development of the Wright YJ65-W-1 "Sapphire" turbojet engine.

XLR27

The Curtiss-Wright XLR27-CW-1 was a proposed four-chamber rocket motor that was to generate a total thrust of 8.400-lbf, 2,100-lbf from each chamber. It was projected to be the supplementary powerplant for the Republic XF-91 Thunderceptor with two of its exhaust nozzles above and two below the turbojet exhaust orifice. Developed under USAAF Secret Project MX-1512, it suffered from progressive problems, and the XLR27 rocket motor program was abandoned.

GARRETT AIRESEARCH F104

Two non-afterburning, axial-flow 5,440-lbf Garrett AiReseach (now Honeywell Aerospace) F104 (Model ATF3-6) turbofanjet engines were used to propel the Northrop Tacit Blue aircraft. Its length is 102 inches, diameter is 33.6 inches, and its dry weight is 1,125 pounds.

GENERAL ELECTRIC AIRCRAFT ENGINES I-A, J31, J33, J35, J47, J53, J73, J79, J85, J87, YJ93, YJ101, YF101 AND F101, F110, F118, F120, F136, F404, XT31

General Electric Aircraft Engines (GEAE) holds the distinction of producing America's first turbojet engine, which propelled America's first jet-powered airplane. The division also designed and built the first production axial-flow type of turbojet engine.

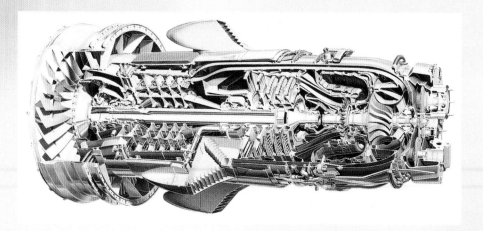

Two Honeywell (formerly AiResearch Garrett) 5,440-lbf Model ATF3-6 turbofanjet engines—military designation F104—propelled the Northrop Tacit Blue air vehicle. Five ATF3-6 type engines (shown here in cutaway) were made available to this once top-secret program. *Honeywell*

I-A

The centrifugal-flow GE Model I-A (a.k.a. Type I, pronounced "eye") was primarily a licensed copy of the Power Jets–designed, Rover-built W.2B, but with several GE modifications.

General Electric was given a Power Jets W.1X engine and drawings for the W.2B. From these, they created its I-A engine, which was first run on April 18, 1942. Two GE I-A engines, producing just 1,250-lbf each, powered the trio of Bell XP-59A Airacomet aircraft.

Further development of the Model I-A led to the Model I-16, which became the J31 series of engines.

J31

The 1,610-lbf J31-GE-1 engine was used by the thirteen service test YP-59A aircraft.

The improved J31-GE-3 engine powered the twenty production P-59A airplanes and developed 1,610-lbf with lower specific fuel consumption.

The 1,150-lbf J31-GE-5 engine gave the thirty production P-59B aircraft their power and allowed them to fly at about 410 miles per hour. The -5 was also refit to the three XP-59A airplanes for ongoing tests.

The J31 is 72 inches long, 41.45 inches in diameter, and has a dry weight of 850 pounds.

J33

The centrifugal-flow General Electric J33 turbojet engine, GE Model I-40 (USAAF AMC Secret Project MX-415), was used to propel several of the aircraft described here. These include the Lockheed XP-80A, YP-80A, XP-80B (XP-80R "Racey"), XFP-80 and XFP-80A, XF-14, Convair XP-81, Bell XF-83, Lockheed YF-94A and YF-94B, and the Convair XF-92A.

The J33 was a development of the GE J31, made larger to produce more power; its power range went from 4,000-lbf to 4,600-lbf. It was produced by both the General Electric Corporation and the Allison Engine Company division of the General Motors Corporation.

The centrifugal-flow, non-afterburning, 4,000-lbf GE J33 (Model I-40) propelled several USAF fighter aircraft including the F-80 series, XP-81, XP-83, XF-92A, and F-94A/-94B. *USAF*

J35

The axial-flow General Electric Aircraft Engines J35 turbojet (Model 7E-TG-180 series) was developed under USAAF AMC Secret Project MX-414 and was employed by a number of bombers and fighters. The J35 was the first production axial-flow turbojet engine produced in the United States. Generally, it is 145 inches long with a diameter of 37.5 inches and dry weight of 2,425 pounds.

XJ35-GE-1

The non-afterburning axial-flow 3,750-lbf XJ35-GE-1 (Model 7E-TG-180-A1) was first in the J35 series of engines, making its first test run in mid-1945. It was a test engine only and not employed by any aircraft.

J35-GE-7

The North American XB-45, Boeing XB-47, and the Martin XB-48 experimental bombers employed the J35-GE-7 (Model TG-180-B1) non-afterburning, axial-flow, 3,750-lbf turbojet engine. In the case of the six-jet XB-47 (the first of two built), only the four inboard engines (housed side by side in the twin-engine nacelles) were -7s.

The -7 was also employed by a number of early fighters, including the two experimental Republic XP-84 fighters.

J31 PRODUCTION LINES AND USE

J31-GE-1	(Model 7E-I-16-A1)	YP-59A
J31-GE-3	(Model 7E-I-16-A3)	YP-59A
J31-GE-5	(Model 7E-I-16-B5)	XP-59A (replaced I-A engines), YP-59A
J31-GE-7	(Model 7E-I-16-B7)	XP-59A (replaced -5 engines), YP-59A

J33 PRODUCTION LINES AND USE

XJ33-GE-1	(Model 7E-I-40-A1)	XP-81
XJ33-GE-5	(Model 7E-I-40-5)	XP-83
XJ33-GE-7	(Model 7E-I-40-A7)	XP-80A
XJ33-GE-11	(Model 7E-I-40-A11)	XFP-80A

J47 PRODUCTION LINE AND USE

J47-GE-17	(Model 7E-TG-190-C17) 6,650-lbf (with afterburner)	YF-86D

J35-GE-9

The non-afterburning, axial-flow, 3,750-lbf J35-GE-9 (Model TG-180-C1) was employed by the experimental XB-45 and XB-47 bombers. The six-jet XB-47 used only two -9s, mounted in the outside single-engine nacelles.

J35-GE-15

The one-off Republic XP-84A was propelled by a single axial-flow, 4,000-lbf J35-GE-15 turbojet engine.

J47

The axial-flow water- and alcohol-injected and/or afterburning GEAE J47 turbojet engine (Model 7E-TG-190 series) was developed under USAF AMC Secret Project MX-827 for numerous USAF bomber and fighter applications.

The XJ47-GE-1 (Model 7E-TG-190-A1) produced 5,820-lbf (with water/alcohol injection) and was first flown on the service test YF-86A (the first of three XF-86s built and given a Y prefix and an A suffix).

The J47-GE-3 (Model 7E-TG-190-A3) was the production version of the XJ47-GE-1, offering 5,820-lbf (with water/alcohol injection). It was used to propel XB-47 number two, which offered a speed of 650 mph.

J53

The afterburning, axial-flow General Electric XJ53-GE-1 (Model 7E-XJ53-GE-1A) turbojet engine was to produce 24,300-lbf (with afterburning) and was developed under USAF ARDC Secret Project MX-908. It was to be the propulsive system for a number of USAF aircraft, including an early Convair XB-58 (two XJ53-GE-X25s) design, which later adopted four GE J79s instead. It was also to power the proposed Douglas Model 1209 supersonic composite bomber and the Convair X-6 nuclear-fueled research airplane, neither of which were ever built.

The similar (though more powerful) XJ53-GE-3 (Model 7E-XJ53-GE-3A) was to develop 27,200-lbf (with afterburning) as a propulsive system for ground- and air-launched cruise-type missiles.

Development of the XJ53 series of Mach 2-rated turbojet engines was cancelled before any examples were test flown.

J73

The General Electric J73 turbojet was developed by General Electric from the earlier J47 engine. Its original USAF designation was J47-21, but the innovative features (variable-inlet guide vanes single-shell combustor case) led to its redesignation as J73.

The advanced J47-21 became the J73-GE-1/-3/-5 engines, which were employed by the Republic YF-84J, North American YF-86H, and production F-86Hs.

An outgrowth of the J73, designated J73-GE-X24A, was developed into the J79.

J79

The afterburner-equipped, axial-flow General Electric J79 (Model CJ805) turbojet engine was used to propel three of the aircraft described here. These included the Lockheed F-104 Starfighter, Convair B-58 Hustler, and McDonnell F-4 Phantom II. It was a Mach 2-plus rated propulsive system that allowed a number of these aircraft to establish world altitude, speed, and time-to-climb records.

The USAF version of the J79 engine was developed under ARDC Secret Project MX-2118. It was an outgrowth of the J73, initially designated XJ73-GE-X24A, then XJ79-GE-1. The XJ79-GE-1 made its first static ground run on June 8, 1954. In this form it produced 14,350-lbf with afterburning.

The J79 is 17.4 feet long, 3.2 feet in diameter, and its dry weight is 3,850 pounds. In its final form, it produces 11,905-lbf without afterburning and 17,835-lbf with afterburning.

Two non-afterburning 9,490-lbf XJ79-GE-13 turbojet engines were used to power the two highly modified Douglas WB-66D airplanes (55-0408 and 55-0410), which were redesignated X-21A for the USAF/NASA laminar flow control (LFC) system flight test program.

J85

The afterburning, axial-flow J85-GE-1 (Model CJ610) turbojet engine was developed by General Electric under USAF ARDC Secret Project MX-2273. A relatively small turbojet engine, it produces as much 5,000-lbf with afterburning. It was the propulsive system for the supersonic Northrop F-5 series of lightweight fighter aircraft.

Early in its development, the J85 was in direct competition with the Fairchild J83, which itself was developed from the Fairchild J44. The Fairchild Engine Division (previously the Ranger Aircraft Engine Division of the Fairchild Engine and Aircraft Corporation) began development of the YJ44-R-1 under USAAF AMC Secret Project MX-996 in 1947. It was used in target drones, missiles, and as jet boosters on several aircraft types.

The Fairchild J83, originally designed to propel the McDonnell ADM-20 Quail air-launched decoy missile, was also considered by Northrop as a propulsion system for its N-156T/N-156F programs. In this competition, though, the General Electric J85 prevailed.

J87

The General Electric J87 was the proposed propulsive system for the WS-125A nuclear-powered bomber (NPB). Then the afterburning, axial-flow J87 (X211) was to be the propulsive system for the continuous airborne missile launching

and low-level penetration system (CAMAL), the Convair NX-2, which replaced the WS-125A NPB program.

The J87 is 41 feet long, 80 inches in diameter, and has a dry weight of 15,745 pounds (turbojet section minus the reactor). It produced 17,300-lbf with afterburning, 13,685-lbf without. In operation onboard the NX-2, two side-by-side mounted J87s were to be fueled by a single XMA-1A nuclear reactor.

YJ93

The YJ93 was created by General Electric to propel both the North American XB-70A Valkyrie and the North American F-108A Rapier. Respectively, these aircraft were developed under USAF ARDC Weapon System 110A and WS-202A.

The YJ93 started life as the General Electric X275, an enlarged version of the J79 turbojet. This evolved to the X279 when Mach 3 cruise speed (in continuous afterburner) became a requirement; it ultimately became the YJ93.

The engine used a special high-temperature JP-6 fuel. The six YJ93-GE-3 turbojet engines that propelled the XB-70A were capable of producing an engine thrust-to-weight ratio of 5:1, allowing for a speed of two thousand miles per hour (approximately Mach 3) at an altitude of seventy thousand feet. The F-108A, though cancelled and never built, was to use two J93-GE-3 turbojet engines to generate its speed of two thousand miles per hour.

The 30,000-lbf-class YJ93-GE-3 is 237 inches long, 52.5 inches in diameter, and has a dry weight of 3,800 pounds.

Prior to the first flight of a XB-70A, this modified B-58A (55-0662)—the third one built and designated NB-58A for these tests—was to be used extensively to evaluate a service test YJ93-GE-3 turbojet engine while in flight. These tests would give future XB-70A test pilots invaluable information on how the J93 would work in practice with the Valkyrie aircraft. The NB-58A/YJ93-GE-3 in-flight test program was cancelled prior to any flights, and only engine run-ups on the ground and taxi tests were performed. *AFTC/HO*

YJ101

The YJ101-GE-100 was the propulsive system for the Northrop YF-17.

In pure turbojets, the afterburner and efflux nozzle is exposed to the turbine's superheated exhaust gases. In a bypass engine (turbofan), while some of the bypass air mixes with the core exhaust and is burned by the afterburner, the remainder is used to cool the engine's external skin and nozzle, eliminating the secondary flow normally needed to achieve this, which in turn considerably reduces complexity, drag, weight, and cost. A further advantage is gained during afterburning in that the bypass air is still relatively rich in oxygen, whereas the core air has already passed through the engine where most of the oxygen is consumed.

The J101 was a physically small engine, 12 feet, 1 inch long (3.68m), with a maximum diameter of 2 feet, 8 ½ inches (0.83m). With three low-pressure and seven high-pressure compressor stages, it achieved a compression pressure ratio of 20:1. It featured just two turbine stages, one high- and one low-pressure, and a variable converging-diverging nozzle. Its static thrust rating was 9,000 lbs (4,082 kg) at full military power, and 15,000 lb (6,800 kg) with full afterburner, which gave it a thrust/weight ratio of 8:1.

The US Navy's acceptance of the Northrop/McDonnell Douglas design for their VFAX program meant that the YF-17's engines would no longer be adequate for its new role, due to the weight increase attributable to the conversion of the lightweight, land-based fighter into a carrier-based, multirole fighter/attack aircraft. The obvious answer was to upgrade the YJ101, as it was now designated.

YF101 AND F101

Four afterburning YF101-GE-100 two-shaft axial-flow turbofanjet engines propelled each of the four Rockwell International B-1A advanced manned strategic aircraft (AMSA). These engines each produced 30,000-lbf, which propelled the second of four B-1A airplanes to Mach 2.22 (1,465 mph) during flight test activities.

The YF101-GE-100 was developed into the afterburning F101-GE-102, which propels the follow-on fleet of B-1B Lancers of the USAF Air Combat Command. The -102 version of the F101 has a diameter of 4.58 feet (55 inches) and a length of 181 inches (15.08 feet); its dry weight is 4,400 pounds. With afterburning it produces 31,000-lbf, which gives the B-1B a maximum speed of Mach 1.25 (830 mph) at fifty thousand feet; it has capacity for 0.92 Mn in very low level operations.

The B-1B isn't as fast as the B-1A because it was designed for low-altitude penetration mission profiles, rather than those for high-altitude penetration. Moreover, since doublesonic speed was no longer required, the B-1A variable-geometry engine air inlets were replaced with fixed engine air inlets on the B-1B.

Eventually, the YF101/F101 bomber engine formed the basis for what became the F110 fighter engine.

F110

Some versions of the F-15 Eagle and F-16C/D Fighting Falcon aircraft are powered by the axial-flow afterburning F110-GE-129 turbofanjet engine, which develops 29,000-lbf with afterburning. It is 181.9 inches long with a maximum diameter of 46.5 inches. It has a bypass ration of 0.96, and its airflow is 270 pounds per second.

F118

The F118-GE-100 is a two-spool, non-afterburning, axial-flow turbofanjet engine that produces 19,000-lbf of energy. It is 101 inches (8.4 feet) long with a diameter of 3.87 feet (46.5 inches); its dry weight is 3,200 pounds.

Four of these engines propel the Northrop Grumman B-2A Spirit stealth bomber to give it a top speed of 0.95 Mn (630 mph) at forty thousand feet. It has a two-stage, low-pressure compressor, and its thrust-to-weight ratio is 5.9:1.

F120

General Electric Aircraft Engines (GEAE) YF120-GE-100 (Model GE37) was an afterburning, axial-flow, turbofanjet engine that could produce 35,000-lbf with afterburning. Two of these propelled the first of two Lockheed/Boeing/General Dynamics YF-22 ATF prototypes, and another two powered the second of two Northrop/McDonnell Douglas YF-23 ATF prototypes.

The F136-GE-100 was the preferred alternative engine for the JSF program, but it was cancelled. *GEAE*

The YF120-GE-100 turbofanjet engine provided an aircraft's supercruise capability, powering YF-22 number one to Mach 1.58 and YF-23 number two to Mach 1.60.

The twin-spool, afterburning YF120 measures 13.9 feet in length, 3.5 feet in diameter, and its dry weight is 4,100.5 pounds.

F136

The afterburning General Electric/Rolls-Royce axial-flow F136 turbofanjet engine is in the 40,000-lbf class (with afterburning) and 25,000-lbf class (without afterburning). It was to be an alternative propulsion system for the Joint Strike Fighter (JSF) program.

F404

The axial-flow General Electric F404-GE-F1D2 turbofanjet engine was developed from the YJ101 and propelled the five prototype YF-117As and subsequent production F-117As.

The non-afterburning F404-GE-F1D2 produces 10,540-lbf. It measures 6.9 feet long with a diameter of 2.9 feet; its dry weight is 1,730 pounds.

XT31

The General Electric axial-flow, 2,300-shaft-horsepower XT3-GE-1 (Model TG-100) turbopropjet engine was developed under USAAF AMC Secret Project MX-464 for use by the Convair XP-81. It was the first turbopropjet engine designed and built in the United States. Its first static test run was in May 1945, and its first flight on an airplane, the XP-81, was on December 21, 1945.

LOCKHEED XJ37

The Lockheed XJ37-LO-1 turbojet engine (Lockheed Model L-1000) was an axial-flow design. The brainchild of Lockheed propulsion engineer Nathan C. Price, it was projected to produce 5,100-lbf.

XJ37

Two J37-LO-1 turbojet engines were to propel the proposed Lockheed Model L-133 interceptor. The L-133 interceptor and the L-1000 engine were both offered to the USAAF. The L-133 was passed over, but the USAAF contracted with Lockheed to further develop its XJ37 (Model L-1000) turbojet engine under AMC Secret Project MX-411.

After Menasco Manufacturing took over on the XJ37 engine project, the model number was changed to L-1000S and USAAF AMC Secret Project MX-913 was issued to the program.

Menasco eventually gave up on its Model L-1000S (XJ37) program, and it was taken over by Wright Aeronautical, who eventually transformed the design into a turbopropjet engine designated XT35-W-1.

MENASCO L-1000S

The Menasco Model L-1000S was the continued development of former Lockheed Model L-1000 (XJ37-LO-1), now under USAF AMC Secret Project MX-913.

Rare period color photo revealing the Lockheed XJ37-LA-1 with propulsive system engineer. Its inventor, Nathan C. Price, is visible at center, near the aft end of the engine. The Menasco Manufacturing Company took over the development of the J37 and later sold it to the Wright Aeronautical Corporation. Write transformed part of the engine into a turbopropjet engine, at first designated as TJ37, then as XT35-W-1 and named "Typhoon." *LM/Code One*

A former powerplant mechanic training aid P&W J48 (Model JT7) on display at the Naval Aviation Museum in Pensacola, Florida. The J48 propelled the North American YF-93A and Lockheed YF-94C. *Photograph by Greg Goebel*

NORTHROP XT37 TURBODYNE

The axial-flow Northrop T-37 Turbodyne was to be a 10,000-shaft-horsepower-class turbopropjet engine to be used on the Northrop B-35 series of flying-wing bombers. The USAAF ordered twenty-five XT37 engines under USAAF contract number W33-038 AC-6218 (13181).

For in-flight engine testing, Northrop was to equip a service test YB-35A (42-102378) with four XT37-NH-1 engines.

The XT37 was designed by Northrop engineer Vladimir H. Pavlecka in 1944. It was built jointly by Northrop and the Joshua Hendy Iron Works of Sunnyvale, California, as the Northrop-Hendy XT37-NH-1 (pusher-type) and the more powerful 10,400 shaft horsepower XT37-NH-3 (tractor-type) under the Northrop N-19 project number, which was in development for several years under USAAF AMC Secret Project MX-562.

Additional development of this engine, which was to spin a pair of four-bladed contra-rotating propellers, was performed by General Electric as the XT37A under the USAAF AMC MX-562-A project number after GE purchased the rights to the design in 1950.

PRATT & WHITNEY AIRCRAFT J48, J57, J58, J75, J91, JT15D, F100, YF119, F135, TF30, TF33

Pratt & Whitney Aircraft (P&WA) has produced some of the most important propulsive systems for many aircraft.

J48

The centrifugal-flow Pratt & Whitney J48 (P&W Model JT7) was developed under license from the Rolls-Royce RB.44 Tay turbojet engine. It was a further

development of the license-built P&W J42 (P&W Model JT6) that was created from the Roll-Royce RB.41 Nene, an earlier centrifugal-flow design.

The J48-P-1 and J48-P-3 turbojet engines were developed under USAF AMC Secret Project MX-1447 and led to the development of the afterburning 8,750-lbf (with afterburning) J48-P-5 used by the Lockheed YF-94C and the J48-P-6, which was employed by the North American YF-93A.

J57

The axial-flow Pratt & Whitney J57 (P&W Model JT3) powered the USAF's first combat aircraft capable of supersonic speed while flying on the straight and level. It was developed under USAF ARDC Secret Project MX-1116 and propelled the Boeing XB-52 and YB-52, Convair YB-60, North American YF-100, McDonnell F-101A and F-101B, and the Convair YF-102 and YF-102A.

The P&W J57 was developed in stages under USAF ARDC Secret Projects MX-1116 for the XJ57-P-1 (Model JT3A) of 9,000-lbf and YJ57-P-3 (Model JT3) of 8,700-lbf for MX-839 or the experimental XB-52 and prototype YB-52 aircraft; MX-1116-A as development of J57; and MX-1116-B for the XJ57-P-5.

The afterburning version of the J57 Turbowasp was the first Mach 1-rated turbojet engine in America. In fact, on May 25, 1953, it propelled the first of two YF-100 service test airplanes to Mach 1.1 (815 mph) in level attitude flight during its very first flight. This marked the first time that a pure combat-type airplane had accomplished this feat.

J58

The Pratt & Whitney J58 (P&W Model JT11D-20) propelled the Lockheed YF-12A to speeds in excess of Mach 3 (2,000 mph). An axial-flow-type, air-inlet, bleed-bypass engine, it produced 32,500-lbf with afterburning. The J58 was also considered for use by the proposed Convair B-58C and by the F-106X version of the Convair Delta Dart.

The J58 measures seventeen feet, ten inches long, four feet, nine inches in diameter, and has a dry weight of about six thousand pounds. At a speed of about 1,600 mph (2.4 Mn), the J58 turbojet engine essentially became a ramjet engine.

A P&W J58 (Model JT11D-20) on display at the Evergreen Air Museum in McMinnville, Oregon. Two Mach 3–rated J58s powered each of the three prototype YF-12As. *Photograph by Greg Goebel*

J75

The Pratt & Whitney J75 (Model JT4) was an afterburning, axial-flow turbojet engine that was developed under USAF ARDC Secret Project MX-1984.

The Pratt & Whitney J75, alongside the General Electric J79, was the first Mach 2–rated engine to be produced in America.

The J75 propelled several USAF experimental and prototype aircraft, including the Republic YF-105B, Convair F-106A, and North American F-107A.

The YJ75-P-3 produced 23,500-lbf with afterburning and propelled the YF-105B aircraft. It measured 21.6 feet in length, 3.6 feet in diameter, and its dry weight was 6,175 pounds.

The J75-P-17 produced 24,500-lbf with afterburning and powered the prototype F-106A aircraft. It measured 19.8 feet in length and 3.7 feet in diameter, with a dry weight of 5,875 pounds.

Schematic illustrating the operation of the J58 engine as the YF-12As would approach and pass Mach 3 speeds. *LM*

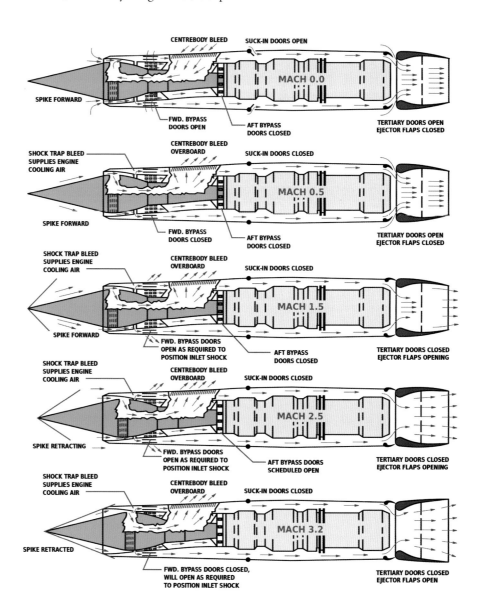

Pratt & Whitney Aircraft

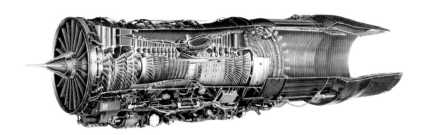

The YJ75-P-9 developed 24,500-lbf with afterburning and was the propulsive system for the prototype F-107A aircraft. Its dimensions and weight were similar to the -3 and -17 versions of the J75.

J91

The Pratt & Whitney J91 (Model JTN9) was to be an alternative nuclear-fueled engine for the WS-125A nuclear, powered bomber.

JT15D

The non-afterburning, axial-flow JT15D-5C turbofanjet engine, built by Pratt & Whitney Canada, was used to propel the Boeing *Bird of Prey* demonstrator airplane. It developed 3,190-lbf, which gave the aircraft a top speed of three hundred mph. It measures 60.5 inches in length, 27 inches in diameter, and weighs about 630 pounds.

F100

The Pratt & Whitney F100-PW-100 (Model JTF22A-25A) was the propulsion system for the prototype F-15A aircraft, producing 23,830-lbf in afterburner. The 23,830-lbf (with afterburning) Pratt & Whitney F100-PW-200 (Model JTF22A-33) is the afterburning, axial-flow, turbofanjet engine that propelled the prototype YF-16 airplanes. The -100 evolved into the -200, then the -220, and finally the -229.

The Pratt & Whitney F100-PW-229 propelled the demonstrator F-15E Strike Eagle, demonstrator F-15SE Silent Eagle, and both the F-16C and F-16D. In afterburner it produces 29,160-lbf and functions as a Mach 2–rated engine. It is 15.9 feet long, 3.875 feet in diameter, and has a dry weight of 3,826 pounds.

YF119

The YF119-PW-100 (Model PW5000) is an afterburning, axial-flow turbofanjet engine that propelled both the YF-22 and YF-23 ATF prototype aircraft (YF-22 is number two, YF-23 is number one).

The YF119 produced 35,000-lbf with afterburning and featured supercruise, whereby the ATF prototypes could fly at Mach 1.5-plus speeds with the use of afterburners.

F135

The F135 was developed from the F119 for exclusive use by the trio of Joint Strike Fighter aircraft. The F135-PW-100 turbofanjet engine is the propulsive system for the F-35A, producing 43,000-lbf (maximum in afterburner) and 28,000-lbf (intermediate power). It measures 18.33 feet long, 3.83 feet in diameter, and has a dry weight of 3,750 pounds.

TF30

The afterburning, axial-flow, Pratt & Whitney TF30-P-1 (Model JTF10A) turbofanjet engine was used to propel the early research, development, test, and evaluation (RDT&E) F-111A aircraft. The troublesome -1 (rife with compressor stalls) gave way to the 18,500-lbf TF30-P-3, which replaced the -1 engines on the F-111A RDT&E aircraft.

TF33

The Pratt & Whitney non-afterburning, axial-flow, P&W TF33-P-3/-103 (Model JT3D) turbofanjet engine is employed by the B-52H and produces up to 17,000-lbf. The TF33 is 20.35 feet long with a diameter of 3.3 feet.

REACTION MOTORS XLR11

Reaction Motors, Inc. (RMI) produced the first rocket motor powerful enough to propel a piloted air vehicle to speeds faster than sound. This historic event took place on October 14, 1947, when the first of three Bell X-1 aircraft, flown by USAF Cap. Charles E. "Chuck" Yeager, exceeded Mach 1 (1.06 Mn attained) while passing through an altitude of forty-two thousand feet above Muroc AFB.

The engine employed by this air vehicle was a 6,000-lbf RMI XLR11-RM-5. The Reaction Motors XLR11 (RMI Model RMI 6000C4) is a four-chamber rocket motor, with each chamber producing 1,500-lbf. The chambers could be fired separately or in groups to provide a maximum of 6,000-lbf.

The XLR11-RM-9 version of the engine, used in combination with a J47 turbojet engine, helped propel the Republic XF-91 Thunderceptor to supersonic speed on December 9, 1952. This represented the first time in US history that a combat-type aircraft flew supersonic in level attitude flight.

The liquid-fuel XLR11-RM-9 was developed under USAAF AMC Secret Project MX-976 and was to be the secondary booster engine, working in concert with the primary afterburning J47 turbojet engine, employed by the XF-91 Thunderceptor.

WESTINGHOUSE ELECTRIC J30, J34, J40, J46

The Aviation Gas Turbine (AGT) Division of the Westinghouse Electric Corporation produced some of the earliest engines used by the turbojet-powered bomber and fighter aircraft. None successful, except for the J30 and J34, which enjoyed limited success.

J30

The axial-flow Westinghouse Electric J30 (Westinghouse Model 19 series) was the first American-designed turbojet engine to run, an event that took place on March 19, 1943.

The J30 was developed under USAAF AMC Secret Project MX-826 as a pair of non-afterburning, 1,365-lbf Model X19B (XJ30-WE-1) engines that propelled the Northrop XP-79B on its first and only flight.

The composite-powered Douglas XB-42A used two non-afterburning XJ30-WE-3 (Model X19XB-2B) turbojet engines, which developed 1,600-lbf. It later used two non-afterburning, 1,700-lbf XJ30-WE-5 (Model X19XB-3) turbojet engines for evaluation of the type for other applications.

Later versions of the non-afterburning J30 were developed under MX-1426, including the 1,560-lbf XJ30-WE-7/-9 and the 1,560-lbf J30-P-7/-9 engines built under license by Pratt & Whitney.

J34

The axial-flow Westinghouse J34 (Model 24C series) turbojet engine was initially developed under USAAF AMC Secret Project MX-825. Used as both a non-

afterburning and afterburning propulsive system, it was employed by no less than eight aircraft: the Lockheed XP-80A, the McDonnell XF-85, the Curtiss XF-87, the McDonnell XF-88 and XF-88A and XF-88B, and the Lockheed XF-90 and XF-90A.

The second of two Lockheed XP-80A airplanes was fitted with a single XJ34-WE-11 (Model X24C-5) turbojet engine to evaluate the Lockheed-made afterburner section for the XP/XF-90 program. This engine produced 4,100-lbf with afterburning.

Four versions of non-afterburning Westinghouse Electric J34 engines (Project MX-1427) were used to propel the four-engine Curtiss XF-87 Blackhawk, which gave it a maximum level flight speed of 600 mph. These included the XJ34-WE-1 (Model X24C-4A), XJ34-WE-5 (Model X24C-4B), XJ34-WE-7 (Model 24C-4B), and the XJ34-WE-9 (Model 24C-4B).

The non-afterburning XJ34-WE-13 (Model X24C-5) engine (Project MX-1174) was used by the XF-88 Voodoo and it produced 3,000-lbf.

The afterburning Model X24C-5 (XJ34-WE-15) engine (Project MX-1174) was employed by both the XF-88A and XF-90A penetration fighter program contestants. In each case, the manufacturers of these aircraft developed their own afterburner sections for this engine. Their maximum military power ratings were 3,150-lbf, but, with afterburning, these engines produced approximately 40 percent more power, or about 4,100-lbf. Using their homemade afterburner sections, their top speeds increased to 690 miles per hour and 665 mph, respectively, in level attitude flight.

The 3,000-lbf (without afterburning) XJ34-WE-3 (Model X24C-4B) turbojet engine propelled the McDonnell XF-85 Goblin. This engine was later replaced by the non-afterburning, 3,000-lbf J34-WE-22 (Model 24C-4B) production-type engine.

The 4,900-lbf (with afterburning) J34-WE-17 (modified Model 24C-8) engine powered the XF-88B tri-motored Voodoo. It too was developed under MX-1174.

J40

The Westinghouse J40 turbojet engine is an axial-flow design, offered as the propulsive system for a number of USAF aircraft, including an early version of the Boeing XB-52. This plane was to incorporate eight afterburning XJ40-WE-6 (Model X40E2) turbojet engines, rated at 7,500-lbf.

The XJ40-WE-6 was developed from the original J40, the afterburning XJ40-WE-1 (Model X40E3A-1C).

J46

The afterburning, axial-flow Westinghouse Electric J46-WE-1 and J46-WE-3 turbojet engines were proposed propulsive systems for several aircraft found, including the proposed production version of the McDonnell F-88 Voodoo and the proposed production version of the Lockheed F-90. It was a scaled-up development of the J34.

The XJ46-WE-1 (Model X24C-10) was to develop up to 4,800-lbf (without afterburning), 6,000-lbf (with afterburning). Developed under USAF AMC Secret Project MX-1169, it was never used in any USAF bomber or fighter airplane because of sustained developmental problems.

The XJ46-WE-3 was also to be developed under MX-1169 for possible USAF application, but as with the XJ46-WE-1, the program was halted.

SUMMARY

The foregoing information on these various propulsive systems, used or planned for use by a variety of experimental and prototype bomber and fighter airframes, offers a brief overview of each powerplant. There is much more to tell, and only a book devoted solely to their respective histories could provide the details they deserve.

None of the aircraft discussed here could have succeeded without the engines that propelled them.

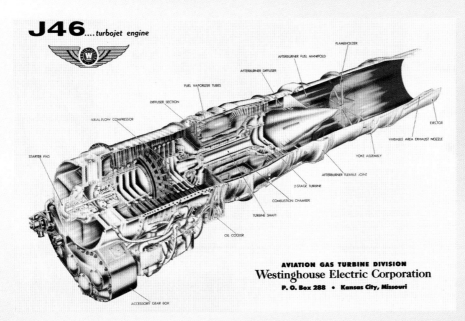

The 6,000-lbf-class Westinghouse J46 turbojet engine (Model 24C series) was considered for use by several aircraft, including the F-88 and F-90—had it gone into production. Instead, the J46 was passed over due to developmental problems. *Westinghouse*

SIX

SUPERSONIC WONDERS AND DOUBLESONIC SENSATIONS

Air power is like poker. A second-best hand is like none
at all—it will cost you dough and win you nothing.
— Gen. George C. Kenney, USAAF

In the course of aviation history, addressing USAF
turbojet-powered combat aircraft in particular, the
need for speed and service functionality in many
different combat missions emerged for specific
types of bomber and fighter aircraft. Many sensa-
tional aircraft appeared between 1955 and 1965,
including those that reached the speed of sound
and others that effortlessly surpassed it.

236

A Glenn L. Martin Company artist's conception of how operational B-68s might have appeared. *GLMAM via Stan Piet*

Several promising aircraft programs had to be sacrificed in order for the USAF to move forward with other, more favorable programs. Some fabulous aircraft appeared during this era and became legends in their own time. It was a glorious time that gave birth to a number of supersonic wonders and doublesonic sensations.

MARTIN XB-68

The Glenn L. Martin Company XB-68 was to be a twin-turbojet-powered, all-weather, medium-class doublesonic tactical bomber assigned to the USAF Tactical Air Command (TAC) as Weapon System 302A.

The USAF ordered engineering data, wind tunnel models, a full-scale engineering mockup, two flying experimental XB-68 airplanes, and a static test airframe from Martin, leading the company's management to dub the program its Model "316" Tactical Bomber. As projected at the time, production B-68 airplanes were to be available for operations in 1962–1965, according with the USAF's principal mission for the B-68: the destruction of ground targets in support of tactical operations.

In its final configuration, the XB-68 was to be propelled by two afterburning, axial-flow, 27,500-lbf (with afterburning) Pratt & Whitney J75 (Model JT4B-21) turbojet engines. Earlier, however, it was to be powered by two afterburning, axial-flow, Curtiss-Wright J67 (Model TJ32C4) turbojet engines. (The USAF cancelled its J67 engine program in 1955.)

Operational B-68s would have had two-man crews, including the pilot and weapons system operator (WSO). The WSO would be responsible for navigation, bombardment, and defensive operations. Each crewman would be seated in his own cockpit on an emergency ejection seat. The cockpits were of a tandem arrangement, with separate openings atop the fuselage for ingress and egress.

The B-68 was of conventional design, optimized to operate in all-weather and at low and high altitudes. It was to have a maximum speed of 2.4 Mn (almost 1,800 mph). Its stainless-steel skin was to absorb temperatures of up to 350°F.

The design of the XB-68 borrowed from Martin's earlier XB-51, featuring an all-moving, T-type horizontal stabilator mounted on top of its vertical stabilizer and rudder assembly. Although it had a tricycle landing gear arrangement rather than a bicycle arrangement (à la XB-51), the airplane would have outrigger landing gears in streamlined housings attached to either wingtip for balance during ground maneuvers.

It featured dive brakes, two tail-mounted deceleration parachutes (drag chutes), in-flight refueling, and rotary-action bomb bay doors.

XB-68 PRODUCTION BREAKDOWN:

| XB-68: | 0-56-427 to 56-450 | twenty-four aircraft on order, cancelled |

ABOVE: A full-scale engineering mockup of the XB-68. By this time, it was designed to be powered by a pair of J75s instead of J67s. *GLMAM via Stan Piet*

LEFT: The experimental XB-68, shown as it appeared during its USAF 689 engineering board mockup inspection stage. *G. De Chiara © 2015*

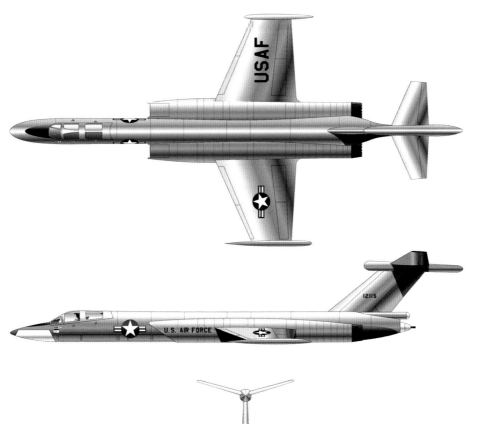

XB-68 SPECIFICATIONS

Propulsive system
Two afterburning, axial-flow, 27,500-lbf (with afterburning) Pratt & Whitney J75 (Model JT4B-21) turbojet engines

Length	105 feet, 9 inches
Height	26 feet, 0 inches
Wingspan	53 feet, 8 inches
Wing area	900 square feet

Maximum speed
2.4 Mn (1,780 mph) (estimated)

Empty weight
49,525 pounds empty (estimated)

Gross takeoff weight
96,600 pounds maximum take-off weight (limited by space), 74,180 pounds for basic mission (subsonic) (estimated)

Payload
One Class C 8,500-pound nuclear bomb or one Class D 3,500-pound nuclear bomb on a rotary bomb carrier

Maximum speed
1,588 mph (2.14 Mn) at 54,700-foot altitude at maximum power

Combat radius
1,250 miles with 3,500-pound (Class D) payload at an average speed of 605 mph for 4.15 hours

Crew
Two (pilot and weapon system operator)

Ceiling
40,950 feet, 100 feet per minute, takeoff weight, military power; 44,800 feet (500 fpm), combat weight, using military power

SUMMARY

While the full-scale engineering mockup was completed and inspected by the USAF, metal was never cut to begin construction on either of the two XB-68 airplanes, as the program had been cancelled before production could begin.

Moreover, in a rather strange turn of events, the designation for this airplane was given to another Martin project, its Titan intercontinental ballistic missile (ICBM), which was also designated XB-68 during the time the USAF was still designating its ballistic missiles with the prefix B ("for bomber"). This ICBM became the SM-68A Titan (SM standing for "strategic missile").

CONVAIR B/RB-58A HUSTLER

The Convair Division of the General Dynamics Corporation built 116 B/RB-58A Hustler airplanes under Weapon System 102A (WS-102A) for the Strategic Air Command (SAC) of the US Air Force at its Fort Worth, Texas, facility between 1956 and 1962. Eight of these were subsequently converted to TB-58As for pilot training and transition purposes. B-58s equipped with ventral reconnaissance pods were designed RB-58A under WS-102L. The premier B-58A made its maiden flight out of Fort Worth on November 11, 1956, with Convair test pilot Beryl Arthur Erickson at the controls.

The doublesonic B-58A was powered by four afterburning, 15,600-lbf (with afterburning) General Electric J79-GE-5A turbojet engines housed in separate underwing pods. It established numerous speed and altitude records and won a number of significant aviation awards and trophies.

Operational B-58As had a three-man crew seated in tandem within their own compartments, comprised of the aircraft commander (pilot), navigator/bombardier (nav/bomb), and defense system operator (DSO).

The Convair B-58 Hustler holds the distinction of being the first (and, currently, only) doublesonic bomber ever produced and put into service with the USAF.

The design of what became the Convair B-58 Hustler was initiated on February 17, 1951, under USAF ARDC Secret Project MX-871. Project MX-871 was identified only as the "GEBO" (*GE*neralized *BO*mber) studies.

As such, Convair (as *Con*solidated *V*ultee *Air*craft would be known), formulated a rather strange two-part aircraft, made up of a bombardment-type air vehicle with an attached pod. The two-part droppable pod was to house a nuclear weapon and fuel.

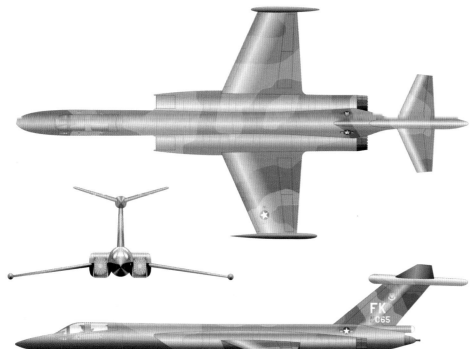

ABOVE, LEFT: A worm's-eye view of prototype B-58A shortly after its rollout on September 4, 1956, showing its massive array of four General Electric J79 turbojet engines. Housed within their respective pod-type engine nacelles, the engines attached under the aircraft's wings with individual aerodynamic pylons. *LM*

ABOVE, RIGHT: Bottom view of prototype B-58A during its first flight. *LM*

LEFT: It's easy to imagine what an operational USAF Tactical Air Command B-68 might have looked like during the Vietnam War—this example is wearing its Southeast Asia camouflage livery. *G. De Chiara © 2015*

Project MX-871 morphed into Project MX-1626, which called for a two-part "land based–refueled" aircraft under USAF contract number AF33 (038)-21250. The USAF Characteristics Summary (CS), dated July 28, 1951, specified a propulsive system of three General Electric J53 turbojet engines, a two-man crew, a maximum fuel capacity of 9,532 US gallons, and an expendable two-part pod containing fuel (35 percent of total fuel), warhead, and guidance equipment; the warhead was to be released as a guided bomb.

The MX-1626 design fell flat and morphed again into Project MX-1964, which was a much improved offering (Model 4) that initially drew the designation XB-58.

MX-1626 SPECIFICATIONS

Length	81.1 feet
Height	22.4 feet
Wingspan	47.3 feet
Combat radius	4,475 miles
Combat range	7,834 miles
Combat speed	1,090 mph
Maximum speed	1,124 mph
Maximum climb rate 9,500 feet per minute; combat weight, maximum power	
Maximum Ceiling 46,400 feet; combat weight, maximum power	
Payload One 6,000-pound nuclear weapon	
Empty weight	38,529 pounds
Maximum takeoff weight	107,250 pounds

On November 18, 1952, the Convair Division of the General Dynamics Corporation in Fort Worth, Texas, was given the green light to proceed with its B-58 program, after which it built two experimental and eleven service test airplanes. All of these were designated B-58A-1-CO (USAF serial numbers 55-660 to 55-672), and thirty-one pods under USAF contract number AF-21230 were approved by the DOD on December 29, 1955. The first two were used as dedicated experimental bombardment airplanes, while the last eleven were used as service test bombardment airplanes. These thirteen preproduction aircraft were never referred to as XB-58 or YB-58 on an official basis, although some references do use this appellation.

In late August 1956, the premier B-58 (55-660) was rolled out of the final assembly area of the Fort Worth factory. After numerous ground tests, including low-, medium-, and high-speed taxi tests, the first B-58A was ready to fly. On November 11, 1956, with General Dynamics' chief engineering test pilot on the B-58 program Beryl A. Erickson in command, the first Hustler made a successful first flight.

On June 29, 1957, with Erickson in command on its twenty-fourth flight, the number one B-58 (55-660) exceeded Mach 2 for the first time; as this was its design speed, it hit 2.03 Mn at 43,250 feet.

Another seventeen initial production B-58s were assigned to the test force, totaling thirty test aircraft, fifteen of which were later brought up to production standard.

The overall success of these thirty test aircraft led to the production of an additional eighty-six Hustler airplanes as follows: thirty service test

Operational B-58s could carry a vast variety of conventional and nuclear arms. *PTH via LM*

ABOVE, LEFT: The first three service test B-58A aircraft rest easy inside a large hangar at GD's facility in Fort Worth. *PTH via LM*

ABOVE, RIGHT: A clear shot of the two-component pod (TCP) during its first takeoff with the B58A. *LM*

LEFT: Three-view of first B-58A as it appeared during its first flight on November 11, 1956. *G. De Chiara © 2015*

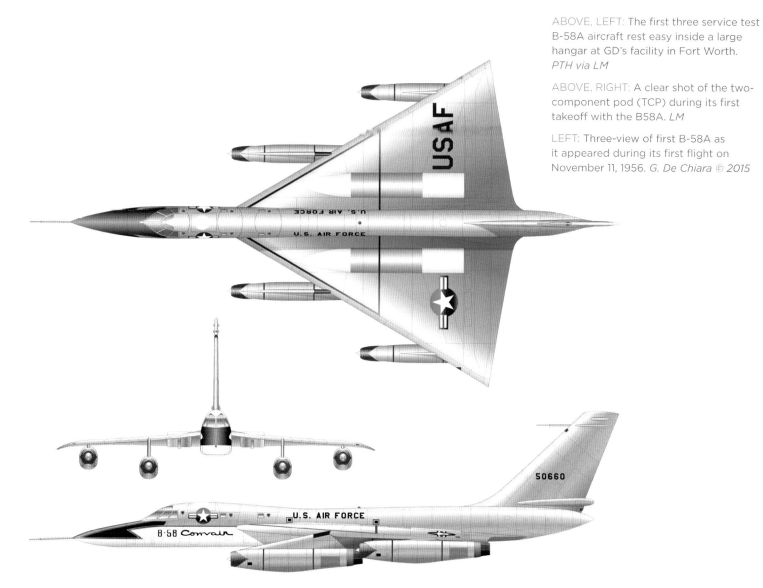

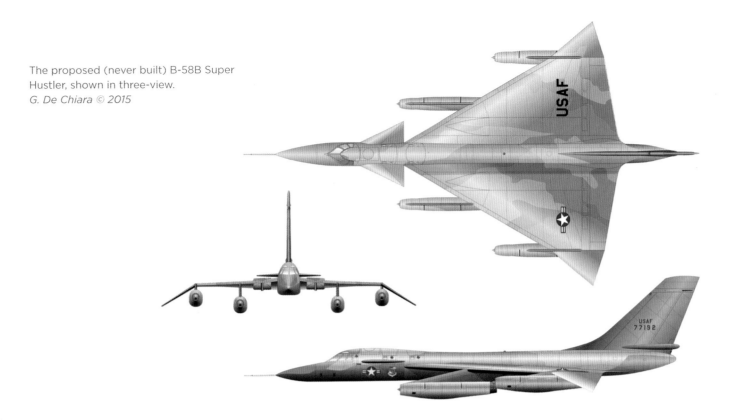

The proposed (never built) B-58B Super Hustler, shown in three-view.
G. De Chiara © 2015

B-58As and eighty-six production B-58As; eight of the service test B-58As were modified to TB-58A.

The B-58, like its F-102 and F-106 counterparts, featured the area rule configuration with its pinched-in fuselage and a delta wing. Its vertical tail, however, was not of a delta shape.

B-58B

One service test B-58B (without the YB prefix) was ordered (USAF serial number 60-1109) to be evaluated as an improved follow-on to the B-58A with conventional weapons as well as nuclear weapons carriage capability. The B-58A had been configured to carry and deliver only nuclear stores.

The B-58B would have had a five-foot fuselage extension, increased gross weight to 186,000 pounds, GE J79-GE-9 engines, increased vertical tail area, strengthened wings and vertical tail, and it was to be the carrier of the proposed Convair FISH/KINGFISH reconnaissance air vehicles—had Convair won the CIA-sponsored Project GUSTO competition.

One early production service test B-58A (USAF serial number 55-0668) was to be modified to become the prototype B-58B. This never came to fruition, however, as the B-58B program was cancelled before any modification work to this particular B-58A was initiated.

B-58C

The B-58C was an unbuilt version of the Hustler. It was to be an enlarged version with more fuel and two or four 32,500-lbf Pratt & Whitney J58 turbojet engines—the same as were used on the Lockheed YF-12A. Design studies were conducted with two and four engine designs: the C model had an estimated top speed approaching Mach 3, a supersonic cruise capability of approximately Mach 2, and a service ceiling of about seventy thousand feet along with the capability of carrying conventional bombs. Convair estimated maximum range at six thousand miles. The B-58C was proposed as a lower-cost alternative to the North American B-70 Valkyrie. As enemy defenses against high-speed, high-altitude penetration bombers improved, the value of the B-58C diminished and the program was cancelled in April 1961.

RB-58A

Nineteen dedicated RB-58A airplanes were ordered under the FY1960 defense budget (USAF serial numbers 60-1130 to 60-1148), but these were cancelled prior to their manufacture.

TB-58A

Eight former service test B-58As were converted to serve as pilot training and transition aircraft, designated TB-58A. A few of these also served as chase aircraft in the XB-70A, YF-12A, and SR-71A flight-test programs.

SUMMARY

The B-58 was a beautiful airplane, especially when it was in the air without its two-part fuel/weapon pod attached. This unsightly pod was critical to its operation, though, so it could not be left off, and it was saddled to the Hustler throughout its tenure in USAF SAC.

The B-58 had a long development time but a very short career in USAF SAC. It first entered service on March 15, 1960; by January 31, 1970, it had been fully retired—just under a ten-year period in operation.

The Convair Division of the General Dynamics Corporation built 116 B/RB-58A Hustler airplanes in Fort Worth, Texas, and twenty-four of these were lost in crashes; two others were damaged in accidents (one returned to flight). In all, ninety-one of the 116 Hustlers built were sent to the scrap yard. Eight of these B/RB-58As were saved, and they are currently on public display.

B/RB-58A SPECIFICATIONS

Propulsive system
Four afterburning, axial-flow, 15,600-lbf (with afterburner) General Electric J79-GE-5A turbojet engines

Length	96 feet, 9 inches
Height	31 feet, 5 inches
Wingspan	56 feet, 10 inches
Wing area	1,542.5 square feet
Empty weight	55,560 pounds
Gross weight	163,000 pounds

Maximum speed
1,320 miles per hour (Mach 2) at 63,100 feet

Maximum range	5,634 miles
Ceiling	63,000 feet

Crew
Three (pilot, navigator/bombardier, and defense system operator)

Armament
One tail-mounted, six-barrel, rotary-action General Electric M61A1 Vulcan 20mm cannon (580 to 1,040 rounds of 20mm ammunition)

Payload
One 6,230-pound W-39 freefall 8.8-foot-long nuclear device

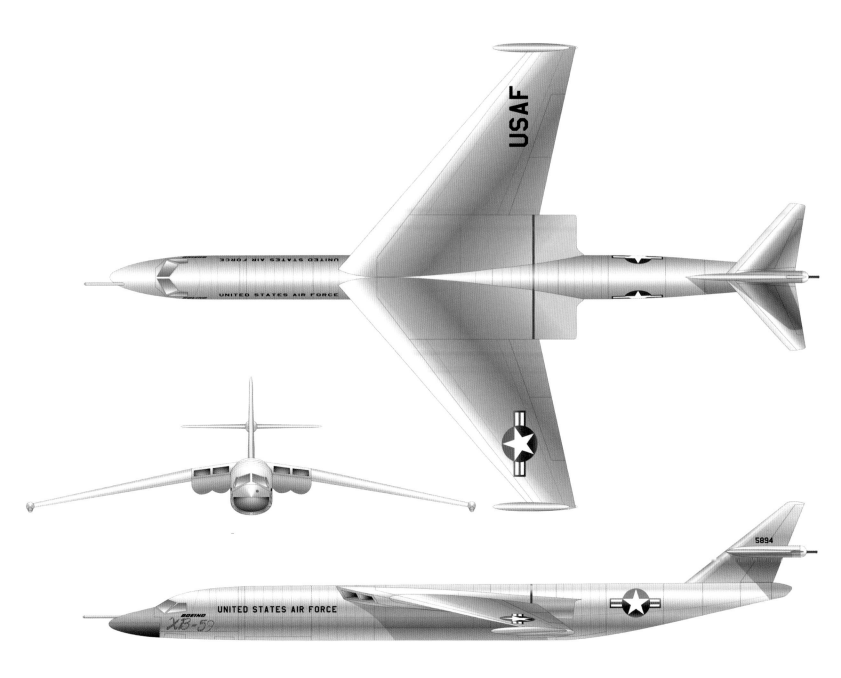

ABOVE: The three-view of the proposed XB-59 (Model 701-299-1) derived from a Boeing-produced desk model in August 1953. This version is believed to be the final configuration. *G. De Chiara © 2015*

BELOW: This left side view of the XB-59 shows how the plane would have looked at rest on its bicycle-type landing gear and wingtip outrigger wheels. *G. De Chiara © 2015*

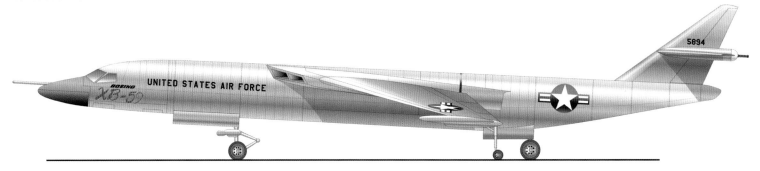

BOEING XB-59

On January 27, 1949, the Boeing Airplane Company was notified that its XB-55 program had been terminated and that its remaining funds would be transferred to its advanced bomber program, which was still in its infancy.

At this time, under USAF AMC Secret Project MX-1712, the Boeing entry was known as Boeing Model 484-405B; it was being developed under USAF contract number AF33 (038)-21388. This contract was issued to Boeing in February 1951, calling for Boeing to supply a full-scale engineering mockup, wind tunnel models, and cost and engineering data.

Much like Project MX-1626 (XB-58) morphed into Project MX-1964, Project MX-1712 (XB-59) morphed into Project MX-1965. As such, the Boeing XB-59 was a direct competitor to the Convair XB-58. The XB-59 (Model 701-333 final configuration) was conceived as a Mach 2-capable medium bomber, to be built as the successor to the Boeing B-47 Stratojet. The first Boeing model number designated XB-59, however, was the Model 701-299-1 design.

Initial production problems with the B-47 were solved, and the XB-55 project was cancelled before any aircraft were built. The development funding from the XB-55 was transferred to the XB-59 and design studies continued. The basic XB-59 was to be powered by four turbojet engines. Both the Pratt & Whitney J75 and General Electric J73-X24A (J79) were used in developmental (paper) studies. The design called for four jets to be mounted inside the inboard wing, two on each side. The landing gear was similar to the tandem main landing gear, with wingtip outrigger design used on both the B-47 and B-52.

As designed, the Boeing XB-59 was to be powered by four afterburning, axial-flow, 14,000-lbf General Electric J73-GE-X24A (later J79) turbojet engines to give it a maximum speed in excess of Mach 2.

On November 18, 1952, Chief of Staff of the USAF Gen. Hoyt S. Vandenberg declared Convair the winner of the competition and all work on the XB-59 ended. The Convair XB-58 (Model 4) would be built.

Neither the XB-58 or XB-59 was anywhere near the prototype construction phase, and the cancellation of the Boeing project was due in part to the expense of funding development for both projects through prototype flight-testing and the growing belief by USAF senior staff officers that high-speed, high-altitude penetration was not a viable offensive strategy during the projected initial operational capability in the early 1960s.

XB-59 SPECIFICATIONS
(FINAL CONFIGURATION)

Propulsive system
Four afterburning, axial-flow, 14,000-lbf (with afterburning) General Electric J73-GE-X24A turbojet engines; this engine became the J79

Length	123 feet, 4 inches
Height	25 feet, 5 inches
Wingspan	81 feet, 4 inches
Empty weight	63,200 pounds
Gross weight	148,000 pounds
Maximum speed	2-plus Mn
Maximum range	2,380 miles
Combat Ceiling	51,000 feet

Payload
One 10,000-pound nuclear device

Armament
One tail-mounted, radar-directed 30mm cannon

SUMMARY

The proposed Boeing XB-59 was an interesting design that most likely would have worked every bit as well as the B-58. Since it was not selected, we will never know.

NUCLEAR-POWERED (NEPA) BOMBER PROGRAM

In 1946, the USAAF began funding the Nuclear Energy for the Propulsion of Aircraft (NEPA) program. On May 26, 1946, the USAAF awarded the Fairchild Engine and Aircraft Corporation contract, which initiated the NEPA project. This Fairchild NEPA project, known as SHAD under USAAF AMC Secret Project MX-821, was described as a "program of developing the use of nuclear energy for aircraft propulsion."

Between 1946 and 1951, a plausible means of fueling a turbojet-powered aircraft with nuclear energy emerged and was studied thoroughly.

On April 30, 1951, the NEPA project was terminated; on the following day, the ANP project was initiated.

ANP PROGRAM

The Manned Aircraft Nuclear Propulsion (ANP) program was based on requirements established by the DOD and was a joint project of the Atomic Energy Commission (AEC) and the USAF for developing a nuclear-powered airplane for military purposes. The US Navy was a minor participant in the program. The AEC was responsible for the nuclear reactor and related shielding, while the USAF was responsible for the remaining parts of the airplane, mainly the turbomachinery (propulsive system), airframe, and auxiliary components.

On March 22, 1955, Headquarters USAF issued General Operational Requirement 81 (GOR 81) for a piloted nuclear-powered intercontinental strategic bombardment weapon system: a nuclear-powered bomber (NPB).

WS-125A

On April 15, 1955, the USAF Air Research and Development Command (ARDC) followed suit and released its System Operational Requirement 18 (SOR 18), which designated the NPB as Weapon System 125A (WS-125A).

The USAF awarded fixed-price redeterminable contracts for studies and investigations for a piloted Nuclear-Powered Intercontinental Strategic Bombardment Weapon System to three airframe contractors: AF 33-(600)-30291 to Boeing, AF 33-(600)-30292 to Convair, and AF 33-(600)-30293 to Lockheed. These studies and investigations were to be considered as part of a design competition leading to the award of development contracts for the NPB under WS-125A.

In November 1955, the USAF directed team-ups of General Electric with Convair and Pratt & Whitney with Lockheed on the WS-125A program.

Boeing had been eliminated from participation in the WS-125A program because it was chosen as a contractor for Weapon System 110A (WS-110A), the chemical-powered bomber (CPB) program.

The NPB air vehicle mockup inspection was to be held sometime in 1958, and the propulsion system manufacturer for the air vehicle would be selected at that time.

First flight with chemical fuel was to be in October 1960, and first flight with nuclear power was to be July 1961. The first inventory airplane was to be available in July 1963, while the first wing of thirty operational aircraft would appear in 1964.

CAMAL

In September 1959, the DOD changed the CAMAL mission from high-speed, high-altitude to subsonic speed (0.8 to 0.9 Mn) at thirty-five thousand feet with a potential life (endurance) of one thousand hours.

The CAMAL concept more believably echoed the "state of the art" of an operational atomic-plane (A-plane), propelled by a nuclear-energy-fueled propulsive system, specifying low-altitude bombing attacks at subsonic speeds.

In 1959, Convair won the right to build the CAMAL airframe, and Lockheed was simultaneously eliminated from the NPB program, its second submission, the proposed TDN L-225, having been rejected.

NB-36H

Earlier, on May 11, 1953, under USAF ARDC Secret Project MX-1589, Convair was authorized to modify a B-36H-20-CF (USAF serial number 51-5712), which became the XB-36H nuclear testbed airplane. As a part of this program (first authorized in 1951), MX-1589-A was assigned to the X-6 nuclear-powered research aircraft, which was to be another modified B-36.

Nicknamed "Crusader," this lone XB-36H (later NB-36H) flew forty-seven times from September 17, 1955, to March 28, 1957. A total of 215 hours flight was accumulated during this time, and the nuclear reactor was operated for eighty-nine hours.

Convair test pilot A. S. Winchell Jr. had made the first flight on the XB-36H. Later, the proposed Convair X-6 was intended to be a follow-on to the NB-36H.

CONVAIR NX-2

In June 1958, the USAF cancelled WS-125A and replaced it with continuous airborne missile launching and low-level penetration

NX-2 CAMAL SPECIFICATIONS
(FINAL CONFIGURATION)

Propulsive system
Three axial-flow afterburning 20,000-lbf (with afterburning) General Electric J87 (Model X211) nuclear-fueled turbojet engines

Length	170 feet, 4 inches
Height	43 feet, 1 inch
Wingspan	134 feet, 2 inches
Wing area	5,000 square feet

NX-2 SPECIFICATIONS
(FINAL CONFIGURATION)

Propulsive system
Three afterburning axial-flow (with afterburning) General Electric XJ87-GE-1 (X211) turbojet engines in twin packs, plus two wing-mounted "booster" turbojet engines

Length	170.4 feet
Height	43.1 feet
Wingspan	134.2 feet
Gross takeoff weight	450,000- plus pounds (estimated)

system (CAMAL). CAMAL more realistically reflected the state-of-the-art for an "atomic airplane," specifying low-altitude bombing attack at subsonic speed. CAMAL would be built by the Convair Division of the General Dynamics Corporation and was designated NX-2 (Convair Model 54).

The Convair NX-2 was to be propelled by three General Electric X211 (later designated J87) engines for its primary propulsion, which would be fed with nuclear fuel via two XMA-1A reactors, one reactor per pair of J87 engines. Two booster turbojet engines, one under either wing in pod-type nacelles, were to provide backup power during the flight-test phase. These booster engines would not be attached to a production NX-2, which was projected to make its first flight in 1963.

The XJ87-GE-1 (X211) engine was completed in September 1959 and put into its test cell to run using conventional fuel. It was at this time that the DOD changed its CAMAL mission objectives to a speed range of 0.8 to 0.9 Mn at thirty-five thousand feet, with a potential endurance of one thousand hours.

The proposed General Electric J87 (Model X211) propulsive system was a *direct* air cycle design that drew air directly through the radioactive core of the reactor. This superheated and compressed air was sent through the engines, where it would violently expand to create the exhaust thrust to move the NX2 forward.

Counter to this design, the proposed Pratt & Whitney XJ91-P-1 (Model JTN9) propulsive system was an *indirect* air cycle system that was to draw reactor heat indirectly through a closed loop and circulate liquid metal from the core of the reactor to heat exchangers, where compressed air was heated to produce thrust.

Several airframe and powerplant contractors subsequently responded to the NPB requirement. The teams of Lockheed/Pratt & Whitney and Convair/General Electric moved forward, with the latter team being the final choice. With this, the Convair X211 program was initiated.

There were a number of programs to test the feasibility of an NPB, including:

MX-1589: Nuclear testbed aircraft (Convair NB-36H)

MX-1589-A: Nuclear-powered research aircraft (Convair X-6)

MX-1627: Low-altitude nuclear-powered bomber studies

In 1956, the USAF decided that the proposed WS-125A NPB was unfeasible as an operational strategic aircraft. Finally, after spending more than $1 billion, the project was cancelled on March 28, 1961.

SUMMARY

The mish-mash of programs associated with proposed nuclear-powered strategic bombardment aircraft lasted some fifteen years and cost just over $1 billion ($1,041,400,000, to be exact).

In the end, the long-running and hapless NPB program had become a passel of requirements, approvals, cancellations, designs, developments, proposals, rejections, studies, and, finally, a colossal waste of time, effort, and money.

Nevertheless, a lot was learned, and it made the use of nuclear energy in the United States safer in the long run. For example, we can thank this abortive program for America's fleet of nuclear-fueled naval vessels.

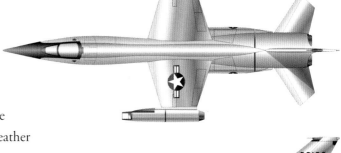

BELL "YF-109" (D-188A)

The Bell Aircraft Corporation signed a joint agreement with the US Air Force and US Navy in 1955 to develop a tactical, all-weather vertical takeoff and landing/short-takeoff-vertical landing (VTOL/STO-VL) Mach 2-plus fighter-bomber that could double as an all-weather area- and point-interceptor interceptor for both services. Bell came up with its D-188 for the US Navy and its D-188A for the USAF under USAF ARDC Project MX-1976.

The USN version (D-188) would most likely have been designated F3L-1, while the USAF version (D-188A) was never officially designated (although the designation F-109 was considered). According to a Bell document modestly entitled "Model Numbers," the US Navy version is listed as follows:

* Model: 2000
* Description: VSTOL Tactical Fighter Airplane
* Project engineer or technical director: D. Wright
* Customer: US Navy
* Authority: US Navy contract NOas 57-836-c

ABOVE: Bell design number D-188A was developed under USAF ARDC Secret Project MX-1976, shown here as it appeared during its full-scale engineering mockup inspection in 1960. G. De Chiara © 2015

BELOW: A phantom view of the Bell Aircraft Corporation D-188A ("YF-109") showing the noteworthy, near-vertical twin-engine placement behind the cockpit. PMB Collection

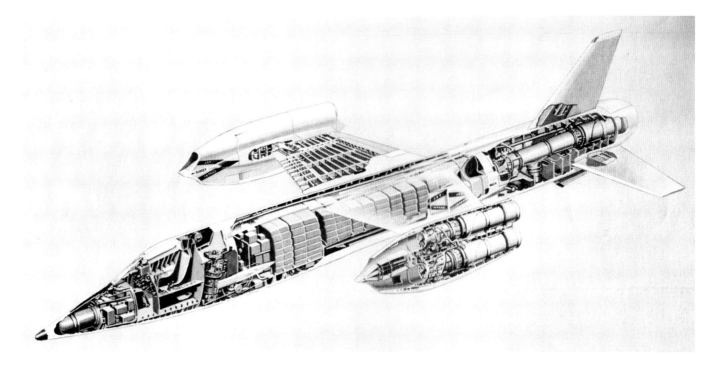

BELL MODEL 2000/ DESIGN NUMBER D-188A (YF-109) SPECIFICATIONS
(AS OF FEBRUARY 6, 1959)

Propulsive system

Six afterburning, axial-flow, 3,360-lbf (maximum power) General Electric J85-GE-5 turbojet engines and two vertically mounted, non-afterburning, axial-flow, 2,455-lbf (military power) GE J85-GE-5 turbojet engines (the use of two non-afterburning, vertically mounted, 2,450-lbf Fairchild J83 turbojet engines instead of two General Electric J85 turbojet engines was likewise considered)

Length	59 feet, 1.8 inches
Height	12 feet, 10.9 inches
Wingspan	23 feet, 9 inches
Empty weight	14,031 pounds
Gross weight	24,883 pounds
Maximum speed	2.3 Mn
Maximum range	1,105 nautical miles
Ceiling	55,000 feet

Armament

Three AIM-9 Sidewinder AAMs, two folding-fin AGM-12 Bullpup AGMs, or one thousand-pound class nuclear device; all weapons were to be carried internally in a 148-inch-long, 30-inch-deep armament bay; external fuel tanks and weapons could also be configured on underwing hard points

Bell attempted several times to acquire a designation for the USAF version of Bell Model 2000. On January 29, 1958, Bell requested that its D-188A be designated YF-109. But on February 14, 1958, Headquarters USAF denied the request.

Bell submitted a second request for redesignation on October 3, 1958, and once again USAF rejected the request, on February 6, 1959.

The reasons were most likely due to the fact that the B version of the McDonnell F-101 Voodoo was to be the F-109 but wound up being designated F-101B. Moreover, the Ryan X-13 was temporarily designated F-109. Thus, with F-109 designations all over the place, Bell received its rejections.

Planning at this time called for the manufacture and flight test of three prototype D-188A airplanes in late 1960 and early 1961, with subsequent production and delivery of operational D-188As in 1962 and after. The D-188A was to be employed by the USAF Tactical Air Command.

A full-scale engineering mockup of the D-188A was built with USAF serial number 59-2109 on its vertical tail. This particular serial number, though, was assigned to a Boeing IM-99A BoMarc surface-to-air interceptor missile. Another serial number (60-2715), said to be associated with the D-188A program, belonged instead to a Martin AGM-12 Bullpup air-to-ground guided missile. Even though three prototypes were to be built, then, no documented issuance of specific serial numbers assigned to these aircraft is available in the record. The D-188A mockup was ready for inspection on February 6, 1959.

As designed, the USAF version of Bell D-188 (the D-188A) was to be propelled by *eight* General Electric axial-flow, 3,850-lbf (with afterburning) J85-GE-5 turbojet engines. Only the fuselage engines were to be equipped with afterburners. The vertically mounted engines would produce 2,600-lbf (without afterburning); a 2.3 Mn speed was projected. Two would have been tail-mounted horizontally, side by side, two would have been mounted side by side vertically in the fuselage aft of the cockpit, two would have been mounted side by side in a nacelle on the left wing tip, and two would have been mounted side by side in a nacelle on the right wingtip. The wing-mounted nacelles/engines would have rotated through a 100-degree arc.

SUMMARY

Had the Bell Model 2000 series of aircraft gone into production, the planes would surely have been unconventional warplanes. Why these USAF/USN D-188 designs were passed over in the spring of 1961 is a mystery. Most likely, their demise can be attributed to the creation of other USN/USAF types at this time that could perform equally as well as or better than the Model 2000's proposed missions; for example, the USN McDonnell F4H (later F-4) Phantom II, which also joined the USAF, could carry more weapons farther and faster than the Model 2000.

Thus, as thought-provoking as Bell's Model 2000 was, it became just another interesting aircraft that was passed over and never built.

McDONNELL DOUGLAS YF-4C (F-110A), YRF-4C (YRF-110A), AND YF-4E PHANTOM II

If the status quo ruled the day, the US Navy would never have adopted a US Air Force fighter plane, and vice versa. As early as November 9, 1959, though, the USAF, its Tactical Air Command (TAC) in particular, opted to take a close look at the USN F4H-1 Phantom II being produced by the McDonnell Aircraft Corporation (McAir) in St. Louis, Missouri.

To begin, in 1961, a program called Operation Highspeed was implemented by the USAF, whereby a USAF Air Defense Command (ADC) Convair F-106A Delta Dart was pitted against a USN F4H-1 Phantom II in a fly-off competition.

This competition was designed to see which type had the best maneuverability, agility, climb rate, energy (speed), and overall range and altitude characteristics. The fly-off ensued and, afterward, it was judged that the Phantom II was superior to the Delta Dart. During the fly-off, the F4H-1 demonstrated superior high-altitude top speed, low-altitude top speed, all-out altitude, climb

F-4C SPECIFICATIONS

Propulsive system
Two afterburning, axial-flow, 17,000-lbf (maximum) General Electric J79-GE-15 turbojet engines

Length	58 feet, 2 inches
Height	16 feet, 6 inches
Wingspan	38 feet, 0 inches
Gross weight	58,000 pounds
Maximum speed	1,400 miles per hour
Maximum range	1,750 miles
Ceiling	59,600 feet

Armament
Up to 16,000 pounds of nuclear or conventional bombs, rockets, missiles, cannon pods

rate, roll rate, radar search range, unrefueled range, maneuverability, and agility. In addition, the Phantom II could carry up to twenty-four 500-pound bombs, whereas the Delta Dart could carry none.

As a result, the USAF almost immediately ordered a single F-110A (USAF serial number 62-12199) for its TAC, which was soon followed by an additional order for two YRF-110A airplanes (USAF serial numbers 62-12200 and 12201). Respectively, according to McAir documentation entitled "McDonnell Model Numbers" and dated July 1, 1974, these became known as McAir Models 98DE and 98DF. This documentation also states the dates these McAir model numbers were assigned: November 13, 1961 (98DE) and January 3, 1961 (98DF). (The date January 3, 1961 is believed to be an error, and should be January 3, 1962.)

On January 24, 1962, two US Navy F4H-1 Phantom II airplanes that featured USAF paint schemes with the designation F-110A lettered on each of their noses was delivered to the USAF Tactical Air Command (TAC). Two USAF colonels, Col. Gordon M. Graham (later Lt. Gen.) and Col. George "Ravin" Laven Jr., picked them up from McDonnell's factory on this day and flew them to Langley AFB, Virginia, for service test evaluations. Respectively, these two modified USN F4H-1 airplanes (USN bureau numbers 149405 and 149406) eventually received USAF serial numbers 62-12168 and 62-12169. To further distinguish these two airplanes from USN F4H-1 Phantom II airplanes, with a suggestion from McDonnell Aircraft, they were unofficially nicknamed "Spectre."

A little more than a year later, on February 11, 1963, USAF Col. Stanton T. Smith Jr., Commander of the 836th Air Division, and USAF Maj. Jerry F. Hogue delivered the first of twenty-seven McDonnell F-4B and RF-4B Phantom II airplanes (twenty-five F-4Bs and two RF-4Bs) to the USAF 4453rd Combat Crew Training Wing (CCTW) at MacDill AFB, Tampa Bay, Florida. These twenty-seven aircraft carried both US Navy and USAF serial numbers.

In the interim, on September 18, 1962, the new US tri-service aircraft designation system went into effect, and the USAF designations F-110A and YRF-110A were replaced by USAF designations F-4C and YRF-4C. In addition, the unofficial name "Spectre" had been disapproved and the USAF F-4C and RF-4C airplanes became officially named Phantom II, just like their USN counterparts.

With this new designation system in place, the single F-110A and the pair of YRF-110A airplanes on order for USAF TAC were officially redesignated F-4C and YRF-4C. Thus, the F-110A and YRF-110A designations had become null and void.

The lone service test F-4C Phantom II (62-12199), the former F-110A, was completed in April 1963. On May 27, 1963, it made its first flight out of Lambert Field, adjacent to the McAir factory, with McAir chief engineering test pilot Robert C. "Bob" Little at its controls. The pair of service YRF-4C Phantom II airplanes
(62-12200 and 62-12201), the former YRF-110A airplanes, soon followed.

The USAF, its TAC in particular, readily adopted the F-4C and RF-4C airplanes and, using FY1963 defense funding, it ordered an initial production batch of 331 F-4Cs and RF-4Cs. Thus, the USAF had officially adopted the USN F-4 Phantom II and had made the plan part of its own fleet, with continued orders for additional variants.

YF-4E PHANTOM II

Aerial combat in the Vietnam War proved that all-missile-armed fighters such as the F-4 Phantom II were at a disadvantage: they should also be armed with guns or cannon for close-in aerial combat. In fact, the Phantom II was the only combat fighter in the air that didn't have guns or cannon. At the behest of the USAF, therefore, it became a priority to arm the aircraft with such weapons; this led to the E version of the Phantom II, initially called F-4E PLUS.

After thorough evaluation of this problem, McDonnell chose to arm the E variant of the Phantom II with a single nose-mounted General Electric M61A1 Vulcan six-barrel rotary-action 20mm cannon. This was offered to the USAF as its Model 98GV-1, becoming Weapon System 327C.

Three service test YF-4E airplanes were evaluated prior to F-4E production. The first service test YF-4E was a modification the first of the two YRF-4C

> ## *Historical Note*
>
> The McDonnell Douglas Corporation was founded on April 28, 1967, when the McDonnell Aircraft Corporation merged with the Douglas Aircraft Company.
>
> The US Navy and US Marine Corps received 1,264 of the Phantom IIs, while the US Air Force received 2,874 of the F-4s. With this, the USAF procured 1,610 more F-4s than did the USN/USMC, the plane's original customer. The Phantom II fighter series was the most produced fighter since World War II, and remains so to this day. Only the F-16 (which is still in production) seems to have a chance to surpass the total number of Phantom IIs produced—more than 4,500 F-16s have been produced as of 2015.

The first of two prototype YRF-4C airplanes (62-12200) takes first flight on August 9, 1963, with William S. "Bill" Ross at the controls; the second prototype (62-12201) would take to the sky about seven weeks later, on September 30, 1963. These were formerly known as the YRF-110A Spectre aircraft. *GSLASM via Mark Nankivil*

airplanes built (USAF serial number 62-12200), which made its first flight on August 7, 1965, at Lambert Field with McDonnell test pilot Joseph R. "Joe" Dobronski under glass. It was powered by a new version of the GE J79, the J79-GE-J1B (eventually emerging as the J79-GE-17), which offered improvements and more power.

YF-4E numbers two and three, a modified F-4C and a modified F-4D (USAF serial numbers 63-7445 and 65-0713), were also equipped with the improved GE J79-GE-J1B engines initially; later they received production J79-GE-17 engines.

The first production F-4E (USAF serial number 66-0284) made its first flight on June 30, 1967, with McDonnell Douglas test pilot Raymond D. "R. D." Hunt and McDonnell Douglas chief systems and radar operator Wayne Wight at the controls.

YF-4C (F-110A), YRF-4C (YRF-110A), AND YF-4E
PRODUCTION BREAKDOWN

YF-4C-9-MC:	62-12168-F4H-1	BuAer 149405 (delivered to Langley AFB, January 24, 1962, with "F-110A" painted on its nose)
YF-4C-9MC:	62-12169, F4H-1	BuAer 149406 (delivered to Langley AFB 24 January 1962 with F-110A painted on its nose)
F-4B-MC (formerly F4H-1):	62-12170 to 62-12196	BuAer Numbers: 150480, 150486, 150493, 150630, 150634, 150643, 150649, 150650, 150652, 150653, 150994, 150997, 150999, 151000, 151004, 151006, 151007, 151009, 151011, 151014, 151016, 151017, 151020, 151021; all reportedly returned to US Navy after USAF evaluation
F-110A-MC:	62-1219	redesignated F-4C-15-MC
YRF-110A-MC:	62-12200 and 62-12201	redesignated YRF-4C
YF-4E:	62-1220	modified from YRF-4C-14-MC, formerly designated YRF-110A
YF-4E:	63-7445	modified from F-4C-17-MC
YF-4E:	65-0713	modified from F-4D-28-MC

The end result of these evaluations was not only good for McDonnell Aircraft/McDonnell Douglas and the US Air Force: it was good for America on the whole, because the F-4 Phantom II not only proved to be a standout combat aircraft in the Vietnam War: it showed itself to be an outstanding war fighter in subsequent combat actions as well. Moreover, the USAF Tactical Air Command eventually procured 2,874 Phantom IIs in four major variants, including the F-4C, RF-4C, F-4D, and F-4E models.

SUMMARY

The F-4 Phantom II program proved to be so successful that the McDonnell Douglas Corporation manufactured 5,057 of them in numerous versions for the US Armed Forces as well for a number of US allies and friends. Another 138 Phantom II airplanes were built in Japan by Mitsubishi for a grand total of 5,195 F-4s built.

Northrop Fang

Edgar O. "Ed" Schmued, designer of the legendary P-51 Mustang when he was employed at North American Aviation, was the co-designer of Northrop Model N-102, dubbed "Fang" by Northrop. The Northrop Preliminary Design Department issued report P.D. 152 on January 19, 1953, which provided the details for the new fighter's design. Schmued's colleague, Welko E. Gasich, doubled as chief engineer on the program. They applied for a US patent on the design on December 5, 1955 (granted on May 14, 1957). As Schmued (assistant chief engineer) and Gasich had designed it between 1952 and 1953, Northrop's Fang was offered to the USAF as a lightweight single-seat, delta-winged fighter-interceptor powered by a single afterburning, axial-flow turbojet engine.

Northrop, realizing the cost advantage of devising a small, lightweight fighter that would meet mission requirements, came up with this surprising design. Northrop's Model N-102 Fang design was begun by Schmued in late 1952, after he joined NoAir, and defined by Gasich. To demonstrate its Fang effort, a full-scale engineering mockup was built as a private venture, without USAF financing. Although Fang never progressed beyond mockup stage, it had a profound influence on Northrop's subsequent fighter aircraft designs.

The full-scale mockup of the single-tail N-102 Fang is a little more impressive than its paper equivalent. *WMOA*

Northrop's N-102 Fang featured a thin delta wing and, in its final form, it was to be powered by a single General Electric J79 afterburning turbojet engine, notwithstanding its design, which had been drawn to accept any of five different turbojet engines. The Fang design featured both a single vertical tail and twin V-shaped "butterfly" vertical tails. The design became an engineering platform for a number of later fighter innovations. For example, it employed a hinged cockpit canopy windshield that could be raised like the hood of a car as a means of accessing instrumentation. It also featured exceptional maintenance accessibility to its interior areas via the use of quick-release, Dzus-type fasteners on its access panels and doors. Additionally, it had provision for rapid powerplant changing.

Northrop's Fang was a small fighter wrapped around a large engine; for this reason, there was not a great deal of room for growth. Though it offered a number of new features, it did not generate enough enthusiasm for orders—however, it provided a foundation on which to build.

The result was the Northrop F-5 series of lightweight fighters.

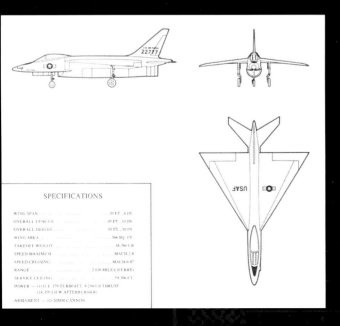

SPECIFICATIONS

WING SPAN	30 FT. 6 IN
OVERALL LENGTH	45 FT. 10 IN
OVERALL HEIGHT	13 FT. 10 IN
WING AREA	366 SQ. FT.
TAKEOFF WEIGHT	18,760 LB
SPEED-MAXIMUM	MACH 2.0
SPEED-CRUISING	MACH 0.87
RANGE	2,010 MILES (FERRY)
SERVICE CEILING	59,300 FT.
POWER — (1) G.E. J79 TURBOJET, 9,290 LB THRUST (13,190 LB W/AFTERBURNER)	
ARMAMENT — (1) 20MM CANNON	

Three-view of Fang with single-tail configuration. *WMOF*

Artist's conception of the Fang during a ground attack mission imagines its swift hit-and-run capabilities. *WMOA*

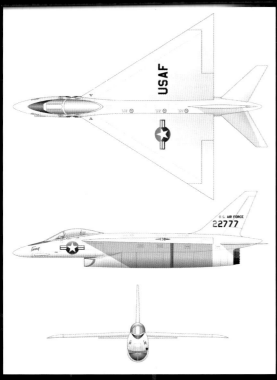

Northrop Fang in three-view as it appeared during its mockup phase. *G. De Chiara © 2015*

N-102 FANG SPECIFICATIONS

Propulsive system
One afterburning, axial-flow General Electric J79 turbojet engine, version to be decided if ordered into production

Length	41.0 feet
Height	10.0 feet
Wingspan	23.0 feet

NORTHROP N-156F, YF-5A (N-156C), F-5A-21, AND YF-5B-21 FREEDOM FIGHTER

In 1949, Oliver Patton Echols (a retired USAF Major General) was elected chairman and chief executive officer of the Northrop Aircraft (NoAir). NoAir founder and President Jack Northrop retired in 1952, at which time Echols became president and general manager. As one of his first official duties, Echols hired Schmued to serve as vice president of engineering. As one of the world's foremost aircraft designers of piston- and turbojet-powered fighters—including the P-51 Mustang, F-86 Sabre, and F-100 Super Sabre, produced by North American Aviation, where he had worked prior to joining NoAir.

Schmued, in turn, hired William F. Ballhaus as chief engineer, having wrested him away from Convair. Ballhaus brought Welco E. Gasich aboard from the Rand Corporation to serve as chief of preliminary design. Ward Dennis, also from the famed Rand think tank, came with Gasich to become head of Northrop's weapon system analysis staff. Rounding out Northrop's new leadership was Thomas V. "Tom" Jones, who would serve as assistant to Ballhaus, ultimately becoming Northrop's president.

These five men—Ballhaus, Schmued, Gasich, Dennis, and Jones—were responsible for generating new military aerospace-related business for Northrop, in particular a new fighter airplane.

Rare color image showing the Northrop N-156F mockup. *WMOA*

Their task was a considerable one. A number of well-known airframe contractors already appeared to have a lock on the lucrative fighter aircraft market. Several were well into the prototype stage on the then new USAF Century series of supersonic/doublesonic fighter-bombers and fighter-interceptors.

Northrop's previous fighter production had been for big, complex, heavy night fighter types with high-volume firepower, namely the P-61 Black Widow (in use during World War II) and the F-89 Scorpion (just entering service but already obsolete). The current trend in military fighter aircraft requirements at the time was for sophisticated types that would fly faster, higher, and farther than previous turbojet-powered fighters. As a consequence, they were larger, more complex, more expensive, and heavier than what had gone before.

While most of the airframe contractors continued to pursue heavier, larger, more complex and expensive airframes, NoAir devised a formula to design a fighter to meet the design criteria of an effective, affordable fighter. Five design goals were set: (1) the airplane must have adequate performance to meet the requirement, but not at the expense of complexity, cost, or safety; (2) the airplane must be capable of performing air-to-air, air-to-ground, and reconnaissance missions without lengthy downtime for change in avionics; (3) the airplane should have two engines for safety and to lower attrition costs; (4) the airplane must have excellent handling characteristics at both low and high speeds and must be able to operate from five-thousand-foot runways; and (5) the airplane must be inexpensive to buy, operate, and support.

Four facts were apparent: (1) the airplane must have supersonic flight capability; (2) the airplane must be lightweight, because cost is directly related to airplane weight; (3) the airplane must have a small frontal area with high thrust-to-weight engines; and (4) engines with sufficient power meeting this description were not known to exist.

The required engine was found, unexpectedly, in mid-1954.

Both General Electric Aircraft Engines and the Fairchild Engine Division of the Fairchild Aircraft Corporation were in competition to develop a small, high-thrust turbojet engine for application on the McDonnell ADM-20 Quail air-launched decoy missile—respectively, the J85 and the J83. But neither type was being developed to incorporate an afterburner section.

When Northrop discovered these two powerplants, it set the wheels in motion to propel its new N-156 fighter with one or the other. The company successfully petitioned GEAE to develop an afterburner section for its J85. Since the J85 had been chosen over the J83 for Quail propulsion, NoAir selected this for its N-156 as well. The engine now existed.

Development of a lightweight supersonic fighter around the new and revolutionary General Electric J85 axial-flow turbojet engine became a priority project within Northrop's Advanced Design department. It would be based on the Northrop YT-38 Talon advanced trainer aircraft, a pair of which had just been ordered by the USAF. The advanced trainer was designated in-house as the N-156T, the fighter as N-156F.

Since the USAF was swamped with its procurement of various Century series fighters, it had little to say about Northrop's N-156F offering. Northrop continued its development with company funds, including the construction of a

N-156F number one (59-4987) demonstrates its weapons-carrying capability. *NMUSAF*

N-156F number one became the prototype F-5A number one (59-4987). It cuts quite a fearsome shadow! *NMUSAF*

N-156F number three (59-4989) splits the sky during a test flight near Rogers Dry Lake. *NMUSAF*

full-scale engineering mockup. It decided that, since the USAF wasn't interested, it would attempt to sell it to US allies.

Allied and friendly nations liked what they saw in the Freedom Fighter, as Northrop had dubbed its N-156F. Interest grew enough that Northrop was forced to receive US Department of Defense authorization to initiate contract negotiations. With authorization granted, Northrop further developed its lightweight fighter for imminent sales abroad; but, because it was a military weapon system, the USAF would oversee the program. For test and evaluation, three N-156F Freedom Fighter prototypes were ordered on February 25, 1958, under the Military Assistance Program (MAP) as USAF SS-420A (Support System 420A).

Since the Northrop T-38A Talon was already in production for the USAF Air Training Command (ATC), and commonality with the N-156F Freedom Fighter was well established, Northrop was able to produce the first prototype in a relatively short time, ceremoniously rolling it out on May 30, 1959—some ten months after YT-38 number one, and thirteen days before YT-38 number two flew.

Flown by Northrop engineering test pilot Lewis A. "Lew" Nelson, N-156F Freedom Fighter prototype number one made its first flight at Edwards AFB on June 30, 1959. As with its two YT-38 stablemates, the aircraft was powered by two interim non-afterburning YJ85-GE-1 turbojet engines. But this didn't matter: Nelson hit Mach 1.04 in a shallow dive. He reported: "Transition to supersonic, as with the YT-38, is so smooth you have to watch the instrumentation to know when it happens."

Northrop's N-156F Freedom Fighter program began rather slowly, but accelerated steadily toward procurement orders. Early in its existence, it won an on-paper F-X competition with the Lockheed F-104 Starfighter and the Chance Vought F8U Crusader. Finally, in April 1962, the USAF ordered seventy-one single-seat and fifteen tandem-seat airplanes, which, under the new designation system, were designated F-5A and F-5B, respectively. The name Freedom Fighter was officially adopted at this time. The N-156F prototypes were designated YF-5A and given a new Northrop Project Number, N-156C. The USAF issued serial numbers 59-4987 to 59-4989 to the three YF-5As and serial number 59-4993 to the static test article, which was designated XF-5A.

In 1965, General Electric notified Northrop that it was at work on an improved version of the J85, which would produce 5,000-lbf with afterburning (up 25 percent).

In order to measure F-5 performance gain with the new engine, General Electric made arrangements with the USAF to modify one tandem-seat F-5B (63-8445) to accommodate the new YJ85-GE-21 engines, which would serve as a flying engine testbed airplane. It was temporarily designated F-5A-21, but quickly redesignated YF-5B-21 for this purpose.

This two-seat F-5B (the eighth example built) was fitted with two afterburning, 5,000-lbf (with afterburning) General Electric YJ85-GE-21A turbojet engines to serve as a prototype for this engine development program; it made its first flight on March 28, 1969, with GEAE jet propulsion chief engineering test pilot John M. Fritz at the controls.

A new top speed of Mach 1.60 (1,080 mph) was attained in subsequent testing. This result gave rise to Northrop's follow-on F-5E and F-5F Tiger II production programs.

More than 2,200 F-5As, F-5Bs, F-5Es, and F-5Fs were built for the USAF as the result of Northrop's bold stand on its N-156 development program. This success led to the development of the proposed F-5G, which became the F-20 Tigershark.

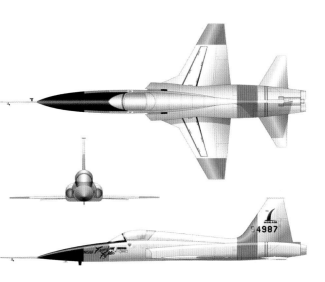

The prototype N-156F (YF-5A number one) in three-view. *G. De Chiara © 2015*

YF-5A SPECIFICATIONS

Propulsive system
Two afterburning, axial-flow, 4,080-lbf (with afterburning) General Electric J85-GE-13 turbojet engines, which replaced the previous YJ85-GE-1 engines

Length	47 feet, 2 inches
Height	13 feet, 6 inches
Wingspan	25 feet, 10 inches
Gross takeoff weight	20,576 pounds
Maximum speed	1.40 Mn (925 mph)
Combat Ceiling	50,700 feet
Maximum range	1,100 miles

Armament
Two M39 20mm cannon, rockets, and missiles

Payload	5,500 pounds of bombs

The rollout of prototype F-5E Tiger II number one (71-1417) wouldn't be complete without its chief engineering test pilot, Hank Chouteau. *NMUSAF*

F-111A
SPECIFICATIONS

Propulsive system
Two afterburning, axial-flow, 18,500-lbf Pratt & Whitney TF30-P-1/P-3 turbofanjet engines

Length	73 feet, 5.5 inches
Height	17 feet, 6 inches

Wingspan
63 feet, 0 inches (extended); 31 feet, 11.5 inches (swept)

Empty weight	46,170 pounds
Gross takeoff weight	82,820 pounds
Maximum speed	2.5-plus Mn
Combat Ceiling	50,000 feet
Maximum range	3,000 miles

Armament
One 20mm M61A1 Vulcan cannon and AAMs

Payload
30,000 pounds of bombs

SUMMARY

The Northrop Aircraft Corporation F-5 Freedom Fighter program proved to be one of the most successful and lucrative fighter production programs of its day. The functional design of the F-5 lent itself to the production of numerous variants and spinoff versions, culminating with the F-20 Tigershark.

The F-5 program began as a private Northrop venture to provide the USAF with a small, light, supersonic aircraft that could double as a fighter and an advanced trainer under Northrop model numbers N-156F and N-156T.

GENERAL DYNAMICS F-111A, FB-111A

After the Republic F-105 Thunderchief had eliminated the North American F-107, it entered into production, then into service with the USAF Tactical Air Command. Although the F-105 was in full-scale production in 1960 (its

What Happened to the F-112 to F-116 and the F-118 Designated Fighter Aircraft?

After September 18, 1962, the DOD didn't *officially* designate any USAF fighter aircraft with digits higher than 111.

However, in a part of the USAF "black world," there are highly classified aircraft programs with aircraft designated F-112 to F-116, and F-118. Still others, with even higher numbers, may exist, though this cannot be confirmed.

As we now know, the F-117, Have Blue, Tacit Blue, and the Bird of Prey aircraft are no longer classified. But the F-112 through F-116 aircraft, and the F-118, remained mysterious—until a number of unclassified documents came to light, starting in 2000. These documents attest to the existence of several formerly classified aircraft test programs that evaluated US, Russian, and Chinese aircraft. Some of these aircraft included:

YF-110B: MiG-21F-13—Have Doughnut
YF-110C: MiG-21F-13/J-7B—Constant Peg
YF-110D: MiG-21F-13—Have Doughnut
YF-110E: Have Phoenix, classified program
YF-110L: Have Phoenix, classified program
YF-110M: Have Phoenix, classified program
YF-112: Su-22
YF-113A: MiG-17F—Have Drill
YF-113B: MiG-23BN—Have Pad
YF-113C: J-5—Have Privilege
YF-113E: MiG-23MS
YF-114C: MiG-17F—Have Ferry
YF-114D: MiG-17PF—Have Glib/Have Idea
YF-116: MiG-25
YF-117D: Tacit Blue
YF-118: MiG-29
YF-118G: Bird of Prey

Vietnam War combat spurs lying in the future), the USAF was already planning to supplement and eventually replace the Thunderchief by 1970. On June 14 1960, therefore, the USAF released SOR-183, which called on the industry to design and develop a replacement for the F-105. Under WS-324A, SOR-183 called for the creation of a tactical fighter experimental (TFX) aircraft. As the framework for a complete weapon system, SOR-183 dictated the following:

- Propulsion by two Model JTF10A-20 Pratt & Whiney TF30-P-1 turbofanjet engines with afterburning
- Two crewmembers, comprised of a pilot and a weapon system operator (WSO)
- Internal weapons carriage for all types of tactical ordnance, including nuclear devices; provision for external weapons carriage and/or external fuel tanks
- Unrefueled range of 3,500 miles, with capability for in-flight refueling
- Provision for a single M61A1 Vulcan 20mm cannon
- Short takeoff and landing (STOL) capability for short and unimproved runway operations; preferably, the TFX would have variable-geometry wings with high-lift devices
- Mach 2.5 speed at best altitude, Mach 1.2 speed at low level

These requirements were demanding, but, if actually built and flown, such a plane would have no equal.

On January 20, 1961, former Senator John F. Kennedy became the thirty-fifth president of the United States. To serve as secretary of defense, he chose Robert S. McNamara. McNamara came to work with the goal of trimming the defense budget while also maintaining America's relatively strong military machine. In a nutshell, his strategy was to create weapons systems that would be adaptable to several branches of the US Armed Forces. For example, he encouraged the

BELOW LEFT: The premier FSD F-111A (63-9766) shines during its October 15, 1964 rollout ceremony in Fort Worth, Texas, with wings extended. *NMUSAF*

BELOW RIGHT: The FSD F-111A in performance, with wings swept aft. Nothing like a little showmanship for the crowd. *NMUSAF*

Weaving among the clouds, GD chief engineering test pilot Dick Johnson controls the first flight of the premier F-111A (63-9766) in December 1964. *NMUSAF*

development of a fighter airplane that could be used equally well by the US Air Force, US Navy, and US Marine Corps. With this drive, and with President Kennedy's backing, he ordered that the TFX be developed under his tri-service policy. His plan seemed solid in theory; in the case of the TFX program, though, it was doomed to fail. By the time the TFX had been adequately developed for USAF use, it had become too heavy to deploy from US Navy aircraft carriers.

Usually, if the USN scorns an aircraft, it will not be procured by the US Marine Corps. For this reason, as we will see, the USAF wound up as the only customer for the TFX.

On October 1, 1961, the USAF released its RFP to the industry. Two months later, on December 6, nine airframe contractors responded to meet the RFP deadline. These included Boeing, McDonnell and Douglas, Lockheed, Republic and Chance Vought, and North American.

General Dynamics, the parent company of Convair, had largely developed aircraft only for the USAF. Since Grumman had developed USN types, the General Dynamic/Grumman team effort was a clever two-pronged strategy to win the TFX competition. Their combined effort was ultimately rewarded—in part, at least.

McDonnell and Douglas, Republic and Chance Vought used a similar ploy.

Initially, none of the airframe contractors came up with a suitable TFX design. Several design competitions were held, until, finally the Boeing TFX had been declared the best choice. On November 24, 1962, though, the team of General Dynamics and Grumman were announced as the winner of the TFX competition, now designated F-111, which resulted in more controversy.

Since every one of the USAF source selection personnel had clearly favored Boeing (its TFX proposal and its Models 818/818N), the assumption throughout the competition period was that it would win the contract. The DOD announcement that General Dynamics would be the prime contractor, with Grumman as the principal contractor, was therefore a complete shock. Only the two airframe contractors, General Dynamics and Grumman, and McNamara showed no surprise. In any event, after congressional hearings, the decision remained firm.

On December 21, 1962, a contract between General Dynamics, Grumman, and the Pentagon became binding. General Dynamics would built eighteen full-scale development F-111As for the USAF and Grumman would build four (later upped to seven) full-scale development F-111Bs for the USN. The first example, a USAF F-111A, was to fly in two years and be delivered six months later.

Weapon System 324A was assigned to the USAF F-111A program and the eighteen FSD airplanes were issued USAF serial numbers 63-9766 to 63-9783.

On October 15, 1964 (sixteen days ahead of schedule), the first F-111A (63-9766) rolled out at General Dynamics' Fort Worth, Texas, facility. By December 20 it was ready for its maiden flight. The next day, it was taken aloft for the first time, piloted by General Dynamics' chief engineering test pilot

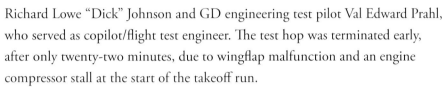

Richard Lowe "Dick" Johnson and GD engineering test pilot Val Edward Prahl, who served as copilot/flight test engineer. The test hop was terminated early, after only twenty-two minutes, due to wingflap malfunction and an engine compressor stall at the start of the takeoff run.

The first FSD F-111As, to be used for research, development, test, and evaluation (RDT&E) purposes were powered by two afterburning Pratt & Whitney TF30-P-1 turbofanjet engines that produced 18,500-lbf (with afterburning). These were replaced with more reliable TF30-P-3 engines as they became available to the growing fleet of eighteen FSD F-111As.

The TF30 engine experienced developmental problems in the early TFX program, causing several crashes. To eliminate these difficulties, changes were made to both the engine air inlet system and to the engine itself. It should be noted that the TF30 was the world's first afterburning turbofanjet engine.

Under WS-324A, the F-111A came equipped with AN/APQ-113 attack radar system, AN/APQ-110 terrain-following radar system, and the AN/APQ-20A inertial bombing and navigation system. It could carry a wide array of various tactical ordnance, including nuclear, within its weapons bay and attached to underwing pylons.

Unofficially named "Aardvark" by those associated with it (due to its long nose section), the F-111 was thought to be the last USAF fighter to carry designation above 100. Then, in 1988, like some long lost ghost from the past, the F-117 materialized.

SUMMARY

The life story of the General Dynamics Corporation F-111 was the result of the DOD-devised TFX program, which was designed to provide fleets of fighters for both the USAF and the US Navy. The program was plagued by fits and starts, on-again, off-again development uncertainties, and political rhetoric. In the end, though, it allowed General Dynamics to manufacture a large number of F-111s for TAC in a number of variants, with several spinoffs, including the FB-111A for SAC.

FB-111A

The General Dynamics Corporation FB-111A was developed under the Weapon System 129A (WS-129A) program to serve as an interim bomber with the USAF Strategic Air Command.

Thanks in large part to the slow development of the Advanced Manned Strategic Aircraft (B-1) program, USAF SAC saw the need to replace some of the aircraft in its fleet of B-52s. The FB-111A was chosen in 1965 to supplement and then replace the B-58, B-52E, and B-52F aircraft. In 1966, the USAF signed a contract to acquire a fleet of FB-111A aircraft: 263 at first, in 1968, but then reduced to seventy-six in 1969.

The prototype, piloted by Val Prahl, flew on July 30, 1967, while the first production FB-111A flew on July 13, 1968. The first six production FB-111As were used as service test aircraft.

In 1968, the UK's RAF cancelled its order for the F-111K, which was then in production. With this, the components for these forty-eight airplanes were diverted to the FB-111A program.

To create the FB-111A, in part, General Dynamics used the longer wings of the cancelled US Navy F-111B to increase its range and underwing stores-carrying capability. The length of the fuselage was increased by two feet, one inch over that of the F-111A to accommodate more fuel, which was increased to 585 gallons internally. Its landing gear was beefed up to compensate for its increased maximum takeoff weight. The FB-111A employed the Triple Plow II engine air inlets to feed their 20,350-lbf (with afterburner) Pratt & Whitney TF30-P-7 turbofanjet engines.

With the advent of the B-1B Lancer, the FB-111A was no longer needed. Beginning in 1989, thirty-four of the planes still in inventory were reconfigured, for tactical rather than strategic operations, becoming the F-111G. When SAC was disestablished in 1992 and the newly formed Air Combat Command (ACC) began operations, these F-111Gs were transferred to the ACC to be used primarily for combat crew training purposes, only to be retired the following year. The FB-111As were retired in 1991.

In 1993, the Royal Australian Air Force (RAAF) bought fifteen F-111Gs to supplement its fleet of F-111Cs. These F-111Gs were retired in 2007, while the remaining RAAF F-111Cs were retired in 2010.

FB-111B

In 1979, General Dynamics proposed its FB-111B (GD in-house designation) to the USAF as an improved FB-111A. As proposed, the FB-111B was to be powered by two General Electric F101 turbofanjet engines with fixed engine air inlets rather than Pratt & Whitney TF30s. Its fuselage was to be lengthened to 88 feet 2½ inches (an increase of 12 feet 8½ inches), and it was to be optimized to carry additional AGM-69A SRAMs on underwing pylons.

General Dynamics proposed the creation of 155 FB-111Bs from converted F-111Ds and FB-111As. This program was deemed too costly, though, and it was terminated before any aircraft could be produced.

FB-111H

The FB-111H was a lengthened version of the FB-111A, proposed in 1977 as an alternative to the Rockwell B-1. It was similar to the aforementioned FB-111B proposal, but would be ten feet longer than the FB-111A, with F101 engines and advanced electronics. The main difference was its enlarged weapons bay, capable of carrying twelve SRAMs internally.

SUMMARY

The FB-111A was largely an interim strategic bomber employed to serve after the retirement of some B-52s and the B-58, before the advent of the B-1. All seventy-six production FB-111A airplanes were delivered to USAF SAC, with the last received on June 30, 1971.

F-111A SPECIFICATIONS

Propulsive system
Two afterburning, axial-flow, 20,350-lbf (maximum) Pratt & Whitney TF30-P-7 turbofanjet engines

Length	75 feet, 6 inches
Height	17 feet, 0 inches
Wingspan	70 feet, 0 inches
Empty weight	47,481 pounds
Gross weight	110,646 pounds
Maximum speed	1,452 miles per hour
Maximum range	6,150 miles
Ceiling	50,000 feet

Crew
Two (pilot, copilot/bombardier/navigator)

Armament
No built-on-board armament was carried

Payload
Six AGM-69A SRAMs and/or six special stores (B-43, B-57, B-61 nuclear weapons), or twenty-four 750-pound class M-117A1 conventional bombs

View of the prototype FB-111A depicting an inflight refueling where all of its fuel tanks—internal and external—are topped off. *NMUSAF*

An FB-111A prototype carries four inert AGM-69A SRAMs during separation tests. *LM/Code One*

TRIPLESONIC MARVELS

If our air forces are never used, they have achieved their finest goal.

— Gen. Nathan F. Twining, USAAF

In early 1942, when turbojet engines and the aircraft to be powered by them first began to take shape in the United States, maximum speeds were projected in the range of 500 to 550 mph. Such speeds would have been 50 to 150 mph faster than the fastest contemporary piston-powered and propeller-driven aircraft, making them more than acceptable for the combat needs of the period.

Later in 1942, the Bell XP-59A Airacomet took wing. The United States' first turbojet-powered airplane, this aircraft barely reached 400 mph, some 50 mph slower than existing US piston-powered fighters. This was completely unacceptable.

The next turbojet-powered fighter appeared in early 1944. The Lockheed XP-80 raised the bar to 500 mph; a vast improvement, but with German combat aircraft such as the Messerschmitt Me262 fighter capable of 500-plus mph, even higher speeds became necessary.

XB-70A number one (AV-1) cruises above the desert with its wingtips full down. *NMUSAF*

Artist's rendering of what a pair of operational F-103s might have looked like during an intercept mission. *Artwork by Jozef Gatial*

The need for speed was not adequately addressed for some time. In fact, it wasn't until after World War II that the USAAF acquired a turbojet-powered fighter fast enough to check enemy fighters. This aircraft, the 600-mph Republic XP-84 Thunderjet, appeared with too little power, too late.

Turbojet engine development moved forward, and even higher speeds were attained. By fall 1947, the North American XP-86 Sabre was able to realize top speeds in excess of 650 mph. Its then unique sweptback flying surfaces, eliminating the once-dreaded compressibility problem, and its afterburning General Electric J47 turbojet engine offered unprecedented performance. Summer 1953 saw a later version of the Sabre surpassing 700 mph, incredible speed for the day. This demonstrated that, once a more powerful engine was available, supersonic speed was now possible.

Such speed had already been reached earlier by specially built rocket-powered, air-launched research aircraft, such as the Bell X-1 on October 14, 1947. Still, speeds on this magnitude were not yet attainable by a turbojet-powered combat-type airplane.

Now designated F-86, the Sabre had evolved into the F-100 Super Sabre, and, in spring 1953, a service test YF-100 made its first flight. Astoundingly, in level attitude flight at an altitude of thirty-five thousand feet, the Super Sabre hit Mach 1.10 (815.6 mph)! This was the first officially recorded supersonic speed flight made by any combat-type of airplane in the world.

Much-improved aircraft designs and better turbojet engines generated even higher speeds and, by summer 1956, a service test Lockheed YF-104A Starfighter, powered by a single afterburning General Electric J79 turbojet engine, became the first combat-type airplane in the world to exceed Mach 2 in level attitude flight.

Several other USAF fighters, such as the Republic F-105B Thunderchief, Convair F-106A Delta Dart—and even a much heavier USAF bomber, the

Convair B-58A Hustler—made these once unthinkable doublesonic speeds commonplace throughout the late-1950s and into the early 1960s. Suddenly, it now seemed that, given enough power, there was no speed limit.

Or was there?

Indeed, there was a speed limit: the heat barrier. While the sound barrier had been broken and surpassed several times over, the heat barrier became a much greater problem. In fact, it is a seemingly insurmountable problem still being addressed today. This limit has held the maximum speed of aircraft to about 3.35 Mn (about 2,480 mph). If an airplane attempts a speed beyond this point, it will begin to melt in midair under the temperatures generated, about 2,000° F.—too hot for the alloys or composites used by current aircraft.

Nevertheless, in late summer 1963, the Lockheed YF-12A became the first USAF fighter to exceed Mach 3.0 (2,000 mph). By fall 1965, the much larger and heavier North American XB-70A Valkyrie achieved triplesonic speed.

From fall 1942 to spring 1953, fewer than eleven years had elapsed from subsonic aircraft to the advent of aircraft capable of supersonic speed. From spring 1953 to summer 1956, in a little more than three years, supersonic speed had advanced to doublesonic speed. Then, from summer 1956 to summer 1963, in just seven years, doublesonic speed had developed into triplesonic speed.

It took twenty-one years for USAF turbojet-powered bombers and fighters to go from speeds around a mere 400 mph to more than 2,000 mph.

A number of triplesonic bombers and fighters were designed for use by the USAF, but only a pair of these triplesonic marvels emerged.

REPUBLIC XF-103

Cancellations, albeit frustrating, are a way of life for all airframe and powerplant contractors and subcontractors. On August 21, 1957, blaming budget cuts and claiming that future aircraft could provide similar information, the USAF simultaneously disappointed the Republic Aviation Corporation, the Curtiss-Wright Corporation, and the Hughes Aircraft Company with its abrupt cancellation of Weapon System 304A. With a single swing of its ax, the USAF

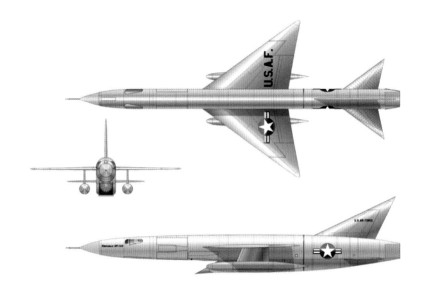

This is the XF-103 as it appeared during its 1954 full-scale engineering mockup inspection. The XF-103—its propulsive system in particular—was ahead of its time, and several other Mach 3—rated engines with greater promise were in development. The first flights of the two XF-103 airplanes were scheduled for March and November 1959. But the XF-103 program was cancelled in August 1957, long before these events were to take place. *G. De Chiara © 2015*

dismantled the programs represented by Republic's Model AP-57, the XF-103; Curtiss-Wright's combined turboramjet engine, comprised of the afterburning XJ67-W-3 turbojet engine and the XRJ55-W-1 ramjet engine; and Hughes's radar and missile fire control, missile and rocket armament package. Surly it exasperated a large number of subcontractors, as well, and it was a major source of frustration for everyone involved. Nevertheless, that's the way things go in the defense industry.

The proposed Republic F-103, in fully missionized form, was to be a single-seat, all-weather, all-missile/rocket-armed area- and point-defense interceptor powered by the Curtiss-Wright dual-cycle turboramjet engine, capable of triplesonic speed at very lofty altitudes. With its proposed armament of Hughes Falcon air-to-air guided missiles and rockets in addition to unguided 2.75-inch-diameter rockets, it was to seek and destroy any known enemy aircraft (whether in existence or in development) via its semi-automatic ground environment (SAGE) system. In this design, its pilot would essentially be the F-103's monitor (during takeoff, climb, descent, and landing), while its onboard SAGE system data link would fire upon the targets automatically.

The SAGE system was online with North American Air Defense Command (NORAD) when the F-103 program was terminated. In the event of an attack, the SAGE system could transmit coded instructions via data link to autopilot equipment onboard manned interceptor aircraft after being scrambled for intercept. Being semi-automatic, by choice, the ADC (Air Defense Command) commander could either have the information relayed by voice to interceptor pilots, or, if the interceptor had the proper equipment, the SAGE system computer itself could transmit instructions automatically to the interceptor aircraft, actually guiding them to the interception and destruction of selected targets. Once initial data had been received, the computer would have already calculated the attacker's direction of travel, speed, altitude, and the actual point of interception. The F-103 was to have this "proper equipment."

Missile away! A pair of operational F-103As pursue the bad guys. One plane fires its starboard side GAR-9X (later AIM-47A). *Artwork by Jozef Gatial*

s a complete weapon system, operational F-103s were to be pure all-weather interceptors capable of operations day or night, in any environment Mother Nature might offer up, short of a full-blown blizzard or hurricane. It was to see targets with either its infrared (heat-detecting) eyeball or its forty-inch-diameter pulse-Doppler radar dish, then fire its heat- and/or radar-guided missiles and/or rockets to kill the threats. All of its unguided rockets, guided missiles, and/or guided rockets were to be carried internally within four bays, two on either side of the fuselage. The unguided rockets were to be carried in a retractable, belly-mounted tray.

Under the direction of Republic's chief designer, Alexander Kartveli, and chief engineer William O'Donnell, the Model AP-57 was proposed to the USAF in January 1951 in Republic's attempt to win the MX-1554 ("1954 interceptor") competition. Although Convair's Model 8 (YF-102) won this competition, Republic's entry drew much interest.

XF-103

The XF-103 was proposed in the same contract competition that resulted in the Convair F-102 under USAF ARDC Project MX-1554. The Republic XF-103, Weapon System 204A, received MX number 1554-A, while its competitor, the Convair F-102, Weapon System 201A, got MX number 1554.

The Republic design, Model AP-57, underwent a design performance evaluation conducted by the Air Material Command on March 27, 1951, significantly early in the program. The results showed an estimated top speed of Mach 2.23 (1,655 mph), which was near the limiting Mach 3.0 quoted to address excessive (estimated) turbine inlet air temperature. The Republic design was ranked eighth out of nine entries for design; the only aircraft scoring lower was the initial Convair F-102 entry. The evaluation follows.

Design Performance Evaluation by the Air Materiel Command

The Republic Model AP-57 [has] a Delta wing swept-back 55 degrees at the leading edge and incorporates a variable incidence feature. The horizontal tail is of the Delta configuration with a sweep-back angle of 60 degrees at the leading edge. The alighting gear was of the tricycle type and retracts into the fuselage. The turbojet was a Wright Aeronautical Corporation XJ67-W-1 powerplant with afterburner. A Ferri-type two-dimensional engine air inlet is used. The thrust of the turbojet engine was limited at high Mach numbers by the allowable turbine inlet temperature. The air was to bypass the engine compressor and turbine, and using the afterburner as a ramjet combustion chamber, the available thrust was greatly increased above 2.0 Mach number.

Six (6) MX-904 missiles and 36 2.75 in. FFA rockets internally stowed were proposed. The missile launching system was complex with six (6) individual tracks and actuating cylinders. Accessibility of all launchers for loading was adequate, though rockets and especially missiles were poorly located in the immediate area of the engine air scoop. Most of the Electronic and Control system were located in a compartment adjacent to the radar equipment. Fuel was contained in five pressurized fuel cells. No self-sealing or armor was proposed for the fuel system.

The USAF canceled the XF-103 development contract on 21 August, 1957 when the aircraft was in the mock-up phase.

Most of the electronic and control system was located in a compartment adjacent to the radar equipment. Accessibility of the radar for checking in the airplane was not too good. A ladder was required for access to all components, unless the entire nose package was removed. This limited the use of test equipment and would have made minor adjustments difficult. The entire collection of rack-mounted components could be removed in a detachable nose section along with the radar antenna and associated equipment for complete replacement or for major service.

The fuel was to be contained in five pressurized fuel cells. No self-sealing or armor was proposed for the fuel system.

The electrical system included a 24–28 volt direct current (DC) system and a three phase four wire 208 volt 400 cycle alternating current (AC) system. The electrical power source proposed would have required the development of an alternator, constant speed drive, and a ram air turbine.

The projected first flights of the two XF-103 airplanes on order were March 1959 and November 1959, respectively. Design was initiated in April 1950; contract was approved September 1951, and mockup inspection was March 1953.

SUMMARY

The design of what became the XF-103 was initiated in April 1950 as part of the 1954 AWI program. The contract was approved in September 1951, and the mockup inspection was held in March 1953. The USAF canceled the XF-103 development contract on August 21, 1957. The estimated first flight dates for the two XF-103s had been earmarked for March 1959 (number one) and November 1959.

It has been reported that USAF serial numbers 53-7824 and 53-7825 were assigned to these two aircraft, but these numbers conflict with numbers assigned to Piper L-21B aircraft.

NORTH AMERICAN XF-108 AND F-108A RAPIER

On September 23, 1959, the USAF abruptly cancelled the North American Aviation F-108A Rapier program. No further explanation was given.

The USAF Air Defense Command had planned to procure as many as 420 F-108 Rapiers to equip fighter-interceptor squadrons located at bases best suited to protect the US borders.

A great deal of speculation preceded the development of these planes, since the Hughes AN/ASG-18 radar and missile/rocket fire control system and Hughes GAR-9 (later designated AIM-47A) development programs were expected to be used in another interceptor that was under development. As it turned out, this speculation proved correct: in February 1964, the existence of the Lockheed YF-12A interceptor was announced by President Lyndon B. Johnson.

XF-103 SPECIFICATIONS
(USAF CHARACTERISTICS SUMMARY, DATED JULY 1, 1957)

Propulsive system
One Wright afterburning, axial-flow YJ67-W-3/XRJ55-W-1 double-cycle turbojet-ramjet engine (the afterburner was to be used independently as a ramjet engine when air bypassed around the turbojet engine); with afterburning, the turbojet engine was to produce 22,100-lbf, the ramjet was to produce 37,400-lbf (maximum)

Maximum speed
Mach number 3-plus (2,000 mph)

Length	81 feet, 9 inches
Height	18 feet, 3 inches
Wingspan	35 feet, 8 inches
Wing area	401 square feet
Empty weight	32,575 pounds (calculated)
Maximum weight	55,780 pounds (calculated)
Ceiling	69,570 feet

Armament
36 unguided 2.75-inch-diameter "Mighty Mouse" FFARS and six Hughes GAR-1 and/or GAR-3 Falcon AAMS

NAA artist rendering of an F-108A depicting its cruising profile. *CW*

This three-view shows a late configuration F-108A study, featuring pre-porthole FCO windows and ventral stabilizers extended circa early 1959. *G. De Chiara © 2015*

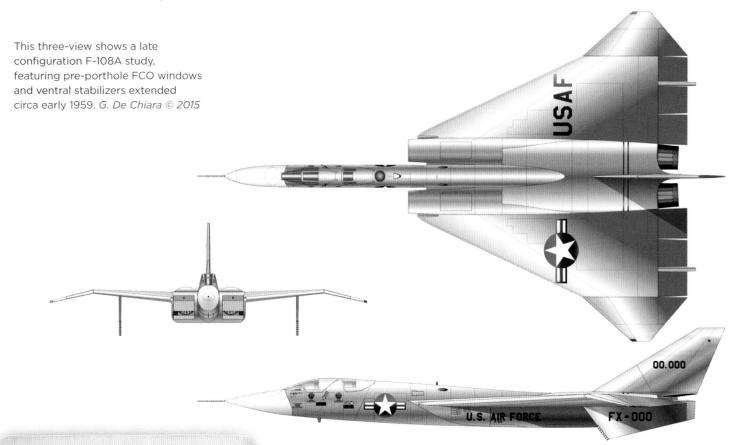

F-108A SPECIFICATIONS

Propulsive system
Two afterburning, axial-flow, 29,300-lbf General Electric J93-GE-3AR (Model 7E) turbojet engines with thrust reversers

Length	89.2 feet
Height	22.1 feet
Wingspan	57.4 feet
Empty weight	50,907 pounds
Gross weight	102,533 pounds

Maximum speed
Mach 3.0 (about 2,000 mph)

Ceiling
77,400 feet (point intercept), 79,400 feet (area intercept)

Maximum range	2,486.8 miles

Crew
Two (pilot, weapons system operator)

Armament
Three Hughes AIM-47A Falcon AAMs

The history of the F-108 traces back to October 6, 1955, when the USAF ARDC issued a general operational requirement (GOR 114) that called for the development of an advanced long-range interceptor, experimental (LRI,X) for its Air Defense Command. It was to be capable of flying and fighting in any environment, armed exclusively with air-to-air missiles and/or rockets.

NAA received a letter contract on June 6, 1956, calling for a LRI with a combat speed of at least 2,000 mph (3-plus Mn) that could operate at seventy thousand feet in all types of weather, day or night. It was to be a complete weapon system (WS-202A) propelled by two afterburning turbojet engines, armed with nuclear warhead air-to-air guided missiles, and controlled by a two-man crew (pilot and weapons system operator).

COMPETITION

North American Aviation wasn't the only airframe contractor to offer LRI designs to the USAF.

Martin initially suggested its Model 302, followed by its Model 308 (Model 302 alternate one) and Model 314 (Model 302 alternate two), before bowing out of the competition.

Full-scale engineering mockup illustrating the round porthole-type window for the FCO. Taken on July 27, 1959, this is the latest known mockup photo, proving that the round FCO window was final. *CW*

McDonnell offered several designs under its Models 109A and 109B, 110A, and 110B, and Model 111A; most of these featured the use of two afterburning Wright J67-W-1 turbojet engines. McAir Model 110A featured the use of three J67-W-1s.

Northrop came up with several offerings of its own. The first of these was its Model N-126 Delta Scorpion. The second was Model N-144, which it dubbed Long-Distance Interceptor. It then offered its Model N-167, which it referred to as a Two Engine Long Range Interceptor.

Republic offered its AP-75, which didn't get further than the drawing board.

It's unclear if some other airframe contractors, possibly Douglas or Lockheed, had also entered the LRI fray, but it is likely that they did.

In any event, it was the offering from NAA that got the nod—to no avail, as it turned out.

SUMMARY

The F-108A was expected to fly in March 1961 and to begin initial operations in January 1963, but the program was cancelled eighteen months prior to its projected first flight.

The abrupt cancellation of the F-108 Rapier program in September 1959 was shocking to say the least, for it had nearly come to fruition as far as its design and mockup development programs were concerned. That is, North American was nearly ready to cut first metal for the assembly of the prototypes, and was only waiting for a go-ahead with the award of a production contract. NAA had satisfied the demands of USAF ADC and it was ready to proceed to the next level. Even though NAA continued its development well into 1960, the USAF's decision stood firm and its cancellation was final.

Why did the USAF cancel the F-108 program? Most aviation historians believe the reason was the advent of Lockheed's family of A-12 aircraft, which produced the YF-12A. This aircraft carried the same radar and missile fire control system and missile armament slated for use by none other aircraft than the Rapier.

YF-12A number one (60-6934) lands after its first flight on August 7, 1963. *LM/Code One*

LOCKHEED YF-12A (AF-12)

Some aircraft are ahead of their time; one such example was the Lockheed YF-12A. This amazing aircraft was a service test, all-weather, all-missile-armed triplesonic interceptor like nothing that had come before it or anything that has come after. It still looks futuristic after more than fifty years.

The Lockheed YF-12A was derived from the Lockheed A-12 program. Following a series of design efforts from 1956 to 1959 to find an advanced manned reconnaissance aircraft to supplement the Lockheed U-2, the Lockheed A-12 design was selected on August 29, 1959. Lockheed received a limited go-ahead on its A-12 project on September 1, and a full-scale go-ahead on the development, construction, and flight testing of twelve A-12 aircraft on January 26, 1960.

The first A-12 (USAF serial number 60-6924) was officially flight-tested on April 30, 1962, by Lockheed test pilot Louis Wellington "Lou" Schalk Jr., who had made an unofficial test hop four days earlier on April 26.

At this time, on an interim basis while awaiting its design Mach 3–rated Pratt & Whitney J58 turbojet engines, the A-12 number was powered by two Mach 2–rated Pratt & Whitney J75 turbojet engines. The J58 didn't become available to the A-12 program until January 1963.

As a matter of course, all airframe contractors design and develop their military airframes to perform not only their primary function, but a host of other functions as well. While the A-12 was created specifically to perform high speed (Mach 3-plus), high-altitude (eighty thousand-plus feet) photographic reconnaissance, its airframe and powerplant combination lent itself to other aircraft requirements. For this reason, Lockheed proposed two (at least) other aircraft types from the A-12: a reconnaissance

strike (RS) bomber as an alternative to the North American RS-70, and an improved manned interceptor (IMI) as an alternative to the North American F-108.

When the USAF abruptly cancelled the F-108 program on September 23, 1959, it simultaneously announced that its proposed Hughes AN/ASG-18 radar and fire control system, along with its Hughes AIM-47 Falcon air-to-air nuclear warhead missile armament package, would be further developed for possible future use. Five years later, on September 30, 1964, the American public found out what that possible future use entailed: it had been successfully adopted within the second derivative of the A-12, the Lockheed YF-12A IMI, which had been covertly flying for some thirteen months. Lockheed test pilot James D. Eastham had performed its first flight on August 7, 1963.

Lockheed received official USAF go-ahead on the development, construction, and flight-testing of three YF-12A aircraft on March 17, 1960, just six months after the demise of the F-108; and the YF-12A incorporated the Rapier's entire radar, infrared, fire control, and missile armament package. These three aircraft were formerly A-12 numbers seven, eight, and nine, and received USAF serial numbers 60-6934 to 60-6936.

Not only was the YF-12A the largest and heaviest interceptor ever built, it was fastest, highest flying, and longest ranging. Moreover, it was the most advanced airplane of its type. It incorporated a two-man crew seated in tandem (pilot and fire control officer) on zero-speed and zero-altitude emergency ejection seats, a large-area delta wing, twin inward-canted (fifteen degrees from the vertical) all-movable vertical tails that doubled as rudders, tricycle landing gear with three wheels and tires on the main legs, and two wheels/tires on the nose gear, a large diameter braking parachute, and two 32,500-lbf (maximum power) Pratt & Whitney turboramjet engines housed with two nacelles at the midpoint of either wing. Another feature was the aircraft's radar cross-section-reducing chines that incorporated the infrared detection and homing units (small, white-colored "eyeballs") faired into the bladelike, squared-off chines on either side of the fuselage.

The YF-12A was a derivative of the super-secret CIA-funded A-12 spy plane, a forerunner of the SR-71 Blackbird.

Three-view of the YF-12A number one (60-6934).
G. De Chiara © 2015

YF-12A SPECIFICATIONS

Crew
Two (pilot and fire control officer)

Propulsive system
Two axial-flow afterburning 34,500-lbf (with afterburning) Pratt & Whitney J58 (Model JT11D-20A) turbojet engines

Length	101 feet, 0 inches
Height	18 feet, 6 inches
Wingspan	55 feet, 7 inches
Wing area	1,795 square feet
Empty weight	60,730 pounds
Gross takeoff weight	140,000 pounds

Maximum speed
Mach 3.35 (2,275 mph)

Ceiling	90,000 feet
Maximum range	3,000 miles

Armament
Three Hughes AIM-47A AAMs; provision was made for a single rotary-action General Electric six-barrel Gatling-type M61A1 Vulcan 20mm cannon housed within the forward bay on the port (left-hand) fuselage chine

AF-12

Once the final design of the A-12 was frozen, it was time to fabricate and assemble all of the materials required to manufacture the twelfth version of the Archangel series of air vehicle designs. So unique was the design of the A-12, it lent its airframe to several other types of aircraft beyond its dedicated reconnaissance platform. In fact, several other aircraft types were offered up, including the AF-12, B-12, and RB-12; the first served as a dedicated interceptor, the second was a strategic bomber, and the third was a reconnaissance bomber. The bomber and reconnaissance bomber versions were offered as alternatives to the B-70 and RB-70, temporarily known as the B-71 and RB-71. The interceptor version was offered to replace the F-106, which could fill the void left by the cancelled F-108.

YF-12A

Lockheed was authorized by both the CIA and USAF to complete three A-12 airframes, numbers seven (Article 127), eight (Article 128), and nine (Article 129) as service test, all-missile-armed interceptors in October 1960 under the highly classified KEDLOCK program. These airframes already had the USAF serial numbers 60-6934 to 60-6936 applied to them, which they retained. At first these were known only as AF-12s, but eventually they received the YF-12A designation.

Skunk Works test pilot James D. "Jim" Eastham (Dutch 52) was appointed chief engineering test pilot on the YF-12A program and, on August 7, 1963, he completed a successful first flight on YF-12A number one (60-6934) at Area 51. Pulling double duty, Eastham was also responsible for writing the pilot's flight manual for the aircraft. On March 13, 1964, Eastham initially flew YF-12A number three (60-6936). In the interim, Eastham had successfully flown YF-12A number two (60-6935) on its first flight (the actual date could not be confirmed).

YF-12A number three (60-6936) crashed to destruction on June 24, 1971, while serving with NASA. The airplane caught fire on its approach to Edwards AFB, caused by the failure of a distressed fuel line. USAF pilot Lt Col. Ronald J. Layton and WSO Maj. William A. Curtis were forced to eject; both survived the ordeal.

SUMMARY

On May 14, 1965, the US Air Force placed a production order for ninety-three F-12Bs for its Air Defense Command (ADC). Secretary of Defense Robert S. McNamara would not release the funding for three consecutive years, however, due to costs incurred from the Vietnam War. Updated intelligence placed a lower priority on defense of the continental United States, so the F-12B was deemed unnecessary. In January 1968, the F-12B program was officially ended.

With this, what might have been the best interceptor on the planet was passed over to become nothing more than an interesting chapter in aviation history.

NORTH AMERICAN XB-70, XB-70A, XB-70B, YB-70, AND XRS-70 VALKYRIE

The North American Aviation B-70 Valkyrie was one of the most elegant and exotic aircraft ever built and flown. Its beauty and grace couldn't save it from its ultimate fate, though, as no production orders were issued to make this plane a reality.

Although numerous examples were to be built under several designations, just two of these unique airplanes were actually built and flown.

The B-70 program began on an official basis in October 1954, when Headquarters USAF issued General

1962 US Tri-Service Aircraft Designation System

On September 18, 1962, the Tri-Service Designation System was put in place to standardize aerospace designations in the US Air Force, US Army, and US Navy (including the US Marine Corps and US Coast Guard). This included aircraft, armament, electronic equipment, missiles, and rockets. This action was by order of the US Department of Defense, effective immediately. It was designed to help eliminate high-digit designations and to alleviate interservice confusion.

For a few examples, as far as fighter aircraft go, the North American FJ-4 Fury became the F-1E Fury, the McDonnell F2H Banshee became the F-2 Banshee, the McDonnell F3H Demon became the F-3 Demon, and the McDonnell F4H Phantom II became the F-4 Phantom II.

Bomber aircraft started anew at B-1A. Other types, (such as attack aircraft) not featured in this work, were likewise redesignated. Fighter aircraft started afresh at F-1A.

Operational Requirement 38 (GOR 38) calling for a manned intercontinental strategic bombardment weapon system that would at first supplement and then replace the B-52, beginning in 1965.

On March 22, 1955, Headquarters USAF released GOR 82, which superseded GOR 38. It too called for a piloted intercontinental strategic bombardment weapon system that would be capable of carrying a twenty-thousand-pound payload of high-yield nuclear weapons, increased to twenty-five thousand pounds by an amendment in September 1955.

The USAF Air Research and Development Command (ARDC), the agency in charge of the program, followed suit and issued its Specific Operational Requirement No. 22 (SOR 22), which identified the new bomber as Weapon System 110A (WS-110A) and established 1963 as the target date for the first wing of thirty operational aircraft.

SOR 22's performance demands were in part for a 0.90 Mn cruise speed and a "maximum possible" speed during a 1,150 mile (1,000 nautical miles) penetration. The ARDC made a revision of SOR 22 on April 15, 1955, that demanded WS-110A's cruise speed should be *no less than* 0.90 Mn, unless a lower speed would result in a significant range increase. Instead of the subsonic speed requirement mentioned in GOR 38, a maximum possible "supersonic" speed within the combat zone was desired. On October 11, 1955, the ARDC amended the revised SOR 22. The amendment set July 1964 as the target date for the first operational wing of bombers. The purpose of this delay was to avoid cost and overall weapons system risks, if at all possible.

In early 1955, Headquarters USAF had already released GOR 96 for a piloted intercontinental reconnaissance weapon system, bound by similar objectives as the previously established bombardment weapon system, known as WS-110A.

The ARDC issued a study requirement of GOR 96 in July 1955 that validated a reconnaissance version of WS-110A. The reconnaissance system was identified as WS-110L. The two systems were combined soon afterward, becoming in the process Weapon System 110A/L (WS-110A/L).

One month earlier, in June 1955, the Air Staff directed that development of WS-110A/L be initiated as soon as possible with a multiple, competitive phase I program. Although six eligible airframe contractors were contacted, only Boeing and North American elected to submit proposals.

On November 8, 1955, the USAF awarded a letter of intent to purchase contracts to both Boeing and North American for phase I development of WS-110A/L. Boeing received $2.6 million, while North American received $1.8 million.

Each contractor had to furnish a design for the required weapon system, engineering data, wind tunnel models, drawings, specifications, reports, and other information, such as cost estimates. They also had to conduct studies and wind tunnel tests, and build a full-scale engineering mockup. The mockup was

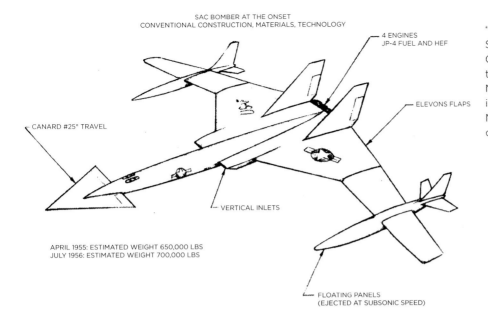

SAC BOMBER AT THE ONSET
CONVENTIONAL CONSTRUCTION, MATERIALS, TECHNOLOGY

4 ENGINES
JP-4 FUEL AND HEF

ELEVONS FLAPS

CANARD #25° TRAVEL

VERTICAL INLETS

APRIL 1955: ESTIMATED WEIGHT 650,000 LBS
JULY 1956: ESTIMATED WEIGHT 700,000 LBS

FLOATING PANELS
(EJECTED AT SUBSONIC SPEED)

"Back to the drawing boards!" exclaimed Strategic Air Command (SAC) Cmdr. Curtis Emerson LeMay when he saw the first designs from both Boeing and North American. "This isn't an airplane, it's a three-ship formation!" This was North American's initial WS-110A offering in 1955–56. *USAF*

to be completed and ready for USAF inspection within two years of the date on which the contractor accepted the contract (no later than mid-April 1958). It was stipulated that the airframe contractor's fees should not exceed $450,000.

The two letter of intent contracts became definitive in 1956. The Boeing contract, AF33 (600) AF-31802, signed on March 15, 1956, specified a total estimated cost of $19.9 million, the North American contract, AF33 (600) AF-31801, signed on April 16, 1956, was for $9.9 million, subject to renegotiation.

The aircraft needed to meet the WS-110A/L requirements demanded advanced airframes and propulsive systems. In mid-March 1956, the demands of WS-110A/L were lofty to say the least. Undaunted, both firms moved forward to achieve their respective goals of creating just what the USAF was wanting.

Ironically, to meet WS-110A/L requirements, both firms came up independently with similar designs, featuring main bodies and two giant, external fuel tanks complete with their own jettisonable wing extensions on either side of the main bodies. These gargantuan three-part designs prompted SAC Cmdr. Curtis Emerson LeMay to order both firms "back to the drawing boards. These aren't airplanes, they're three-ship formations!" They were unacceptable designs, and LeMay and the USAF wanted no part of them.

By the end of February 1957, it had become feasible that WS-110A/L could be an all-supersonic cruise aircraft instead of a split-mission subsonic-cruise/supersonic-dash aircraft. The key was to employ a high energy fuel (HEF) called boron, an innovation that led to chemical-powered bomber (CPB) nomenclature.

On September 18, 1957, the USAF gave Boeing and North American the new system characteristics established for the competition. These characteristics called for a maximum speed range of Mach 3.00 to 3.20, target altitudes from seventy thousand to seventy-five thousand feet, unrefueled range of 6,100 to 10,500 miles, and a gross weight between 475,000 and 490,000 pounds.

Created around 1960, this NAA artist concept imagines an operational B-70 taking off, complete with SAC markings. *CW*

XB-70

On December 23, 1957, the USAF selected the North American WS-110A/L design over that of Boeing. The USAF awarded North American a new phase I contract, USAF contract number AF33 (600) AF-36599, on January, 24 1958. This phase I contract called for fifteen test aircraft and an operational wing of thirty aircraft by 1965.

In February 1958, the USAF cancelled the development of WS-110L, which had been part of WS-110A since March 1956. At the same time, WS-110A was designated XB-70.

A development engineering inspection and a full-scale engineering mockup review were held on March 2 and 30, 1959, respectively.

On August 11, 1959, the DOD cancelled the HEF program and the so-called CPB program disappeared. As it turned out, in laboratory tests, HEF would be highly detrimental to the turbojet engines using the substance for fuel.

When first ordered, the airplane had been designated XB-70. Several years before the first flight, however, on September 18, 1962, it was redesignated XB-70A (adding the suffix A) as part of the then new DOD aircraft, missile, and rocket designation system.

XB-70A

Two XB-70A airplanes were built (USAF serial numbers 62-0001 and 62-0207) under NAA Charge Number NA-278.

B-70

By December 1, 1959, the USAF planned to procure a single B-70 under NAA Charge Number NA-239, and its first flight was estimated to take place some twenty-five months later, in January 1962.

XB-70B

One XB-70B or Air Vehicle Three (AV-3) was to be built and flight-tested, then delivered in FY1965 (USAF serial number 62-0208). Unlike its XB-70A counterparts, the XB-70B (also NA-278) was to have a four-man crew: pilot, copilot, navigator/bombardier, and flight observer. Its estimated first flight date was December 1964. But this would never be, as the XB-70B was cancelled just as it had entered into its early stages of assembly.

YB-70

In August 1960, the USAF sought to procure eleven service test YB-70 aircraft under NAA Charge Number NA-274, with deliveries beginning in FY1965.

The YB-70 was to have the uprated 29,500-lbf (with afterburning) J93-GE-3 turbojet engines due in part to its higher 190,895-pound empty weight and gross takeoff weight of 554,609 pounds.

Its projected payloads were to be special (nuclear) stores: one twenty-five-thousand-pound Class A bomb, two ten-thousand-pound Class B bombs, two Class C bombs, or four Class D bombs.

It was later decided that one, not eleven, prototype YB-70 aircraft would be procured, but the program was cancelled before even this one example could be built.

XRS-70

In March 1962, the USAF proposed a new version of the B-70 that could serve SAC as a reconnaissance strike aircraft, carrying and delivering rocket-powered, air-to-ground nuclear armed guided missiles as its primary mission. For this specific mission, it was designated RS-70. Its secondary mission was to offer reconnaissance, whereby it would fly back over targets that had already been attacked and then restrike if necessary.

The USAF, in conjunction with North American Aviation, proposed an initial buy of sixty RS-70 aircraft, to be operational in FY1969. A second batch of 150 RS-70s was to be delivered in the following year. Each airplane was to cost $50 million.

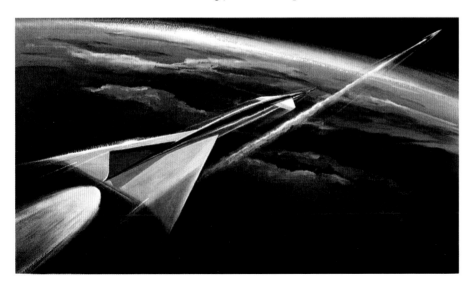

An early artist conception of the proposed but never built RS-70 depicts a missile launch aimed at a sub-space enemy target in the highest reaches of the atmosphere. The RS-70 was designed to pursue the target after missile impact and confirm that the target had been destroyed; if not, it was meant to restrike and destroy the target. *CW*

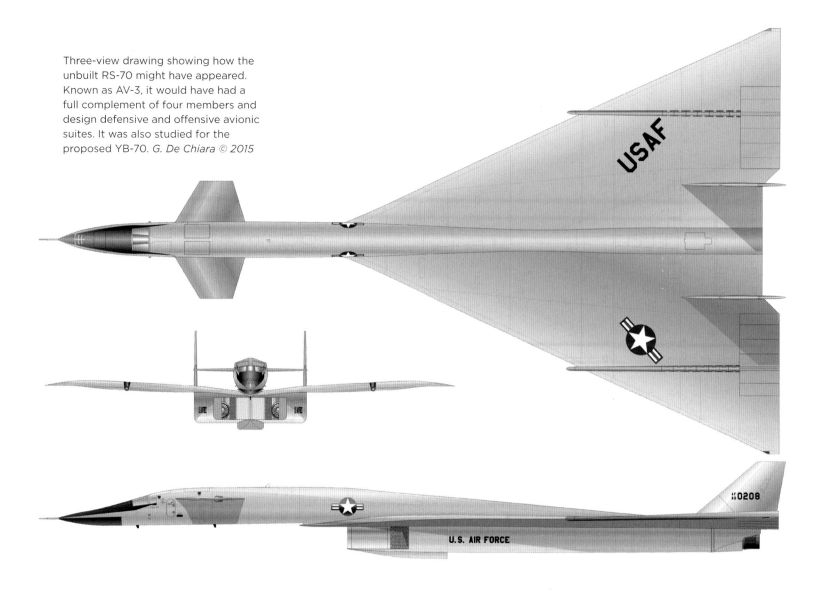

Three-view drawing showing how the unbuilt RS-70 might have appeared. Known as AV-3, it would have had a full complement of four members and design defensive and offensive avionic suites. It was also studied for the proposed YB-70. *G. De Chiara © 2015*

Three test aircraft were scheduled for production at first, then the number was upped to five test aircraft, with the last two built with the full array of equipment found in a production RS-70. The test aircraft were designated XRS-70.

Only $52.9 million was allocated to the XRS-70 program, which wasn't enough to build all five aircraft. Whether the three aircraft could be built with available resources became a moot point, because in July 1963 the entire RS-70 program was axed. The reason for this action became clear after the existence

B-70
PRODUCTION BREAKDOWN

XB-70A-1-NA:	1-62-0001	formerly designated XB-70
XB-70A-2-NA:	1 -62-0207	formerly designated XB-70; built with five-degree dihedral wings
XB-70B-NA:	0-62-0208	cancelled before completion

of the Lockheed A-12 series of triplesonic aircraft was made known, culminating with the public announcement of the SR-71A.

SUMMARY

On February 4, 1969, NASA pilot Fitzhugh L. "Fitz" Fulton and USAF pilot USAF Lt. Col. (later Col.) Emil "Ted" Sturmthal flew the surviving XB-70A from Edwards AFB to Wright-Patterson AFB in Dayton, Ohio, where it was scheduled to go on permanent display at the US Air Force Museum (now the National Museum of the United States Air Force). The flight took three hours and eighteen minutes, after which a normal landing concluded the 1,880 mile trip. It was a subsonic flight during which only 0.91 Mn (600 mph) was attained; the maximum altitude reached was thirty-three thousand feet. It was the eighty-third flight of XB-70A number one, the 129th flight overall.

In an interview after the flight, Lieutenant Colonel Sturmthal said: "I'd do anything to keep the B-70 in the air except pay for it myself." Since the B-70 program was estimated to have cost $1.5 billion, and each flight $11.6 million, Lieutenant Colonel Sturmthal's post-flight comment is understandable.

The on-again, off-again B-70 program was plagued with inconsistencies and uncertainties throughout its fifteen year lifetime (1954 to 1969). Yet the program produced a triplesonic twosome that performed as advertised. As of 2015, more than forty-five years after XB-70A number one made her last landing, the North American XB-70A Valkyrie aircraft remain the largest and heaviest aircraft to have ever flown at three times the speed of sound.

XB-70A SPECIFICATIONS

Propulsive system	
Six afterburning, axial-flow, 28,000-lbf (with afterburning) General Electric YJ93-GE-3 turbojet engines	
Length	185 feet, 8 inches
Height	30 feet, 7 inches
Wingspan	105 feet, 0 inches
Empty weight	230,876 pounds
Max takeoff weight	542,049 pounds
Maximum speed	
1,980.5 mph between the altitudes of 65,000 to 71,750 feet (3.0-plus Mn) with a minimum afterburner power setting	
Combat Ceiling	76,950 feet
Combat range	4,561.68 miles
Armament	None
Payload	Test equipment only

Portrayal of the final WS-110 (B-70) offering from Boeing, its Model 804. *Artwork by Jozef Gatial*

A sunny day greets XB-70A number two (AV-2) during its rollout ceremony on May 29, 1965. *NMUSAF*

Interesting Proposals

During the course of searching for all-weather, strategic, medium-, and long-range fighters and interceptors, a number of US airframe contractors offered thought-provoking proposals.

Boeing, Douglas, Martin, Northrop, North American Aviation, McDonnell, and Republic were among many companies that submitted proposals. Here are some of the more intriguing models:

BOEING MODEL 712 LONG-RANGE INTERCEPTOR

The Boeing Model 712 was its proposed long-range interceptor to meet USAF WS-202A's specifications. It measured 101.31 feet long, 22.20 feet high, with a wingspan of 55.34 feet. It had tricycle landing gear, a two-man cockpit, and it was to be powered by two Wright Aeronautical J67-W-1 turbojet engines.

DOUGLAS MODEL 1355 LONG-RANGE INTERCEPTOR

The proposed Model 1355 was Douglas' version of a long-range interceptor. As proposed on August 12, 1954, the prototype was to be powered by two axial-flow, 21,500-lbf (with afterburning) Wright Aeronautical YJ67-W-1 turbojet engines, one under either wing in pod-type nacelles. It was to be 87.3 feet long and 20.1 feet high, with a wingspan of 48 feet.

It was to be manned by two crewmembers (pilot and radar operator) and armed with eight AIM-4 Falcon AAMs and forty-eight 2.75-inch FFARs. The weapons would have been carried in two bays on retractable launchers: the FFARs in forward bay, Falcons in aft bay. Its empty weight was to be 41,612 pounds, with a maximum weight of 69,786 pounds.

MARTIN COMPANY MODEL 302 LONG-RANGE INTERCEPTOR

Martin's proposed long-range interceptor was its Model 302, a four-engine design with two of its engines housed side by side in its tail section, the other two mounted in pod-type nacelles, one under either wing. All four propulsion units were to be afterburning Wright Aeronautical J67 turbojet engines.

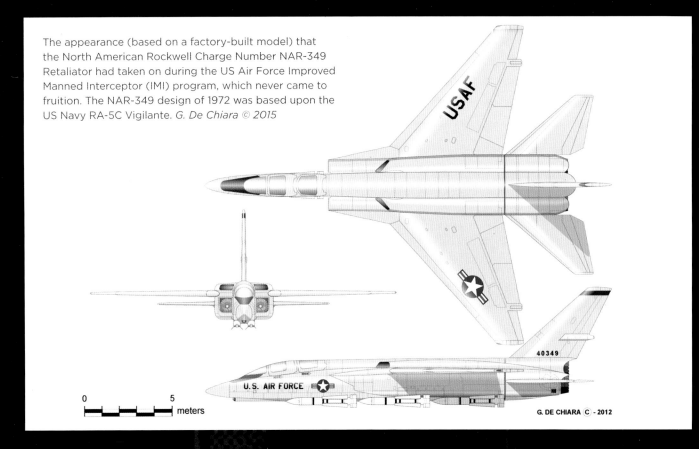

The appearance (based on a factory-built model) that the North American Rockwell Charge Number NAR-349 Retaliator had taken on during the US Air Force Improved Manned Interceptor (IMI) program, which never came to fruition. The NAR-349 design of 1972 was based upon the US Navy RA-5C Vigilante. *G. De Chiara © 2015*

Its proposed armament, carried internally in two weapon bays, was to be forty-eight folding-fin aerial rockets housed in a retractable tray in its forward bay and eight Falcon air-to-air missiles housed in its aft bay.

It was a two-place (seated in tandem) high T-tail design with sweptback flying surfaces, with bicycle-type landing gear and wingtip outriggers for taxiing balance, similar to Martin's XB-48 and XB-51 aircraft.

The Model 302 evolved into Models 308 and 314, which were referred to in-house as Alternate Number 1 for 302, Alternate Number 2 for 302, respectively.

MCDONNELL MODELS 109A/109B, 110A/110B, AND 111A LONG-RANGE INTERCEPTOR

The Model 109A, initiated on June 17, 1954, was a two-place aircraft similar to the F-101A *Voodoo* with a length of 89.8 feet, an aspect ratio of 4.5, and a wing with an area of six hundred square feet. It was to be powered by two Wright Aeronautical J67-W-1 turbojet engines and armed with six Hughes Falcon missiles or McDonnell Model 103E 141-inch long missiles. It was to have the Hughes LRI radar and fire control system. The Model 109B begun on August 18, 1954, was to be the same as Model 109A, but with a length of 96.5 feet and a much-reduced wing area of 78 square feet.

Begun on July 14, 1954, the Model 110A was a two-place aircraft with a length of 86.8 feet, an aspect ratio of 5.5, and a wing area of one thousand square feet. It was to be propelled by three J67-W-1 turbojet engines, armed with eight Falcon missiles and forty-eight folding-fin 2.75-inch rockets or three AUW rockets (McAir nomenclature; the AUW rocket type mentioned here is unknown to this writer, but it may be the type that became the Douglas AIR-2 Genie or the Hughes AIM-47 Falcon, both of which had nuclear warheads) and forty-eight 2.75-inch rockets, equipped with the Hughes LRI radar and fire control system. The Model 110B, started on August 18, 1954, was a two-place aircraft with a length of 79.0 feet, an aspect ratio of 4.31, and a wing area of 819 square feet. It was to be powered by two J67-W-1 turbojet engines, and it was to use the armament, radar, and fire control system of the Model 110A.

Initiated on October 26, 1954, the Model 111A was to be a two-place aircraft with a length of 80.6 feet, aspect ratio of 6.0, and a wing area of eight hundred square feet. It was to be powered by two J67-W-1 turbojet engines and retained the armament, radar, and fire control system proposed for the Model 110A.

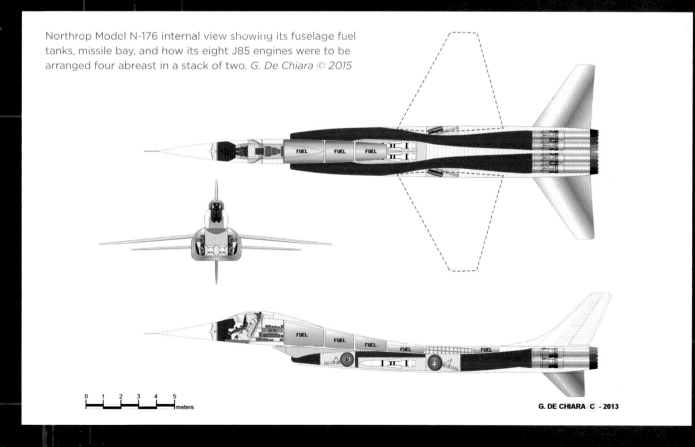

Northrop Model N-176 internal view showing its fuselage fuel tanks, missile bay, and how its eight J85 engines were to be arranged four abreast in a stack of two. *G. De Chiara © 2015*

G. DE CHIARA (C) - 2013

Northrop N-176 soars into action as depicted here in this accurate representation. *Artwork by Jozef Gatial*

NORTHROP PROJECT NUMBERS N-126/-126B PERIMETER DEFENSE DELTA SCORPION

According to an official Northrop Aircraft document entitled "NorAir Project Designations," the Project N-126 (Model Specification NS-107) identified an aircraft called the "Perimeter Defense Delta Scorpion." Project N-126B (NA-108) called for the same. There is no further explanation of either project.

NORTHROP PROJECT NUMBER N-132 STRATEGIC FIGHTER

The Northrop Project N-132 Strategic Fighter program received USAF contract number AF33 (616)-2066, but no further information is extant in the "NorAir Project Designations" document.

NORTHROP PROJECT NUMBER N-144 LONG-RANGE INTERCEPTOR

The "NorAir Project Designations" document describes Project N-144 (Model Specification NS-109) only as a "Long-Range Interceptor (LRI-GOR)." Here, LRI means "long-range interceptor," GOR means "general operational requirements."

NORTHROP PROJECT NUMBER N-149 LONG-RANGE INTERCEPTOR WITH XJ79 ENGINES

Northrop Project Number N-149 (Model Specification NS-110) is mentioned as LRI (XJ79 engines) in the "NorAir Project Designations" document.

NORTHROP PROJECT NUMBER N-164 LONG-RANGE INTERCEPTOR (PRELIMINARY DESIGN DRAWING P.D. 2689)

The "NorAir Project Designations" document lists Project N-164 as a long-range interceptor (Preliminary Drawing P.D. 2689) and calls for a full-scale engineering mockup (Model Specification NS-123) to be built for inspection under USAF contract number AF33 (600)-31229.

In the same document, Northrop Project Number N-167A calls for an LRI 1/10th scale wind tunnel model to be built and tested in the NACA Ames facility in Sunnyvale, California.

NORTHROP PROJECT NUMBER N-176 MEDIUM-RANGE INTERCEPTOR

Northrop Project Number N-176 was to be a medium-range interceptor, according to the "NorAir Project Designations" document. It was to be powered by "two engines," with

The last look at XB-70A number two while she was still a complete and whole airplane. Soon after this image was freeze-framed on June 8, 1966, the orange-tailed NASA F-104N (NASA 813) flying just off her right side collided with her. Both aircraft crashed to destruction, killing NASA pilot Joe Walker and XB-70A copilot Carl Cross. Al White, pilot of the XB-70A, ejected and survived the ordeal, but he was badly injured upon landing and never flew again. *NMUSAF*

With its twin vertical tails sheared off, the wounded *Valkyrie* flew on for a short time before it veered off course and entered into a flat spin; it spiraled all the way to the ground. Piloted by Joe Walker, the F-104N exploded and disintegrated immediately as it fell away to the ground below. Walker was killed instantly. *NMUSAF*

each being a quad-pack of General Electric afterburning J85 turbojet engines, for a total of eight engines.

NORTHROP PROJECT NUMBER N-184 MEDIUM-RANGE INTERCEPTOR LIGHTWEIGHT

The "NorAir Project Designations" document lists Northrop Project Number N-184 simply as a medium-range interceptor lightweight, with no other information.

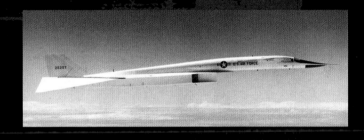

Taking wing: a right-side view of XB-70A number two showing its wingtips lowered to their mid-down twenty-five-degree deflection angle. *NMUSAF*

BELOW: With its six J93s pouring black smoke, XB-70A number one departs the main runway at Edwards AFB during a rapid landing gear retraction test, around mid-1965. *NMUSAF*

NORTH AMERICAN ROCKWELL NR-349 "RETALIATOR" IMPROVED MANNED INTERCEPTOR

The North American Rockwell NR-349 "Retaliator" was to be a USAF version of the USN RA-5C Vigilante, initially propelled by two (then three) afterburning, axial-flow, 17,860-lbf (with afterburning) General Electric J79-GE-10 turbojet engines. The earlier twin-J79-powered offering featured a single North American Rocketdyne XLR46-NA-2 liquid-fueled rocket motor to boost into high-altitude speed. The tri-motor J79-powered version was to be armed with six Hughes AIM-54 Phoenix AAMs carried externally on ventral pylons.

REPUBLIC MODEL AP-75 LONG-RANGE INTERCEPTOR

The Model AP-75 was Republic's bid to build a long-range interceptor under WS-202A. Very few details are available on this proposed LRI.

EIGHT

ADVANCING THE STATE OF THE ART

So when President Carter cancelled the B-1A production program in mid-1977 we of course knew why he had taken such drastic action. Simply put, it was that new technology called "stealth" which allowed an aircraft, whether it be an attack, bomber, or fighter, to be nearly invisible to radar. But in 1977 this advanced technology was as secret as the Manhattan Project in World War II had been, if not more so. It was a matter of national security—in the first degree.

—Walter A. "Walt" Spivak, chief engineer

When American airpower first entered the Vietnam War, the USAF sent large numbers of its turbojet-powered bombers and fighters into battle. Eventually, these included the B-52, B-57, B-66, F-100, F-101, F-102, F-104, F-105, F-111, F-4, and the F-5. With this array of bomber and fighter aircraft, all three of the major combat commands of the USAF—the Aerospace Defense Command (ADC), Strategic Air Command (SAC), and the Tactical Air Command (TAC)—were represented between

McAir chief engineering test pilot Irv Burrows makes a successful first flight in the prototype F-15A number one (71-0280) at Edwards AFB on July 27, 1972. During the flight, Burrows left the gear down and extended the speed brake. The speed brake was later enlarged for improved aerodynamic braking. *NMUSAF*

1962 and 1973. It was a costly air war for America, with 1,737 combat losses during that long-running conflict.

The aforementioned bombers and fighters served well in the Vietnam War, but some hard lessons were learned. The main takeaway was that all-missile-armed fighters also needed machine guns or cannon so that they could continue close-in air-to-air battles after their missiles were gone.

The result was a need for more advanced, more capable bombers and fighters, a requirement that spawned the USAF AMSA, FX, and LWF programs, which in turn gave birth to the B-1, F-15, and F-16. In part, this bomber and these two fighters were responsible for advancing the state of the art.

McDONNELL DOUGLAS F-15A EAGLE, BOEING (McDONNELL DOUGLAS) F-15E STRIKE EAGLE, BOEING F-15SE SILENT EAGLE

With the announcement on April 23, 1991, that the Lockheed/Boeing/General Dynamics YF-22 had won the advanced tactical fighter (ATF) competition and would soon enter into its engineering manufacturing and development (EMD) phase, the working days of the McDonnell Douglas F-15 Eagle began to wane; this plane's retirement was looming on the horizon. But what a fantastic career it had had.

The McDonnell Douglas Corporation F-15 Eagle series of aircraft have more than earned their keep. Until the advent of the F-22 Raptor, the F-15 was regarded as the best air superiority fighter in the world. Moreover, it has never been defeated in combat, and boasts an outstanding kill ratio of 105 to 0!

The history of the F-15, manufactured by the McDonnell Douglas Corporation in St. Louis, Missouri, traces back some fifty years to 1964. At the time of its birth, the idea of a pure air superiority fighter—one that could secure the airspace over a battlefield and allow ground troops and attack aircraft to operate without threat of hostile air attack—began to play in the minds of USAF planners. This new aircraft, the first dedicated to the air superiority role since the F-86 Sabre of the late 1940s, was to achieve absolute air superiority by defeating the advancing enemy threat in any all types of aerial combat. Thus, the FX (fighter experimental) concept was born.

Beginning in November 1965, the USAF Aeronautical Systems Center (ASC) (formerly Aeronautical Systems Division (ASD)) initiated a series of conceptual studies on this new breed of fighter plane. That action was followed by further evaluation studies, which were completed in September 1966; FX concept formulation was completed in August 1967; and FX point design studies were finished in June 1968. On September 12, 1968, the proposed FX was officially designated F-15. (On April 28, 1967, the Douglas Aircraft Company had merged with the McDonnell Aircraft Corporation to form the McDonnell Douglas Corporation.)

The Day-Glo adorned prototype F-15A streaks across the sky over Edwards AFB during its first flight. *NMUSAF*

On September 30, 1968, the RFP on the F-15 was released to eight airframe contractors: Boeing, Fairchild-Republic, General Dynamics, Grumman, Lockheed, McDonnell Douglas, North American, and Northrop. The RFP called for, in part, the following:

- Single-man cockpit and weapon system
- High engine thrust-to-weight ratio, close to, or exceeding, unity (the number one)
- Twin-engine arrangement, each engine to be rated in the 25,000-to-30,000-lbf
- Long ferry range, to Europe without stopover or in-flight refueling
- Low-loading wing, optimized for Mach 0.90 buffet-free performance
- 360-degree visibility from the cockpit
- Maximum gross takeoff weight of forty thousand pounds for the air superiority role
- Long-range pulse-Doppler radar with look-down, shoot-down capability
- Maximum speed of Mach 2.5 at best altitude; 1.20 Mn at low level

Furthermore, armament would be comprised of a single rotary-barrel, 20mm cannon, four AIM-7 Sparrow, and four AIM-9 Sidewinder AAMs.

Four airframe contractors submitted F-15 proposals, and three of them—Fairchild-Republic, North American, and McDonnell Douglas—received contracts on December 30, 1968, to finalize their respective F-15 proposals, due on June 30, 1969.

Following a nearly six-month evaluation period on the F-15 proposals, USAF Secretary Robert C. Seamans Jr. announced on December 23, 1969, that McDonnell Douglas would build the F-15.

On January 2, 1970, a contract was signed for the procurement of the McDonnell Douglas Model 199B, the F-15 under USAF contract number AF-70-0300. The initial order was for twenty full-scale development (FSD) airplanes; eighteen single-seat F-15As; and two tandem-seat TF-15As (later redesignated F-15Bs). No Y for service test prefix was assigned to these twenty

F-15A SPECIFICATIONS

Propulsive system
Two afterburning, axial-flow, 25,000-lbf Pratt & Whitney F100-PW-100 turbofanjet engines

Length	63 feet, 9.75 inches
Height	18 feet, 7.5 inches
Wingspan	42 feet, 9.75 inches
Empty weight	26,500 pounds
Gross weight	41,500 pounds
Maximum speed	2.5-plus Mn
Ceiling	50,000 feet
Maximum range	3,000 miles

Armament
One M61A1 Vulcan 20mm cannon; four AIM-7 Sparrow AAMs; four AIM-9 Sidewinder AAMs

---

airplanes. Respectively, these aircraft were assigned USAF serial numbers 71-280 to 71-289 and 71-113 to 71-120 (F-15As); 71-290 and 71-291 (F-15Bs).

The successful design of the McDonnell Douglas F-15, now officially named Eagle, was engineered primarily by F-15 general manager Donald "Don" Malvern, chief engineer George S. Graff, and Robert C. "Bob" Little. Little, the former chief engineering test pilot, later became a corporate vice president at McDonnell Douglas. The final proposal was made up of 37,500 pages of data, and its product had consumed some 2.5 million man-hours.

After careful examination of available and soon-to-be-available powerplants, McDonnell Douglas selected the upcoming 25,000-lbf, afterburner-equipped Pratt & Whitney F100-PW-100 turbofanjet engine (Model JTF22). This engine was selected in February 1970.

The Pratt & Whitney F100 produced 25 percent more engine thrust per pound of engine weight than the best engines at the time. With a gross takeoff weight in the forty-thousand-pound range, two F100s gave the F-15 some 50,000-lbf of propulsion. Thus, for the first time, an aircraft had a high-enough engine thrust-to-weight ratio that it could actually accelerate while going straight up!

The first FSD F-15A airplane rolled out on June 26, 1972, and it was subsequently transported to Edwards AFB via a Lockheed C-5A Galaxy transport airplane. After a series of ground evaluations—including low-, medium-, and high-speed taxi runs—it was ready for flight.

F-15A/B FULL-SCALE DEVELOPMENT AIRCRAFT
PRODUCTION BREAKDOWN

F-15A-1-MC:	2-71-080 and 71-0281	
F-15A-2-MC:	3-71-0282 to 71-0284	
F-15A-3-MC:	2-71-0285 and 71-0286	
F-15A-4-MC:	3-71-0287 to 71-0289	
F-15B-3-MC:	1-71-0290	formerly designated TF-15A-3-MC
F-15B-4-MC:	1-71-0291	formerly designated TF-15A-4-MC
F-15A-5-MC:	4-72-0113 to 72-0116	
F-15A-6-MC:	4-72-0117 to 72-0120	

The amazing Streak Eagle was a modified F-15A (72-0119) used to establish a number of time-to-climb records, many of which still stand today. In 1975, from January 16 to February 1, the Streak Eagle set eight records. During record number eight, it reached an altitude of 98,425 feet in just 3 minutes, 27.8 seconds, from brake release at takeoff, "coasting" to nearly 103,000 feet before its descent. *NMUSAF*

Then on July 27, 1972, McDonnell Douglas chief engineering test pilot Irving L. "Irv" Burrows controlled its maiden flight: a fifty-minute test hop, during which the airplane attained a speed of 320 mph at twelve thousand feet.

Subsequent flight-testing and armament evaluations found the Eagle to be everything required, and more. It proved to be the most maneuverable and agile fighter-interceptor airplane built to date.

As a complete weapons system (System Management Code 328A), the F-15A Eagle (now retired) incorporated the Hughes Aircraft high-frequency, pulse-Doppler AN/APG-63 attack radar (the Hughes AN/APG-70 unit was employed by later F-15 variants), and carried four Sparrow missiles, four Sidewinder missiles, and a single 20mm cannon as its basic air-to-air armament package.

Later versions of the Eagle carry the AIM-120 AMRAAM. In addition, all versions of the Eagle can carry a vast assortment of air-to-ground weapons, including nuclear. As of 2015, more than 1,100 F-15A, F-15B, F-15C, F-15D Eagle, and F-15E Strike Eagle aircraft have been produced, with the F-15E still in production.

BOEING (McDONNELL DOUGLAS) F-15E STRIKE EAGLE

The F-15E Strike Eagle was McDonnell Douglas's answer to the competing General Dynamics F-16XL, and it was the F-15E that was ordered into production. No Y for service test prefix was assigned.

To initiate the development of the F-15E, McDonnell Douglas modified the second of the two F-15B full-scale development airplanes built (USAF serial number 71-0291) to serve as a demonstrator. This aircraft, known as the advanced fighter capability demonstrator, made its first flight on July 8, 1980.

In March 1981, the USAF announced the Enhanced Tactical Fighter (ETF) program to procure a replacement for the F-111. The program was later renamed the Dual-Role Fighter (DRF) competition. The concept envisioned an aircraft capable of launching deep interdiction missions without requiring additional

F-15E
SPECIFICATIONS

Propulsive system
Two afterburning, axial-flow, 25,000-lbf (with afterburning) Pratt & Whitney F100-PW-220 turbofanjet engines or 29,000-lbf (with afterburning) F100-PW-229 turbofanjet engines

Length	63.8 feet
Height	18.5 feet
Wingspan	42.8 feet
Empty weight	37,500 pounds
Gross takeoff weight	81,000 pounds
Maximum speed	2.5-plus Mn (1,875 mph)
Maximum range	2,400 miles
Combat Ceiling	60,000 feet

Armament
One 20mm multi-barrel M61A2 Vulcan cannon mounted internally with five hundred rounds of ammunition; four AIM-9 Sidewinder missiles and four AIM-120 AMRAAM, or eight AIM-120 AMRAAM missiles; any air-to-surface weapon in the Air Force inventory (nuclear and conventional)

support by fighter escort or jamming. General Dynamics submitted the F-16XL, while McDonnell Douglas submitted the F-15E.

The DRF evaluation program ran from 1981 through April 30, 1983, during which the F-15E logged more than two hundred flights, demonstrated takeoff weights of more than seventy-five thousand pounds, and validated sixteen different weapons-carrying configurations.

The single-engine F-16XL was a promising design that, with its radically redesigned cranked-delta wing, greatly boosted performance; if selected, the single- and two-seat versions were to be designated F-16E and F-16F, respectively. On February 24, 1984, the USAF chose the F-15E.

The main reasons behind the USAF's decision were the lower development cost of the F-15E compared with that of the F-16XL ($270 million versus $470 million), twin-engine redundancy, and the future growth potential of the F-15E.

Two tandem-seat F-15D-41-MC airplanes (USAF serial numbers 86-0183 and 86-0184) were modified to serve as the F-15E Strike Eagle demonstration airplanes. These in turn became F-15E-41-MC airplanes.

The first of these made its first flight at St. Louis on December 11, 1986, with McDonnell Douglas test pilot Gary L. Jennings at the controls. The fight lasted seventy-five minutes and Jennings flew it to 0.9 Mn at forty thousand feet.

Since that first flight nearly thirty years ago, time and again the F-15E Strike Eagle has more than proved its value in combat.

Moreover, remembering the Penetration Fighter program of the late 1940s, it seems the USAF has finally found what it had been looking for way back then: a bona fide penetration fighter.

BOEING F-15SE SILENT EAGLE

The F-15SE (SE, suffix for Silent Eagle) was offered to the USAF and friendly nations as a lower cost alternative to acquiring higher cost stealth fighters.

To create the F-15SE demonstrator, Boeing modified the first F-15E-41-MC Strike Eagle built (USAF serial number 86-0183) by adding two outward canted vertical tails, internal missile/rocket-carrying conformal housings, special radar-absorbing coatings, and other refinements such as the Raytheon APG-8 Active Electronically Scanned Array (AESA) radar system. It was first shown by Boeing on March 17, 2009.

On July 8, 2010, Boeing F-15 chief test pilot Daniel A. "Dan" Draeger flew the inaugural flight on the F-15 Silent Eagle demonstrator

(known as F-15E1) from Lambert St. Louis International Airport in St. Louis, Missouri, an eighty-minute test hop, during which Draeger opened and closed the left-hand conformal weapons bay (CWB).

"Our intent was to verify all systems are operational in a flight environment. This flawless flight allows us to move into the next phase. In the next couple of weeks, we will ferry F-15E1 to the test range and launch an AIM-120," said Draeger. He added, "Everything about the flight went according to plan. We saw nothing unusual during the flight, and we cleared the desired flight envelope needed to fire the missile at the test range; that is pretty much unheard of on a first flight."

Boeing attempted but failed to export its Silent Eagle to Israel, Japan, Saudi Arabia, and South Korea.

SUMMARY

Boeing merged with McDonnell Douglas on August 1, 1997, but retained its name, the Boeing Airplane Company. Since the F-15 was still in production, it became the Boeing F-15 after the merger.

The F-15 Eagle series of air superiority fighter aircraft has played an integral role in the arena of aerial combat for some forty years, not only for the United States but for the air forces of allied and friendly nations.

What began as the FX program in 1965 to supplement and replace the F-4 Phantom II evolved into a successful fighter-interceptor production program, whereby McDonnell Douglas/Boeing has built and continues to build a superior multi-role aircraft: the F-15E Strike Eagle.

ABOVE: The F-16's official name is Fighting Falcon, but its pilots prefer the unofficial name Viper. Also known as the "Electric Jet," the prototype two-seat F-16B (75-0751) is shown here; its first flight was on August 8, 1977. *USAF*

RIGHT: The first of two prototype YF-16s (70-1567) made an inadvertent takeoff on January 20, 1974, during a high-speed taxi test, whereby some damage was incurred during its first landing. The *official* first flight came on February 2, 1974, at Edwards AFB, with GD chief engineering test pilot Phil Oestricher at the controls, as shown here. *USAF*

GENERAL DYNAMICS YF-16, F-16/79, F-16/101, AND F-16XL FIGHTING FALCON

At the beginning of the 1970s, under what it called the Lightweight Fighter program (LWF), the USAF initiated a study into energy maneuverability and the emerging technologies that might be used for future fighter aircraft. In 1971, the USAF opted to test the feasibility of these concepts by seeking the actual development of prototype LWF aircraft. The program had three main goals: (1) to explore the advantages and uses of these and other emerging technologies; (2) to reduce the technical and cost risk potential of full-scale development and production programs; and (3) to provide the Department of Defense with options readily available to military aerospace needs.

The General Dynamic Corporation F-16 Fighting Falcon program continues to be one of the most successful fighter production programs ever.

The Grumman Aerospace Corporation won the right to produce a pair of forward swept-wing (FSW) aircraft to test the FSW concept. This was GD's FSW offering based upon the F-16A. *LM/Code One*

YF-16

The first service test YF-16 was ceremoniously rolled out on December 13, 1973. It was transported to Edwards AFB via a Lockheed C-5A Galaxy on January 7, 1974. The Flight Test Division of General Dynamics had earlier selected Phillip F. "Phil" Oestricher to be chief engineering test pilot on the YF-16 flight test program.

On January 20, 1974, during a high speed taxi test, the airplane inadvertently lifted off to make what was later dubbed "the unintentional first flight." Oestricher made a quick tight circle flight and immediately landed the anxious airplane. It sustained slight damage to its right-hand stabilator, which required some repair. The *intentional* first flight, flown by Oestricher, took place on February, 2, 1974, lasting ninety minutes without any major squawks.

YF-16 number two arrived at Edwards AFB on February 27, 1974, and made its first flight, with GD experimental test pilot Neil R. Anderson in control, on May 9, 1974.

It was decided to demonstrate the YF-16 aircraft at the 1975 Paris Air Show using YF-16 number two. On May 8, 1975, after making a practice demonstration flight at Fort Worth, one of the main landing gear assemblies on YF-16 number two refused to extend for landing. It was decided to retract the extended nose gear and the one extended main gear to attempt a wheels-up belly landing. Pilot Anderson flew it around until it was nearly out of fuel, then successfully greased it onto the grass alongside of the main runway in what's been called a perfect belly landing, with very little damage; he walked away unscathed.

"When it finally stopped, he jumped out of the cockpit and ran about 50 yards and threw his helmet on the ground," USAF Capt. (later Brig. Gen.) Robert L. "Bob" Efferson said. "You don't belly [land] in modern-day fighters, but he did it. He did a remarkable job." Efferson was deputy commander of operations with the 301st Tactical Fighter Wing based at Carswell AFB at the time.

The first flight of the single-seat prototype F-16XL (75-0749) occurred on July 20, 1982, with GD test pilot James McKinney at the controls. The F-16XL aircraft competed against and lost to the F-15E, which became the Strike Eagle. *NMUSAF*

F-16/79

General Dynamics' test pilot James A. "Jim" McKinney initially flight-tested the F-16/79 on October 29, 1980, at GD's Fort Worth, Texas facility.

F-16/101

The first flight of the F-16/101 came on December 17, 1980, with Jim McKinney under glass.

F-16XL

Two F-16XLs were built as conversions from full-scale development aircraft. The number one aircraft (S/N 75-0749) was a single-seat version, while number two (S/N 75-0747) was a two-seat version. The aircraft were flight-tested from mid-1982 through 1985 before being put into storage. The number one aircraft was pulled from storage in 1989 and modified for test work with NASA for studies on supersonic laminar flow and sonic boom research (in conjunction with an SR-71). The second aircraft was returned to service at the Dryden Flight Research Center in 1992 and took over the supersonic laminar flow project from aircraft number one.

The F-16XL is a radically different version of the standard F-16. The most obvious design change is the "cranked-arrow" double delta wing. The aircraft is also more than four feet longer than the production F-16A. The XL was initially intended as a multi-role strike aircraft competing against the F-15E Strike Eagle. The F-16XL never entered production, and only the two prototypes were built.

The F-16XL program began life in February 1980 as the proposed Supersonic Cruise and Maneuvering Prototype (SCAMP).

The first F-16XL (USAF serial number 75-0749) made a successful first flight on July 3, 1982, with General Dynamics' test pilot Jim McKinney at the controls. The number two F-16XL (USAF serial number 75-0747), previously the third FSD F-16A, was first flown on October 29, 1982, by General Dynamics test pilot Alex V. Wolfe with Jim McKinney in the back seat.

An F-16 flaunts the vast array of ordnance—conventional and nuclear—it can carry and deliver. *NMUSAF*

SUMMARY

Today's F-16 Fighting Falcon is a far cry from the YF-16 of yesterday, in looks and in performance.

NORTHROP YF-17 COBRA

Although it was terminated on January 13, 1975, as far as the USAF was concerned, the history of the Northrop Aircraft Corporation YF-17 Cobra lightweight air combat fighter service test aircraft traces back to August 30, 1965. This was an era when Northrop's bestseller, the F-5, was rolling off the production line at a steady rate, a time when Northrop was free to turn its attention toward future fighter design efforts. Indeed, Northrop was ready to look toward the F-5's future, in the 1970s and beyond, when all versions of its F-5 would have to be supplemented and ultimately replaced. This foresight gave birth to Northrop's privately funded F-5X program, which came under the guidance of project director Leon F. "Lee" Begin Jr. This project (Northrop Project Number N-300) spawned the proposed P-530 Cobra.

The YF-16 (foreground) and YF-17 fly together in formation. The YF-16 prevailed in the air combat fighter (ACF) competition, but the YF-17s did everything they were asked to do and more. The design evolved into the F/A-18 Hornet series of fighters for the US Navy. *NMUSAF*

The twin-jet YF-17 was capable of flying supersonic without using its afterburners. It first accomplished this feat on June 11, 1974, on just its second test flight. The YF-17s were powered by two 14,400-lbf GE YJ101-GE-100 turbofanjet engines. The second YF-17 (72-1570) first flew on August 21, 1974. *NMUSAF*

Northrop's proposed P-530 Cobra was to be a single-seat, high-performance fighter designed with emphasis on the air superiority role. As projected, it could also be effectively used for close air support, interdiction, reconnaissance, and high-speed intercept missions. It was an entirely new advanced fighter, designed to meet the defense needs of the United States' allies in the late-1970s through the year 2000. Its maximum speed was in the Mach 2.0 range; to attain that kind of performance, it would employ a pair of specifically tailored afterburning turbojet engines.

Under its core-engine concept, General Electric had initiated development of its Model GE1 two years earlier. One version of this powerplant, using its baseline GE1 core, was the GE1/J1A2 turbojet engine that soon evolved into its Model GE15 (J101 as later designated), an advanced turbojet engine that would produce 15,000-lbf with afterburning and would have a high engine-thrust-to-weight ratio. This was most impressive because the F-5's then latest J85-GE-21 only produced 5,000-lbf with afterburning. By comparison, in other words, the J101 would produce yet another 10,000-lbf without being much larger and heavier than the J85. It was the engine of choice, and it would power the P-530 Cobra (formerly F-5X).

By mid-1966, Northrop's P-530 Cobra team had designed a twin-engine, single-seat fighter configuration optimized to employ a pair of afterburning GE YJ101-GE-100 turbojet engines delivering a total power output of 30,000-lbf. Its unique design featured twin outward-angled vertical fins; a large-area, all-movable stabilator; a large, 360-degree cockpit canopy for excellent all-around visibility; and long-chord leading-edge extensions (LEX) running horizontally from the wing's leading edges to a point even with the front of the cockpit canopy. along either side of the fuselage. These LEX strakes gave the P-530 the appearance of a hooded cobra, poised to strike, which lent the P-530 its nickname.

With in-house financing, Northrop built a full-scale model of its P-530 Cobra. And, with DOD and USAF permission, Northrop showed it to potential customers among the United States' allies. The overwhelmingly favorable response to this privately funded venture surprised Northrop, as did the attention given to another of their proposed designs. This lightweight, high-performance air combat fighter would offer performance that would likely be superior to any current USAF fighter in service—and it was likewise noticed by the USAF.

As development of Northrop's P-530 and GE's GE15 proceeded, the F-15 (F-X) competition went to McDonnell Douglas on December 23, 1969, when the USAF selected its entry over those submitted by Northrop and others.

A bit of a lull followed, during which the USAF came up with its Lightweight Fighter (LWF) program in early 1971. Suddenly, based on its P-530 Cobra design, Northrop had an airframe and powerplant combination well suited to the LWF program. Northrop chose to enter the LWF competition with a derivative of its Cobra; to this end, a number of projects were initiated under Project Director Michael Gerald "Jerry" Huben.

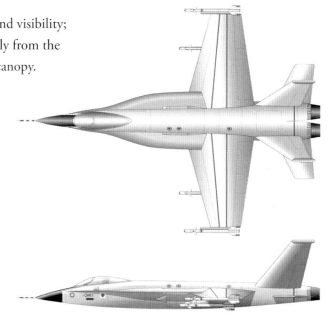

The P-530 Cobra configuration (according to mockup) in three-view. *G. De Chiara © 2015*

The two YF-17 prototypes flew 288 test flights, totaling 345.5 hours. The YF-17 attained a top speed of Mach 1.95, a peak load factor of 9.4 g, and a maximum altitude of over fifty thousand feet. It could attain a sustained thirty-four-degree angle of attack in level flight, and sixty-three degrees in a climb at fifty-eight miles per hour. *NMUSAF*

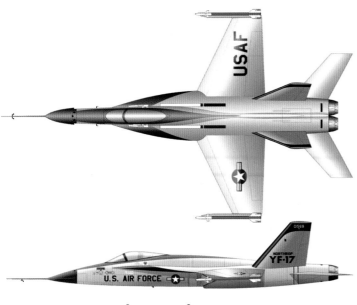

The first of the two prototype YF-17 aircraft (P-600) as it appeared on its first flight, June 9, 1974. *G. De Chiara © 2015*

YF-17 SPECIFICATIONS

Propulsive system
Two afterburning, axial-flow, 15,000-lbf class General Electric Aviation YJ101-GE-100 (Model GE15) turbojet engines

Length	56 feet, 0 inches
Height	16 feet, 6 inches
Wingspan	35 feet, 0 inches
Empty weight	15,000 pounds
Gross weight	23,000 pounds
Maximum speed	2-plus Mn
Maximum range	2,990 miles
Ceiling	50,000 feet

Armament
One M61A1 Vulcan 20mm, six-barrel rotary-action cannon; two AIM-9 Sidewinder AAMs

The first of these, Northrop Project Number N-314, was begun on May 2, 1971. The goal was to produce an advanced energy/maneuverable (E/M) fighter powered by two J101 (GE15) engines. By May 19, this effort had evolved into the baseline E/M concept, Northrop Project Number N-315. On this same day, four additional projects were begun under the direction of Huben:

- Project Number N-316: an advanced E/M fighter with a pair of cheek-type engine air inlets to feed a single Pratt & Whitney F100 turbofanjet engine
- Project Number N-317: an advanced E/M fighter powered by a single J101 (GE15) turbojet engine
- Project Number N-318: an advanced E/M fighter powered by two non-afterburning J101 (GE15) turbojet engines
- Project Number N-319: an advanced E/M fighter with a one-engine air inlet to feed a single P&W F100 turbofanjet engine

By September 28, 1971, Project Number N-321, the proposed P-610 LWF concept had emerged. This lighter version of the P-530 Cobra soon became known as preliminary design P-600. Northrop assigned Walter E. "Walt" Fellers as P-600 project director. It would be Northrop's LWF competition entry, and, when the RPF for an LWF was made public on January 8, 1972, Northrop was nearly ready to respond with its Cobra design.

As noted above, Northrop and General Dynamics were awarded similar contracts in April 1972 to proceed with their respective LWF development programs. Northrop's entry, still unofficially known as Cobra, was designated YF-17.

For its LWF role, the YF-17, like the YF-16, sported wingtip air-to-air missile rails for AIM-9 Sidewinders, a single six-barrel rotary action General Electric M61A1 (formerly T-171) Vulcan 20mm cannon, and three ordnance attachment points.

SUMMARY

Although no production F-17s were ordered and built, the aircraft's excellent design lent itself to the creation of what became the outstanding McDonnell Douglas F/A-18A/B/C/D Hornet, the Boeing (formerly McDonnell Douglas) F/A-18E/F Super Hornet, and the Boeing EA-18G Growler series of fighter and electronic attack aircraft.

Credit goes to the Northrop design team, headed by Project Engineer Walter Edwin "Walt" Fellers, which created the YF-17 service test aircraft.

ROCKWELL INTERNATIONAL B-1A

The Rockwell International Corporation B-1A evolved from a flock of prior programs initiated with the purpose of finding a suitable replacement for the B-52.

These programs had included the Subsonic Low Altitude Bomber (SLAB), Extended Range Strike Aircraft (ERSA), Low-Altitude Manned Penetrator (LAMP), Advanced Manned Precision Strike System (AMPSS), and, finally, the Advanced Manned Strategic Aircraft (AMSA) program, which spawned the B-1A program (as it was designated in April 1969).

Rockwell and Boeing became the finalists on the AMSA program and fought head to head to win the production contract. General Dynamics was eliminated.

In this arrangement, Rockwell International was to build 244 B-1A airplanes for the USAF SAC under Weapon System 139A (WS-139A).

On December 8, 1969, North American Rockwell (NAR) and General Electric Aircraft Engines (GEAE) were announced as the winners of the airframe and powerplant contracts for the B-1A AMSA program.

North American Rockwell was to build two structural test airframes and five flyable evaluation aircraft, while General Electric was to build forty engines. In 1971, these projected numbers were cut to one structural test article and three flyable evaluation aircraft (USAF serial numbers 74-0158 to 74-0160).

The first flight of B-1A number one occurred on December 23, 1974, with chief test pilot Charlie Bock at the controls. It flew from Palmdale to Edwards AFB. *NMUSAF*

AIR SUPERIORITY FIGHTER PROGRAM DESIGNS FROM LOCKHEED-CALIFORNIA COMPANY (CALAC)

X-1:	CL 1250-3-P&WA STF371 turbofanjet	baseline configuration
X-2:	CL 1250-4-P&WA STF371 turbofanjet	baseline configuration
X-3:	CL 1250-7-P&WA STF371 turbofanjet	baseline configuration
X-4:	CL 1250-6-P&WA STF371 turbofanjet	baseline configuration
X-5:	CL 1250-6A-P&WA STF371 turbofanjet	baseline configuration
Y-1:	CL 1250-9-P&WA STF371 turbofanjet	candidate configuration
Y-2:	CL 1250-10-P&WA STF371 turbofanjet	candidate configuration
Y-3:	CL 1250-8-P&WA STF471 turbofanjet	candidate configuration
Z-1:	CL 1250-11-P&WA STJ353 turbojet	candidate configuration

Final Report for the Period November 1, 1969, to July 31, 1972, under USAF contract number F33657-70-C-0511 for Project Number 668A

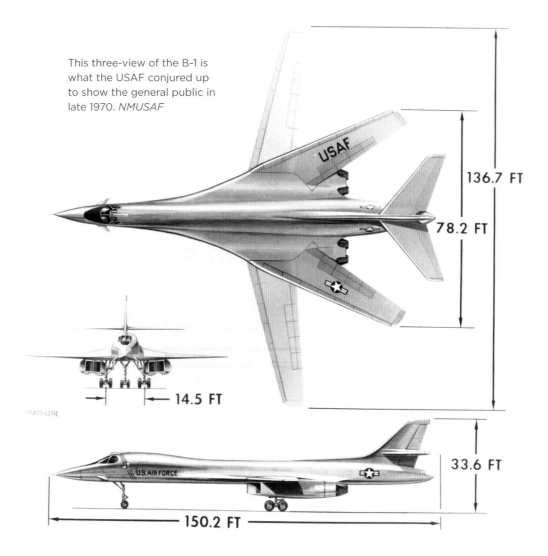

This three-view of the B-1 is what the USAF conjured up to show the general public in late 1970. *NMUSAF*

136.7 FT

78.2 FT

14.5 FT

33.6 FT

150.2 FT

A fourth evaluation B-1A airplane, built to production specification, was subsequently ordered in FY-76 (USAF serial number 76-0174).

The first flight of the premier B-1A was to occur in April 1974 and Initial Operational Capability (IOC) was to be met in 1979. At the time of the order, this program called for the production of the four flyable evaluation airplanes and 240 production B-1A aircraft.

Although NAR's headquarters was located in Inglewood, a suburb of Los Angeles, California, the B-1A evaluation aircraft were built at its facility in Palmdale, about 100 miles northeast of LA. This was the same place the XB-70A Valkyrie aircraft had been assembled some ten years earlier.

Rockwell produced four B-1A airplanes at its Palmdale, California, facility at USAF Plant 42 between the years 1974 and 1976. The first of these rolled out on October 26, 1974, and made its first flight from Palmdale to Edwards AFB on December 23, 1974. The B-1A number one completed 138 flights, while B-1A number two had its first flight on June 14, 1976—*after* number three, due to ground test delays. B-1A number three itself flew for the first time on March 26, 1976, while B-1A number four's first flight took place on February 14, 1979, after which it completed seventy flight tests.

B-1B TEST PROGRAM

Two of the four B-1A airplanes—numbers two and four (USAF serial numbers 74-0159 and 76-0174)—were modified and used extensively in what was called the B-1B Test Program.

The modifications to B-1A number two included the installation of B-1B flight control system components, while the B-1A number four received modified avionics.

The first B-1B Test Program airplane (74-0159) began its flight-test program on March 23, 1983. The second B-1B Test Program airplane (76-0174) initiated its flight-test program on July 30, 1984.

During the B-1B Test Program, on August 29, 1984, B-1A number two crashed to destruction east of Boron, California, some twenty-two miles northeast of Edwards AFB. While aloft at a relatively low altitude (between three thousand and four thousand feet), a Rockwell chief test pilot, Tommie Douglas "Doug" Benefield, neglected to transfer the fuel tanks at the correct time.

The B-1A number one rolls out on
October 26, 1974. It was assembled
and debuted from the same building in
which the two XB-70A airplanes were
built. *NMUSAF*

B-1A number three (74-0160) cruises effortlessly on a test flight. The first three aircraft had the four-seat emergency ejection capsule, while B-1A-4 came with four individual ejection seats. *NMUSAF*

B-1A SPECIFICATIONS

Propulsive system
Four afterburning, axial-flow, 30,000-lbf (with afterburning) General Electric YF101-GE-100 turbofanjet engines

Length	150.2 feet
Height	33.6 feet

Wingspan
136.7 feet (extended); 78.2 feet (sweptback)

Empty weight	190,000 pounds
Gross takeoff weight	389,300 pounds

Maximum speed
2.22 Mn (attained by B-1A evaluation aircraft number two on October 5, 1978)

Maximum range	5,300 miles (unrefueled)
Combat Ceiling	50,000-plus feet
Armament	None

Payload
115,000 pounds (75,000 pounds internal; 40,000 pounds external)

This error led to a shift in the plane's center of gravity, enough to make the airplane uncontrollable.

The three crewmen successfully ejected, rocketing up and away from the out-of-control aircraft in its escape capsule. Tossed about inside the capsule, pilot USAF Maj. Richard V. Reynolds and flight test engineer USAF Capt. Otto J. Waniczek suffered chest and back injuries upon hitting the ground, while copilot Benefield died from the injuries he sustained. Upon later inspection, it was discovered that the escape capsule parachutes had not deployed fully, which led to the module's hard impact on the desert floor in a right-hand, nose-low attitude; its fall was therefore not adequately cushioned by the built-in inflation bladders attached to the bottom of the capsule.

SUMMARY

The success of the long-running B-1A flight test program proved the value of the aircraft as a dedicated weapons system. Nevertheless, its production was halted by President Jimmy Carter.

B-1A
PRODUCTION BREAKDOWN

B-1A:	1-74-0158	retired in April 1981 to Lowry AFB after 138 flights
B-1A:	1--74-0159	modified to serve in the B-1B Test Program; crashed to destruction on August 29, 1984
B-1A:	1-74-0160	sent to Lowry AFB as ground trainer designated GB-1A
B-1A:	1-76-0174	equipped with four conventional emergency ejection seats; modified to serve in the B-1B Test Program
B-1A:	0-	was to serve as a fifth B-1A, but was cancelled prior to completion; parts of it were used to make the first almost totally "hand-built" B-1B\

F-20 number one streaks across the blue during its first flight in its original factory livery on August 30, 1982. Northrop built three F-20s with factory numbers G.G. 1001, GI. 1001, and GI. 1002, respectively. F-20 number two (82-0063) is the only surviving F-20. It sits in its 1983 Paris Air Show aggressor livery, sporting its 340 identity number. *NMUSAF*

To paraphrase an old adage, "You can't keep a good airplane down." This holds true for the B-1 program, since the B-1A program was resurrected by President Ronald Reagan in 1981 as the B-1B program.

B-1A number one (74-0158) was retired to Lowry AFB in Denver, Colorado, where it was taken apart to serve as a weapons trainer. B-1A number three (74-0160) was retired to Lowry AFB as well, where it served as a ground trainer designated GB-1A; it is now located within the Wings Over the Rockies Air and Space Museum located at the former Lowry AFB. B-1A number four (76-0174) is located at the Strategic Air and Space Museum in Ashland, Nebraska.

NORTHROP F-20A (F-5G) TIGERSHARK

The Northrop Aircraft Corporation F-20 Tigershark program was a privately funded effort on the part of Northrop to sell an alternative to the F-16 fighter to the USAF, USN, and US allies and friendly foreign nations.

F-5G/F-20

The F-20's design began as an advanced version of the F-5E Tiger II, which sported temporary in-house designation F-5G. Northrop employees Robert R. Sandusky Jr., Victor C. Brock, and Lionel O. Gay served as its designers.

In its final configuration (Northrop Model N-317), this plane featured a single afterburning, axial-flow, 17,000-lbf (with afterburning) General Electric YF404-GE-100 turbofanjet engine for its propulsive system.

F-20A

Three preproduction-like F-20 Tigershark demonstration airplanes were built and, although they weren't destined for USAF procurement, they were issued USAF serial numbers 82-0062 to 82-0064; the DOD designation F-20 was issued to them in November 1982.

Controversy mounted after the Tigershark received its F-20 designation, since this meant that the designation F-19 had been bypassed. Rumor had it that the F-19 was a super-secret stealth attack and/or fighter aircraft being secretly tested at Area 51 on Nevada's Groom Lake. This story was true, but the plane in question wasn't the F-19—it was the F-117.

In addition to their USAF serial numbers, the F-20s sported civilian registration numbers N4416T, N3986B, and N4467I respectively.

A fourth F-20 (82-0065) had been under construction in November 1986, but the Tigershark program was cancelled before it could be completed.

Thirty-two months after program's go-ahead, the first F-20A rolled out one month early, on August 1, 1982. It made its first flight on August 30, 1982, with Northrop test pilot Russell J. "Russ" Scott at the controls. The plane achieved Mach 1.04 during its inaugural flight.

Rolled out on July 25, 1983, F-20 number two made its first flight on August 26, 1983, piloted by Northrop test pilot Darrell E. Cornell. F-20 number three was first flown on May 12, 1984, with Cornell once again at the controls.

Sadly, two Northrop test pilots perished in the crashes of the first two of the three F-20s built. F-20 number one crashed to destruction on October 10, 1984, at Suwon Air Base in Suwon, South Korea, during a Republic of Korea Air Force demonstration flight, killing Cornell. F-20 number two crashed to destruction the following year, on May 14, 1985, at Goose Bay, Newfoundland/Labrador while preparing for the 1985 Paris Air Show in France killing, David B. "Dave" Barnes. Both crashes were attributed to g-induced loss of consciousness (G-LOC), whereby both pilots became unconscious due to the excessive g-forces encountered; the aircraft were deemed fault free.

The third and only surviving F-20 is on display at the California Science Center in Los Angeles, California.

SUMMARY

The ten-year-long F-20 program didn't end in Northrop's favor, as no orders for the Tigershark were forthcoming from any branch's air arm. All who required similar capacity found that the improved version of the General Dynamics F-16 served well enough. Thus, on November 17, 1986, Northrop ended its work on its long-running F-20 Tigershark program, with an estimated total cost of $1.2 billion.

By the time this program ended, Northrop was heavily involved in another effort, called the Advanced Tactical Fighter (ATF) program. Moreover (and still

F-20
PRODUCTION BREAKDOWN

F-20-NO:
3-82-0062 to 82-0064
82-0065 was cancelled

mostly secret), it was also involved in the ATB (Senior Ice) program, whereby Jack Northrop's dream of an operational flying-wing bomber would become a reality.

BOEING B-1B AND B-1R LANCER

The Boeing B-1B Lancer didn't have any dedicated X (experimental) or Y (prototype/service test) aircraft, but the General Electric/USAF/ Rockwell B-1B Combined Test Force (CTF) used the first nine B-1B aircraft manufactured for these purposes.

The B-1B evolved from a long line of strategic bombardment aircraft proposals. Its creation was the culmination of thousands of design studies that had preceded it, then followed by four flying prototypes designated B-1A built by NAR under design number D481.

The program to produce 240 operational B-1A airplanes was axed by President James Earl "Jimmy" Carter Jr. on June 30, 1977, when he cited, in part:

> The principal reason for my decision is the high unit cost, which has grown from original estimates of $30 million to $100 million per aircraft. I've decided to produce the ALCM [Air Launched Cruise Missile], which is less expensive. Also, the ALCM, with an RCS [radar cross section] smaller by an order of magnitude, is considered less vulnerable to SAM [surface to air missile] defenses than the B-1.

His successor, President Ronald Wilson Reagan, entered office on January 20, 1981. One of his first actions was to keep a campaign promise and reinstate the B-1 program by ordering 100 production B-1B airplanes through the Department of Defense.

B-1B

On October 2, 1981, President Ronald W. Reagan authorized the production of 100 B-1 strategic bombers designated B-1B. The B-1B was the result of the USAF Long Range Combat Aircraft (LRCA) program.

Since the two modified B-1As had served largely as B-1B prototypes, there were no prototype B-1B airplanes as such: all 100 were production airplanes. However, for USAF familiarization, the first nine production B-1Bs were used as evaluation aircraft at Edwards AFB.

The first B-1B (USAF serial number 82-0001) was ceremonially rolled out in Palmdale, California, on September 4, 1984. It made its first flight on October 18, 1984, with the following

Prototype F-20 number one as it appeared in mid-1983 during its USAF demonstration flights. *G. De Chiara © 2015*

F-20 SPECIFICATIONS

Propulsive system
One afterburning, axial-flow, 17,000-lbf General Electric YF404-GE-100 turbofanjet engine

Length	46 feet, 6 inches
Height	13 feet, 10 inches

Wingspan
(with launcher rails): 26 feet, 8 inches

Empty weight	7,890 pounds
Gross weight	28,000 pounds
Maximum speed	2-plus Mn
Maximum range	2,160 miles
Combat Ceiling	55,000 feet

Armament
Two M39 nose-mounted 20mm cannon, air-to-air/air-to-ground missiles and rockets

Payload
Up to 9,000 pounds of bombs

ABOVE: The sole surviving F-20 on display at the California Science Center. Its progenitor was the T-38 Talon, which hangs above it. *Photograph by Darkest Tree*

TOP: F-20 number one fires a Hughes AGM-85 Maverick air-to-ground missile. The F-20s proved they could carry and deliver a vast array of ordnance—bombs, missiles, and rockets. *NMUSAF*

crewmembers: Rockwell B-1 Division chief test pilot Mervin L. Evenson, copilot USAF Capt. D. E. Hamilton, USAF Lt. Col. Leroy B. Schroeder, and USAF Maj. S. A. Henry.

USAF SAC received its first B-1B (82-0001) on June 29, 1985, exactly thirty years after the first B-52 was delivered to SAC. That aircraft, the first B-1B built, was delivered to Dyess AFB in Abilene, Texas, and was appropriately named the *Star of Abilene*. The second B-1B (83-0065), the Lancer, was to be delivered on this date, but it had been grounded at Offutt AFB near Omaha, Nebraska, because of engine foreign object damage (FOD) and the first B-1B took its place. The B-1B achieved its IOC (Initial Operational Capability) on October 1, 1986.

The first operational B-1B Lancer (85-0073), named *The Wings of Freedom*, was delivered to the 28th Bomb Wing at Ellsworth AFB, South Dakota, on January 21 1987. Gen. John T. Chain Jr., Strategic Air Command commander, delivered it. After he landed, the general made remarks that now seem prophetic:

> Too often, people think of SAC bombers . . . only as carriers of nuclear weapons. We have to change that perception. The B-52 and B-1 have an enormous capability to contribute to this nation's conventional capability . . . their firepower is awesome. They could strike a terrorist-type target anywhere in the world flying nonstop from the United States and return with air refueling.

On May 2, 1988, the hundredth and last B-1B was delivered.

On March 15, 1990, the B-1B was officially and appropriately named Lancer, due in part to its primary role of being a penetration bomber. The B-1A never received a nickname or an official name. To those associated with the B-1B Lancer, the nickname Bone ("B-one") is most often heard. B-1B pilots simply refer to it as "my jet."

Boeing purchased NAR on December 6, 1996, creating the Boeing North American Division of the Boeing Airplane Company. At this time, then, the Rockwell International Corporation, North American Aircraft Division B-1B Lancer, became the Boeing North American B-1B Lancer.

Since this purchase and merger, the Boeing B-1B Lancer has become the backbone bomber of the three fleets of USAF strategic bombers, the other two being the B-52H and B-2A fleets.

B-1R ("Boner")

The B-1R was a proposed upgrade of existing B-1B aircraft. The B-1R (R for "regional") would be fitted with advanced radars, air-to-air missiles, and four 35,000-lbf Pratt & Whitney F119 engines. This variant would have a top speed of Mach 2.2, but with 20 percent less range.

Existing external hard points would be modified to allow multiple conventional weapons to be carried, increasing overall load out. For air-to-air defense, an Active Electronically Scanned Array (AESA) radar would be added and some existing hard points modified to carry air-to-air missiles. If needed, the B-1R could have escaped from unfavorable air-to-air encounters with its Mach 2-plus speed.

Alternative "regional bomber" proposals included the Lockheed Martin FB-22 and Northrop Grumman FB-23.

SUMMARY

The B-1B Lancer has more than adequately proven its multi-role capabilities during combat—and it continues to do so. Whether it's a high-, medium-, or low-level attack, the Bone continues to deliver precise bombardments.

B-1B SPECIFICATIONS

Propulsive system	
Four afterburning, axial-flow, 30,780-lbf (with afterburning) General Electric F101-GE-102 turbofanjet engines	
Length	145.76 feet
Height	34.02 feet
Wingspan	
136.68 feet (15 degrees, wing forward), 78.23 feet (67.5 degrees, wing aft)	
Empty weight	186,807 pounds
Gross takeoff weight	477,000 pounds
Maximum speed	
1.2 Mn at altitude; 0.85 Mn at 200 feet	
Maximum range	6,675 miles (unrefueled)
Combat Ceiling	50,000 feet
Armament	None

Payload
133,800 pounds-75,000 pounds (internal); 58,800 pounds (external), up to eighty-four 500-pound Mk-82 or twenty-four 2,000-pound Mk-84 general-purpose bombs; up to eighty-four 500-pound Mk-62 or eight 2,000-pound Mk-65 Quick Strike naval mines; thirty cluster munitions (CBU-87, -89, -97) or thirty Wind-Corrected Munitions Dispensers (CBU-103, -104, -105); up to twenty-four 2,000-pound GBU-31 or fifteen 500-pound GBU-38 Joint Direct Attack Munitions; up to twenty-four AGM-158A Joint Air-to-Surface Standoff Missiles; GBU-54 laser-guided Joint Direct Attack Munition; 1,000-pound and 2,000-pound GPS-guided JDAM weapons

B-1B
PRODUCTION (EVALUATION AIRCRAFT) BREAKDOWN

B-1B:	1-82-0001	Lot I (aircraft 1 only)
B-1B:	0-82-0002	cancelled
B-1B:	7-83-0065 to 83-0071	Lot II (aircraft 2 thru 8)

McAir chief engineering test pilot Irv Burrows makes a successful first flight in the prototype F-15A number one (71-0280) at Edwards AFB on July 27, 1972. During the flight, Burrows left the gear down and extended the speed brake. The speed brake was later enlarged for improved aerodynamic braking. *NMUSAF*

NINE

STEALTH BOMBERS AND FIGHTERS

We painted them [F-117As] black because the TAC [Tactical Air Command] commander wanted them black. The Skunk Works plays by the Golden Rule: he who has the gold sets the rules! If the general had wanted pink, we'd have painted them pink.

—Ben Rich

In 1974, the Defense Advanced Research Projects Agency (DARPA), an arm of the US Department of Defense, created a confidential project code-named "Harvey" (named after the invisible white rabbit from the movie of the same name, starring James M. "Jimmy" Stewart). DOD created Project Harvey to discover how to build air vehicles with very low observability so that they could not be detected by radar—in other words, they would be "invisible" to an adversary.

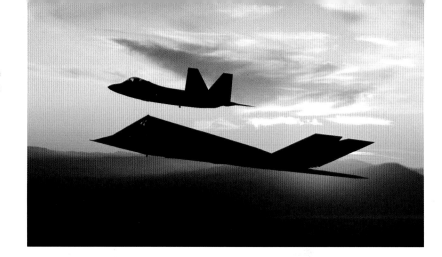

An F-117 (foreground) and an F-22 fly together in formation. The F-22 has replaced the F-117 as the USAF's stealth fighter. *LM/Code One – photo by Andy Wolfe*

DARPA awarded five $1 million contracts to Fairchild Republic, Northrop, McDonnell Douglas, General Dynamics, and Grumman. Their job was to come up with the least observable air vehicle possible. The winner would move forward and receive a contract to build and flight-test two demonstrator aircraft.

Lockheed, a leader in manufacturing air vehicles with low radar signatures, had been passed over. Under the direction of then new Skunk Works president Benjamin Robert "Ben" Rich, the company lobbied unsuccessfully to get one of the contracts. Rich then turned to Lockheed management for funding that would support the program with in-house funds, up to $10 million; the request was approved. Rich and the other Skunk Works geniuses set about creating the winning air vehicle.

With the eighties in full swing, the USAF knew it had to look beyond the decade to a time when fielding more advanced combat aircraft, bombers, and fighters that were virtually undetectable by radar; technology better known today as "stealth." This was the genesis of today's several fleets of stealth bombers and fighters.

SUMMARY

The pair of Have Blue experimental survivable testbed (XST) prototypes played a significant role in the creation and success of the F-117 stealth fighter. Although both were lost in crashes, the eighty-eight test flights between them demonstrated that near-invisible aircraft could be put into service and operate without detection. The success of the F-117 Nighthawk program offered further proof.

LOCKHEED YF-117A NIGHTHAWK

The Lockheed Aircraft Corporation F-117A Nighthawk was the first operational stealth airplane in the world. In fact, until the advent of the B-2 stealth bomber, it was the world's only stealth aircraft. Although the existence of such a warplane had been rumored for several years, as the F-19 the existence of the F-117A wasn't made public until

HAVE BLUE SPECIFICATIONS

Propulsive system	
Two non-afterburning, axial-flow, 2,950-lbf General Electric J85-GE-4A turbojet engines	
Length	38 feet, 0 inches
Height	7 feet, 6 inches
Wingspan	22 feet, 6 inches
Empty weight	8,950 pounds
Gross weight	12,000 pounds
Maximum speed	0.80 Mn (about 600 mph)

Lockheed Have Blue

The Have Blue program was begun in November 1975 under a joint USAF and DARPA contract with Lockheed and its Skunk Works to build and fly two XST air vehicles. The first example (Have Blue number one, HB-1), the aerodynamic testbed, was initially flight-tested at Area 51 by Lockheed test pilot William C. "Bill" Park Jr. on December 1, 1977. The second example (HB-2) was first flown on July 20, 1978, at Area 51 by USAF Lt. Col. Norman Kenneth "Ken" Dyson.

The XST aircraft were assembled in Building 82 at Lockheed's Burbank, California, facility before they were transferred to Groom Lake, Nevada, within the Area 51 complex; there they began their respective flight tests. The highly classified Have Blue program ended in November 1978 when Lockheed got the go-ahead on the Senior Trend program; flight-testing on HB-2 continued as well.

Both aircraft were lost in separate crashes. Have Blue number one crashed to destruction on May 4, 1978, on its thirty-sixth test flight; HB-2 was lost on July 11, 1979, on its fifty-second test flight. Both Park and Dyson survived these crashes, and the remains of the aircraft were buried somewhere within the Area 51 complex. Lockheed applied factory serial numbers 1001 and 1002 to their two Have Blue XST air vehicles.

Have Blue number two (Lockheed factory serial number 1002) is shown here shortly after it was built and painted in its light blue livery. To help avoid suspicion on this highly classified program, both Have Blue aircraft were powered by a pair of rebuilt non-afterburning, 2,950-lbf General Electric J85-GE-4A turbojet engines borrowed from the US Navy T-2B Buckeye trainer aircraft engine inventory. Its classified RCS (radar cross section) was far lower than that of the Northrop pole model, and it won the competition *LM/Code One*

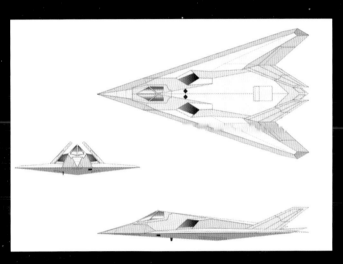

This three-view of Have Blue number one (Lockheed factory serial number 1001) depicts its early appearance before receiving its patched camouflage livery over its original light gray attire. *G. De Chiara © 2015*

November 10, 1988, when Pentagon spokesman James Daniel "Dan" Howard announced the Lockheed F-117A during a press briefing. Along with his announcement came a release of a low-quality photograph of the F-117A that largely distorted its actual appearance. It didn't even have an official name at the time, and was known only as the "Black Jet" to those working on its program.

Developed and produced for the USAF Tactical Air Command (TAC) by Lockheed Advanced Development Company (better known as the Skunk Works), the F-117A Nighthawk was classified as a strike fighter; it was designed from the outset to utilize very-low-observables (that is, "stealth") technologies.

YF-117A (79-10783) makes an early test hop at Area 51. Five prototype F-117As were built, with USAF serial numbers 79-10780 through 79-10784 assigned to them. They were primarily piloted by Lockheed Skunk Works test pilots Harold C. "Hal" Farley and David L. "Dave" Ferguson, and USAF Lt. Col. "Skip" Anderson. They were built in Burbank, California, using conventional aluminum, steel, and titanium alloys for their airframes. Their special shapes and exterior coatings provided their extremely low observable (stealth) requirements. *LM via Denny Lombard*

No X or Y versions of the F-117A were built. Rather, two XST aircraft under the Have Blue program, along with five full-scale development (FSD) F-117A aircraft under the Senior Trend program. These aircraft were used in part for aerodynamics, stealth, and airborne tactical warfare evaluations. (The five FSD F-117A airplanes were retroactively designated YF-117A to reflect upon their service test duties.)

These five FSD F-117A airplanes were ordered for service test duties (Lockheed factory serial numbers 780 to 785). The airplanes were later redesignated YF-117A to better describe their primary service test duties, called Scorpion 1 through Scorpion 5 by those closely associated with them, or known collectively as the "Baja Scorpions." Their first flights are as follows:

Scorpion 1 (780)—June 18, 1981, Harold C. "Hal" Farley Jr. (Bandit 117)
Scorpion 2 (781)—September 24, 1981, David L. "Dave" Ferguson (Bandit 105)
Scorpion 3 (782)—December 18, 1981, Thomas A. "Tom" Morgenfeld (Bandit 101)
Scorpion 4 (783)—June 7, 1982, (flown after Scorpion 5) Tom Morgenfeld
Scorpion 5 (784)—April 10, 1982, (flown before Scorpion 4) Robert Lester "Bob" Riedenauer (Bandit 103)

(When pilots made their first flight on an F-117A or other classified aircraft, they earned "Bandit" status and were issued a corresponding number.)

The first production F-117A (785) attempted to make its first flight on April 20, 1982, out of Area 51, with Skunk Works test pilot Robert L. "Bob" Riedenauer at the controls. Unfortunately, its computerized fly-by-wire flight control system

had somehow been improperly connected; when Riedenauer rotated for takeoff, the airplane immediately went out of control, flipped over onto its back, and crashed upside-down on the ground. Mr. Riedenauer survived, but was severely injured; he would never fly again. This airplane was not accepted by the USAF, and it wasn't counted in the total of fifty-nine production F-117As built and delivered.

On August 23, 1982, the third production F-117A (787) became the first production F-117A to be accepted by the USAF. Its first flight date was July 20, 1982. The first flight date of the second production F-117A (786) (as the first F-117A to fly) was July 15, 1982. Hal Farley, chief engineering test pilot, served as pilot on both of these first flights.

As F-117A production increased, the fleet was assigned to a specifically built air base, complete with individual hangers for each airplane. This once top secret facility is called Tonopah Test Range (Area 10), located about seventy miles northwest of Area 51 and thirty miles southeast of Tonopah, Nevada.

What Happened to the F-19?

Long assumed to be the designation issued to the stealth fighter that materialized as the Lockheed F-117A, the designation F-19 has only been applied to a pair of model airplane kits produced by Testors as the *F-19 Stealth Fighter* (released in 1986) and by Monogram as the F-19A Spectre (released in 1987).

The Monogram kit mimics the airplane shown in a two-page advertisement spread in a magazine purchased by Loral Defense Systems of Akron, Ohio.

Later, a two-page advertisement spread in a magazine appeared that showed the Northrop/Loral F-19A Spectre. The ad stated:

> The Northrop/Loral **F-19A Spectre** pursuit fighter is the next-generation of stealth and speed in flight. With a radar cross-section smaller than a duck in flight, an IR heat signature nearly as low as the heat background, and a noise level so quiet, it won't be heard until the engagement is over, the F-19A is invisible to everything except the human eye. Beauty is in

the eye of the beholder, and the F-19A succeeds in all cases.

At the bottom of ad is an address:

> NORTHROP LORAL Making Advanced Technology Work 1840 Century Park East, Los Angeles, Calif. 90067 USA

This address corresponds with the headquarters address for the former Northrop Aircraft Corporation. A Northrop/Loral F-19A Spectre lineview drawing has surfaced that appears to be legitimate; though it may have been produced as a simple modeler's guide.

Finally, several years ago, a satellite image of the USAF storage and scrap yard known as the 309th Aerospace Maintenance and Regeneration Group (309 AMARG) unit within the Davis-Monthan AFB, Arizona, complex revealed what look like two Northrop/Loral F-19A Spectre aircraft, parked side by side.

Whether such aircraft ever existed is not officially admitted, nor is it documented. If it was ever created, its background and service test history remain a mystery.

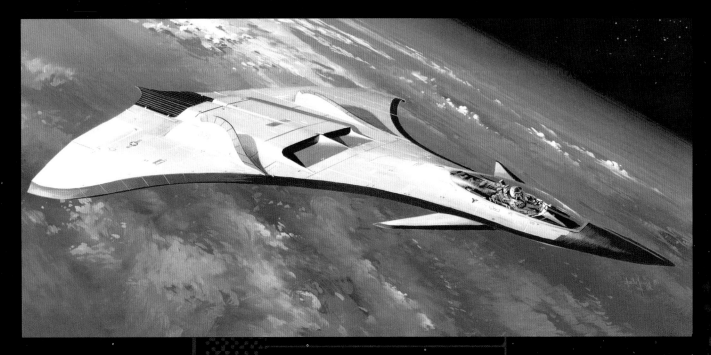

Artist's conception of F-19A Spectre with two short, inward-canted vertical tails, canard foreplanes, and downward-curved wingtips. *Photograph courtesy of the Loral Corporation*

SUMMARY

With the advent of the operational F-117A Nighthawk in October 1983, the USAF Tactical Air Command now possessed a dedicated stealth combat aircraft that could carry and deliver two two-thousand-pound laser-guided (later satellite-guided) JDAM bombs before escaping enemy territory undetected.

On April 22, 2008, the fleet of F-117s was retired and the planes were returned to Tonopah Test Range to be placed in extended storage.

Finally, it should be noted that the F-117A's official name is Night Hawk, but the Nighthawk spelling is most common for this plane.

NORTHROP GRUMMAN B-2A SPIRIT

The Northrop Grumman Corporation B-2A Spirit is a flying wing in the truest sense of the expression. With no fuselage, and no horizontal or vertical tail planes, it is an all-flying-wing aircraft, period. Everything that makes it work is housed within its sweptback wing structure.

The creation of what became the B-2A stealth bomber evolved from the numerous semi-flying-wing aircraft that preceded it, many of which were manufactured by the Northrop Aircraft Corporation from designs produced by that company's founder, John Knudsen "Jack" Northrop.

The B-2 came from the USAF Advanced Technology Bomber (ATB) program initiated with RFPs issued to the industry in September 1980. Four airframe contractors partnered to address the ATB RFP: Northrop, which had teamed with Boeing, and Lockheed, which had teamed with Rockwell.

The Northrop/Boeing entry received the name Senior Ice, while the Lockheed/Rockwell entry got the name Senior Peg.

The proposed Lockheed/Rockwell Senior Peg entry looked much like an enlarged F-117, about the size of a B-58 Hustler, with four embedded, non-afterburning turbofan engines and a two-man cockpit.

The USAF artist's conception of the advanced technology bomber (ATB), officially designated B-2A at the time. Except for the obvious absence of engine exhaust outlets, the final product looked very much like this rendering, which was released to the public in April 1988; the highly anticipated rollout occurred on November 22, 1988, at Northrop's Palmdale, California, facility at USAF Plant 42. It made its first flight on July 17, 1989, nearly eight full months later. *USAF*

F-117A SPECIFICATIONS

Propulsive system
Two non-afterburning, axial-flow, 10,600-lbf General Electric F404-GE-F1D2 turbofanjet engines

Length	65 feet, 11 inches
Height	12 feet, 9.5 inches
Wingspan	43 feet, 4 inches
Empty weight	29,500 pounds
Gross weight	52,500 pounds
Maximum speed	0.92 Mn (617 mph)
Maximum range	1,070 miles
Ceiling	45,000 feet
Armament	None

Payload
Two 2,000-pound precision guided bombs

The B-2A Spirit is downright ominous during final approach for landing. Its Wingspan is 172 feet—exactly the same as that of the YB-49 of 1947. The B-2A features two side-by-side weapon bays. *USAF photo by Airman First Class Michael S. Dorus*

The proposed Northrop/Boeing Senior Ice competitor also had four non-afterburning turbofan engines housed within its body and a two-man crew, but it wasn't faceted like the Senior Peg entry; instead, Senior Ice featured smooth, rounded surfaces.

The USAF ATB selection board found Senior Ice superior to Senior Peg, and decided to move forward on October 20, 1981, when it was announced that Northrop/Boeing would build the ATB. The airplane was officially designated B-2 at this time.

Northrop was awarded the ATB (B-2) contract in November 1981. The contract covered the manufacture of two structural test airframes, one flying prototype, and five evaluation aircraft. The initial plan called for the production of 127 B-2 production airplanes, in addition to the five evaluation airplanes—six counting the prototype—which would be brought up to operational specification.

B-2A

Following its highly visible rollout ceremony in Palmdale, California, the premier B-2A was towed to the flight line for a series of system tests in preparation for the distinctive aircraft's first flight. The extensive ground tests were thorough, taking more than seven months to complete, which is an extraordinarily long time for aircraft flight testing of this kind.

Before any new air vehicle is cleared to fly for the first time, it must first undergo a series of low-, medium-, and high-speed taxi runs in order to make sure that its avionics, brakes, nose wheel steering, flight control surfaces, propulsive system, and all of its other equipment are working properly. In its first series of taxi runs on July 10, 1989, it reached a maximum ground speed of 103 miles per hour. A number of follow-on taxi tests ensued, and, six days later, the unique aircraft was given the green light for flight by USAF B-2 program director, Brig. Gen. Richard M. Scofield.

At 6:37 a.m. on July 17, 1989, the world's first—and, to date, only—stealth bomber rolled down Runway 04 for 4,500 feet before it rotated, lifted off, and flew up and away from Palmdale on its maiden flight. At the controls sat Northrop B-2 Division chief engineering test pilot Bruce J. Hinds and Col. Richard S. Couch, USAF director of the B-2 Combined Test Force (CTF). Hinds was in the left-hand seat and served as aircraft commander, while Colonel Couch was seated in the right-hand seat.

According to plan, the B-2's landing gear remained down and locked during the two-hour, twelve-minute flight. Throughout, as the all-wing air vehicle climbed to a maximum altitude of ten thousand feet, it reached a maximum speed of 218.6 mph (190 knots) while the test pilots performed a series of functional checks of its basic systems. Everything went well.

Then, at 8:29 a.m., after a long, slow approach, the first flight crew brought the boomerang-like airplane in for a perfect landing on the main fifteen-thousand-foot long Runway 22 at Edwards AFB. After the flight, following a lengthy and detailed inspection of all aircraft systems (which lasted nearly a month), the airplane was cleared for its next journey into the skies above southern California. The premier B-2 and its four General Electric F118 turbofanjet engines had performed remarkably well.

The Northrop Aircraft Corporation merged with the Grumman Aerospace Corporation in April 1994 to form the Northrop Grumman Corporation. This merger happened while the B-2 fleet was still being built and prepared for delivery to the USAF Air Combat Command (ACC).

The first operational B-2A (88-0329), Spirit of Missouri, was delivered to the USAF Air Combat Command on December 17, 1993, the ninetieth anniversary of the Wright brothers' first flight. A true flying wing, the Spirit can carry a payload of more than forty thousand pounds. The B-2A was originally ordered for SAC. It became part of the Air Combat Command (ACC) when SAC was disbanded in 1992, and it's now a part of the Air Force Global Strike Command (AFGSC), which was activated in 2009. The B-2A is operated by a pilot and mission commander (co-pilot). On long missions, a third relief pilot joins the crew. The B-2A has a maximum speed of Mach 0.95, just below the speed of sound. Four non-afterburning 17,300-lbf General Electric F118-GE-100 turbofanjet engines propel the aircraft. The B-2A has a maximum ceiling of fifty thousand feet. The B-2A has an unrefueled range of six-thousand-plus nautical miles and more than ten thousand nautical miles with one aerial refueling. *USAF photo by Staff Sgt. Bennie J. Davis III*

B-2A
PRODUCTION (PROTOTYPE AND FSD) BREAKDOWN:

999:	static test article	
1000:	static test article	
1001:	AV-1, B-2A, 82-1066–July 17, 1989	FSD-1
1002:	AV-2, B-2A, 82-1067–October 19, 1990	FSD-2
1003:	AV-3, B-2A, 82-1068–June 18, 1991	FSD-3
1004:	AV-4, B-2A, 82-1069–April 17, 1992	FSD-4
1005:	AV-5, B-2A, 82-1070–October 5, 1992	FSD-5
1006:	AV-6, B-2A, 82-1071–February 2, 1993	FSD-6

B-2A SPECIFICATIONS

Propulsive system
Four non-afterburning, axial-flow, 17,300-lbf General Electric F118-GE-100 turbofanjet engines

Length	69 feet, 0 inches
Height	17 feet, 0 inches
Wingspan	172 feet, 0 inches
Wing area	5,140 square feet
Empty weight	158,000 pounds
Gross weight	336,500 pounds

Maximum speed
High subsonic (Mach 0.9-plus)

Maximum range
6,500 miles (without aerial refueling)

Combat Ceiling	50,000 feet

Payload
Two Boeing 30,000-pound GBU-57A/B Massive Ordnance Penetrator (MOP) "bunker-buster" bombs, each with a 5,300-pound high-explosive warhead

TACIT BLUE (WHALE) SPECIFICATIONS

Propulsive system
Two non-afterburning, axial-flow, 5,050-lbf Garrett AiResearch (now Honeywell) Model ATF-6-1C turbofanjet engines

Length	55 feet, 10 inches
Height	10 feet, 7 inches
Wingspan	48 feet, 2 inches
Wing area	5,140 square feet
Gross weight	30,000 pounds
Ceiling	25,000 to 30,000 feet
Maximum speed	300 mph
Crew	One (pilot only)
Armament	None

In addition to the two static test air vehicles, one flying prototype, and five evaluation airplanes, Northrop/Northrop Grumman produced fifteen production B-2A Spirit strategic bombardment aircraft between 1993 to 1999.

SUMMARY

The B-2A Spirit is solely operated by the 509th Bomb Wing at Whiteman AFB, Missouri. Once part of USAF Air Combat Command, it is now assigned to the Air Force Global Strike Command, established on August 7, 2009.

The Northrop Grumman B-2A Spirit is designed with a pure flying-wing configuration, exactly what Jack Northrop had envisioned from the outset during his quest for such a design.

From his 1929 Flying Wing airplane to his YRB-49A Flying Wing Bomber and flying wing airliner proposal of the late 1940s and early 1950s, Northrop had championed no less than ten semi-flying-wing designs that flew before the B-2. These include the 1929 Flying Wing, N-1M, N-9M, XB-35, YB-35, XP-56, XP-79, XP-79B, YB-49, and the YRB-49A.

When the twenty-first and last B-2A (actually the first B-2A built and flown) was delivered to the USAF, it was named the *Spirit of America*. As patriotic as this was, it might have been more appropriate to name it the "Spirit of Jack Northrop."

LOCKHEED/BOEING/GENERAL DYNAMICS YF-22 LIGHTNING II

The Lockheed Martin F-22 Raptor was born from the Advanced Tactical Fighter (ATF) program first envisaged during the early eighties to supplement, then replace, the F-15 Eagle. The initial requirement for this was a total buy of 750 ATFs, though this amount was reduced several times: first to 648, then 448, 339, 277, 183, and, finally 187 (adding four due to several aircraft losses).

Out of seven contenders, Headquarters USAF selected two primary airframe contractors: Lockheed and Northrop. Both of these firms joined forces with principal contractors Boeing and General Dynamics (Lockheed) and principal contractor McDonnell Douglas (Northrop). Each team was contracted to produce two service test aircraft, each respectively designated YF-22 and YF-23. Since two powerplant contractors—Pratt & Whitney and General Electric—were contending to produce the propulsive system for the winning ATF design, each firm was contracted to produce service test examples of their engines

Northrop Tacit Blue

The Tacit Blue program was devised to investigate the feasibility of fielding a stealthy aircraft for close-up battlefield surveillance without detection, featuring curved surfaces rather than faceted surfaces (as employed on the F-117 stealth fighter). The term Battlefield Surveillance Aircraft Experimental (BSAX) was applied to the Tacit Blue program.

With the USAF on board, DARPA selected the Northrop Aircraft Corporation in late 1978 to build and fly the Tacit Blue low-observable (stealth) demonstration aircraft. This made sense, as Northrop was gearing up to build its Advanced Technology Bomber, the B-2, with curved surfaces as well; Tacit Blue would serve as an important learning tool. The primary goal of this program was to produce an aircraft with the lowest radar cross-section (radar return signal) as possible. Two Tacit Blue air vehicles were ordered, but flight-test funds were only allotted to one of them: the second air vehicle was to serve as a backup if the first example failed.

Front view of the Tacit Blue air vehicle showing its platypus-like nose, cabin windows detail, and dorsal engine air inlet on centerline. A remarkable aircraft. *NMUSAF*

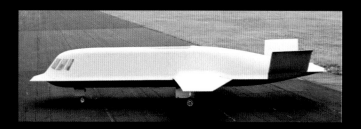

This port-side view of the Northrop Tacit Blue BSAX demonstrator shows its landing gear arrangement and mixture of curved, rounded, and angled exterior surfaces. Its large crew cabin windows are noteworthy. *NMUSAF*

Northrop appointed Richard G. "Dick" Thomas as their chief test pilot for the Tacit Blue program and, on February 5, 1982, he made a successful first flight at Groom Lake, Nevada, in Area 51. With its low radar return signal, Tacit Blue quickly demonstrated that such an aircraft could operate close to battlefield forward lines without detection by enemy radar, in addition to showing that it could continuously monitor enemy forces behind the lines and provide targeting information to battlefield commanders.

Tacit Blue, nicknamed "Whale" because of its "killer whale looks," flew for three years. Its 135th and last flight took place on February 14, 1985, ended the Tacit Blue program. In addition to Thomas (who flew seventy of the 135 flights), four USAF test pilots also flew the Tacit Blue: Maj. Daniel R. "Dan" Vanderhorst (who flew 17 of its 135 flights), Lt. Col. Donald A. "Don" Cornell, Lt. Col. Norman Kenneth "Ken" Dyson, and Lt. Col. John R. "Russ" Easter.

On April 30, 1996, the Pentagon announced the existence of the aircraft and the Tacit Blue program. The airplane was donated to the National Museum of the USAF in Dayton, Ohio, in May 1996, where it can be seen today. No information is currently available for what became of the second Tacit Blue air vehicle, if in fact it was ever built.

The Tacit Blue air vehicle did in fact demonstrate that curved surfaces could offer useful low-observable (stealth) capability for combat aircraft. It helped Northrop designers develop and produce its B-2 stealth bomber, which more than once has proved of tremendous value in combat.

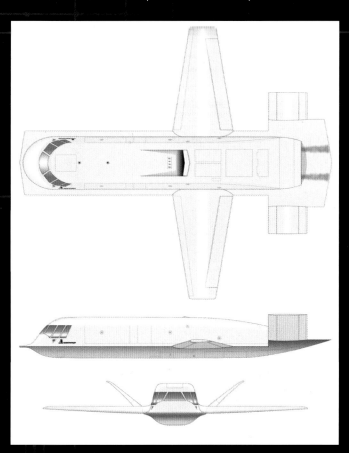

Three-view of the Tacit Blue air vehicle illustrating its beluga-whale-like appearance, recessed dorsal engine air inlet, stubby straight wings, and outward-canted all-movable stabilators (combined horizontal stabilizers and elevators). Note its outward-curved vertical tail tips. *G. De Chiara © 2015*

The two prototype YF-22s first flew together on December 11, 1990. Powered by a General Electric YF120 turbofanjet engine, the YF-22 in the background was equipped with a large canister that housed a parachute for spin tests. *LM/Code One*

respectively designated YF119 and YF120. One YF-22 and one YF-23 would be powered by the YF119, while one YF-22 and one YF-23 would be powered by the YF120. The USAF would therefore have four choices; since it was a winner-take-all competition, the winner would be the preferred airframe and powerplant combination, while the loser would go home.

From this intense competition came the Pratt & Whitney YF119-powered Lockheed-Boeing-General Dynamics YF-22 winner, considered to have "clearly better capability with lower cost, thereby providing the Air Force with a true best value," according to Secretary of the Air Force Dr. Donald B. Rice on April 23, 1991, who announced the winning combination.

Lockheed (now Lockheed Martin) subsequently received an Engineering, Manufacturing, and Development (EMD) contract to produce seven single-seat F-22As and a pair of two-seat F-22B pilot training and transition airplanes. This contract was later amended and the two tandem-seat F-22Bs were cancelled and replaced by an additional two single-seat F-22As, for a total of nine F-22A EMD airplanes.

Pratt & Whitney Aircraft Engines were also given the green light to produce F119 EMD engines for the nine EMD F-22As plus spares. The F119 axial-flow turbofanjet engine is sixteen feet, eleven inches long, with a diameter of forty-six inches, weighing in at 3,900 pounds. It is a twin-spool, counter-rotating, low-aspect ratio engine with three-stage, low-pressure/six-stage high-pressure compressor. It has an annular combustor and offers a 7.95:1 thrust-to-weight ratio.

In a comment to this writer, the late Lt. Gen. David J. "Dave" McCloud (call sign "Marshall") said: "All fighter pilots must have a killer mentality. With the F-22 they'll be very well armed." General McCloud was a part of the ATF selection board that found the YF119-powered entry from Lockheed superior to either one of the two Northrop contenders.

Historical Note

Lieutenant General McCloud was a former member of the Joint Chiefs of Staff as director of force structure, resources, and assessment. At the time of his death, on July 26, 1998, he was the commander of Alaska Command, 11th Air Force, overseeing all military forces in Alaska. A Vietnam War combat veteran, he had flown a wide range of classified, experimental, foreign fighters (such as Red Eagles, as Bandit 6), and many other combat aircraft, primarily fighters, including the F-117A Nighthawk, as Bandit 201.

GE-powered PAV-1 on a test flight shows the type's short, sporty-looking profile. *LM/Code One*

LOCKHEED MARTIN F-22A, F-22B, AND FB-22 RAPTOR

The US Air Force ATF program began in earnest when the teams of Lockheed/Boeing/General Dynamics and Northrop/McDonnell Douglas each received contracts to produce two ATF Prototype Air Vehicles (PAV).

F-22A

The Lockheed Martin Corporation F-22A Raptor fleet of air dominance fighters came from the EMD program, whereby nine single-seat EMD F-22A airplanes were built and thoroughly evaluated. These nine EMD airplanes were ordered under contract number F33657-97-C-0030.

It took a number of years from EMD contract award (six, to be exact) to finally work out design issues and complete the first EMD airplane, which was ceremonially rolled out on April 9, 1997, as the *Spirit of America*. In another five months, it was airborne for the first time. Nevertheless Raptor 01, as the first EMD F-22A (USAF serial number 91-001) was called, successfully entered into flight test on September 27, 1997, with USAF test pilot veteran and F-22 chief test pilot Alfred Paul Metz at the controls. (Metz had also made the first flight on the premier Northrop-McDonnell Douglas YF-23, as Northrop ATF chief test pilot, thereby becoming the only pilot to fly both ATF types of aircraft.)

Meanwhile, the Lockheed Aircraft Corporation had purchased the aircraft manufacturing division of the General Dynamics Corporation

YF-22 SPECIFICATIONS

Propulsive system
(PAV-1): Two afterburning, axial-flow, 35,000-lbf General Electric YF120-GE-100 turbofanjet engines

Propulsive system
(PAV-2): Two afterburning, axial-flow, 35,000-lbf YF119-PW-100 turbofanjet engines

in Fort Worth, Texas, in 1993 and merged with the Martin Marietta Aerospace Corporation in Marietta, Georgia, to become the Lockheed Martin Corporation in 1995. This eliminated ATF principal contractor General Dynamics, leaving only Boeing as an ATF principal contractor. With the Lockheed-Martin Marietta merger, Lockheed subsequently moved its corporate headquarters from Burbank, California, to Bethesda, Maryland.

The eight other EMD F-22As followed suit and, by December 30, 2002, they had all flown. Subsequent testing of these aircraft found the design to be exceptional; additional production contracts were forthcoming through FY2010.

The EMD phase continued on through December 27, 2005, by which time twelve combat-ready F-22As had been delivered. During the EMD phase, the nine EMD F-22As flew 3,496 flights, totaling more than 7,600 flight hours. More than twenty-six thousand flight envelope expansion test points and 3,500 avionics mission test points were met during these flight tests. The EMD F-22 test fleet continued to fly from the Air Force Flight Test Center at Edwards AFB, California, after the EMD phase ended.

F/A-22A

For more than three years, from September 17, 2002, to December 15, 2005, the designation F/A-22A superseded the F-22A designation. This was not popular among USAF personnel, as the prefix F/A stood for "fighter/attack," and put the USAF Raptor into the same classification as the USN F/A-18A/B/C/D Hornet and F/A-18E/F Super Hornet aircraft.

F-22B

In the initial EMD order, provisions were made for two two-seat F-22B pilot training and transition aircraft, in addition to seven single-seat F-22As. These two tandem-seat F-22Bs were canceled, however, and all nine EMD aircraft were built as single-seat F-22As.

Although a second seat was to be provided in the F-22B, its exterior dimensions were identical to those of the F-22A.

FB-22 STRIKE RAPTOR

The proposed Lockheed Martin FB-22 "interim bomber," which was slated to be available for operations in 2018, was an offshoot of Lockheed Martin's earlier X-44 MANTA ("multi-axis, no-tail aircraft") joint USAF/NASA program, which never came to fruition. The X-44 MANTA air vehicle was to be created from a modified EMD F-22A and was projected to take wing in 2007. As proposed, the X-44 MANTA air vehicle was to be a stealthy, tailless aircraft with a delta wing and thrust-vectoring engine exhaust nozzles. The proposed X-44 MANTA program was cancelled before the airplane could be built.

In 2002, the USAF devised its "interim bomber" program, and Lockheed Martin began its studies to create this "2018 bomber," based closely on its earlier modified F-22A design, the cancelled X-44 MANTA.

At first the proposed FB-22 featured a delta wing and no tail, similar to the X-44; this configuration changed to an even larger delta wing, featuring twin outward-canted, vertical tails. It also had a longer, wider fuselage for carrying more fuel for increased range and more internal weapons. The FB-22 design could carry thirty small-diameter bombs (SDBs), which weigh 250 pounds each, compared with the F-22A's payload of eight 250-pound SDBs.

The proposed Lockheed Martin FB-22 was put forward as a candidate for the USAF's requirement for an "interim bomber," with strategic capabilities to become operational by 2018. This would serve as a stopgap design until the entry into service of a future bomber, planned for 2037. However, no orders for the FB-22 were forthcoming.

SUMMARY

The last F-22A Raptor to be built (USAF serial number 10-195, Lockheed number 4195) was rolled off the production line on December 13, 2011. The aircraft made its first flight on March 15, 2012, with F-22 chief test pilot Bret "Lowkey" Luedke at the controls, and it was delivered to the USAF on May 2, 2012. Three days later, it was flown about 8 ½ hours from Georgia to Joint Base Elmendorf-Richardson, Alaska, by Lt. Col. Paul "Max" Moga, where it joined the 525th Fighter Squadron of the 3rd Wing, 11th Air Force, on May 5, 2012. Lieutenant Colonel Moga, then commander of the 525th FS, was formerly one of the elite F-22 demonstration pilots.

The F-22 was declared operational on December 15, 2005. It wasn't until nearly nine years later that it was first used in combat, on

F-22A SPECIFICATIONS

Propulsive system
Two afterburning, axial-flow, 35,000-lbf Pratt & Whitney F119-PW-100 turbofanjet engines with two dimensional-thrust vectoring exhaust nozzles

Length	62 feet, 0 inches
Height	16 feet, 7 inches
Wingspan	44 feet, 6 inches
Empty weight	43,340 pounds
Gross takeoff weight	83,500 pounds

Maximum speed
Mach 2.2-plus (Mn 1.5-plus in supercruise)

Maximum range
Unlimited with aerial refueling

Ceiling	60,000-plus feet

Armament
One 20mm M61A2 six-barrel rotary-action Gatling-type cannon; six AIM-120 AMRAAM missiles; two AIM-9 missiles; two thousand-pound GBU-32 JDAMs; eight 250-pound GBU-39 SDBs

The designs of the two ATF competitors were "worlds apart," as shown here in this side-by-side comparison. *LM/Code One*

SERVICE TEST ATF PAV YF-22 AND EMD F-22A

PRODUCTION BREAKDOWN:

3996:	YF-22-87-700 (Civil Registration No. N22YF)	GE YF120-powered
3997:	YF-22-87-701 (Civil Registration No. N22YX)	P&W YF119-powered
3998:	EMD Radar Cross Section (RCS) static test airframe (full-scale pole-mounted model)	
3999:	EMD static structural loads test airframe	
4000:	EMD static fatigue test airframe	
4001:	F-22A-91-001	
4002:	F-22A-91-002	
4003:	F-22A-91-003	
4004:	F-22A-91-004	
4005:	F-22A-91-005	
4006:	F-22A-91-006	
4007:	F-22A-91-007	
4008:	F-22A-91-008	
4009:	F-22A-91-009	

September 22, 2014, during Operation Inherent Resolve. It reportedly attacked ground targets, as the Lockheed F-117 Nighthawk would have, had the stealth fighter not been replaced on April 22, 2008, by the F-22. It has reportedly targeted Islamic State militant headquarter facilities, among other targets of opportunity, in northern Syria and Iraq, with two thousand-pound GBU-32 joint direct attack munitions, which it carries internally.

NORTHROP/McDONNELL DOUGLAS YF-23 "SPIDER"/"GRAY GHOST" AND FB-23

Appearing before its ATF rival, the first of two Northrop/McDonnell Douglas service test YF-23 prototype air vehicles (PAV-1) was publically unveiled at Edwards AFB on June 22, 1990. This was a first of sorts at the Air Force Flight Test Center (AFFTC), as no new aircraft had been first shown at Edwards before or after the Lockheed YF-12A in 1964. During the rollout ceremony, it was announced that this new ATF was equipped with a pair of Pratt & Whitney YF119-PW-100 turbofanjet engines, while the second YF-23 (PAV-2) would have two General Electric YF120-GE-100 turbofanjet engines.

On October 31, 1986, the USAF announced it had awarded the Northrop Aircraft Corporation a prime contract to build two service test ATF aircraft. Under the $818 million contract, Northrop would serve as the prime contractor to construct and demonstrate-validate (dem-val) two prototype air vehicles (PAV), designated YF-23. It was at this time that Northrop began a fifty-month-long dem-val process of its ATF design. By contract, it would construct two YF-23s for flight testing, employing both the Pratt & Whitney YF119 and the General Electric YF120 ATF propulsive systems.

As Northrop's teammate, the McDonnell Douglas Corporation would be the principal contractor on the YF-23 ATF program, and the two firms would share equally in work, expense, and profit.

PAV-1 and PAV-2 (foreground) fly in loose formation. During the demonstration/validation flights, PAV-1 flew thirty-four flights, totaling forty-three hours, while PAV-2 flew sixteen flights, totaling twenty-two hours. The final flight was flown on December 18, 1990, by PAV-2. Painted in two-tone compass ghost grey, PAV-2 flew for the first time on October 26, 1990, with Jim Sandberg under glass. *NMUSAF*

The USAF selected the two winning ATF contractors after an eight-month-long evaluation of seven ATF entries from Boeing, General Dynamics, Grumman, Lockheed, McDonnell Douglas, Northrop, and Rockwell International.

After the USAF selected Lockheed and Northrop to serve as the ATF prime contractors, prearranged teaming agreements came into effect between the competing airframe contractors. Lockheed teamed with Boeing and General Dynamics, while Northrop teamed with McDonnell Douglas, thereby eliminating Grumman and Rockwell from further ATF work.

Following a series of taxi tests to rotation speed that had begun on July 7, 1990, the first YF-23 successfully completed its first flight on August 27, 1990. It was flown by Northrop ATF chief engineering test pilot Alfred P. "Paul" Metz. Functional tests of its subsystems were performed during the fifty-minute test hop. The airplane reached an altitude of twenty-five thousand feet, attaining a top speed of Mach 0.70 (about 520 mph) before returning to Edwards.

After this first flight, Metz said: "This plane looks good, and I can assure you, it flies well." He added:

> The airplane flies very clean, much cleaner than we expected. I was using considerably less power than I expected, and the chase airplanes were on afterburners just to stay with me—with my landing gear down. It appears to have a tremendous amount of excess thrust, and that's exactly what we wanted with this airplane.

Supercruise was a primary function on the ATF prototype air vehicles, which were capable of cruising supersonically at speeds up to Mach 1.6 without afterburning, as has been reported (the actual speed remains classified). This allows operational F-22As to reach their targets much faster, using less fuel than the best previous fighter aircraft, such as the F-15. Both prototype ATF test engines successfully demonstrated supercruise. YF-23 number one hit Mach 1.43 (1,060 mph) on September 18, 1990, making it the first of the four prototype ATFs to supercruise.

Powered by two General Electric YF120-GE-100 turbofanjet engines, the number two YF-23 made its first flight on October 26, 1990, with Northrop test pilot James R. "Jim" Sandberg under its canopy. The flight lasted forty-four minutes, and PAV-2 reached an altitude of fifteen thousand feet and a speed of Mach 0.485 (360 mph). There was some difficulty with the main landing gear, but otherwise the flight was completed without a hitch.

On November 2, 1990, the USAF released its final RFP to end the dem-val phase and move on to the full scale development (FSD) phase. At this time, it was announced that ATF source selection evaluations would begin after December 31, 1990, and that the winner of the ATF contest would be selected in mid-1991.

As of November 15, 1990, Northrop's two YF-23s had flown twenty-two times, for a total of thirty-three hours. Five pilots—two from Northrop, one from McDonnell Douglas, and two from the USAF—had flown the two aircraft.

The configuration of the YF-23 is matchless, featuring pod-like engine air inlets (one on either side of the fuselage) mounted ventrally; trapezoidal-shaped flying surfaces; fuselage strakes that run from wing apex to wing apex all around the forward fuselage section; and twin large-area, all-movable outward-canted ruddervators (combined rudders and elevators). These last features double as horizontal and vertical stabilizers, thanks to their outward-cant angle of forty-five degrees.

The YF-23 appears to be optimized for high lift and low drag, which allows for high-speed and long-range performance. As with the F-15 Eagle, its designers hoped to supplant and replace the engines of the YF-23 for an engine that could produce more lbf than pounds of aircraft weight. For this reason, it was meant to accelerate while in a near-vertical or almost a ninety-degree climb.

For supercruise, the YF-23 employed a variation of compression lift. It rode atop its own supersonic shock wave, much like a speedboat riding atop its step. In fact, the bottom of its fuselage and nose resemble the bottom of a boat.

From the outset, the ATF was designed to fly and fight in advanced radar networks and dense surface-to-air missile environments anywhere in the world. For its air superiority, it uses first-look, first-shoot capability to destroy any type of enemy fighter, anytime, in day or night, fair or foul weather. To accomplish this, in part, the ATF incorporates the following attributes:

- Very-low-observable (stealth) technologies; a very low radar cross-section (RCS), low infrared (IR) or heat signature, and low-energy avionics emissions
- Construction with advanced composite materials to increase airframe strength while reducing airframe weight
- High agility and maneuverability via the fly-by-computer flight control system
- Advanced multifunction cockpit displays
- High engine-thrust-to-weight ratio propulsive system, with two-dimensional convergent/divergent exhaust nozzles, or thrust-vectoring exhaust nozzles
- Adequate armament

The ATF armament package was comprised of one multi-barrel Vulcan 20mm cannon, four short-range AIM-9 Sidewinders, and four medium-range AIM-120 Slammer AAMs. The missiles would be carried internally and launched from hydraulically operated weapons racks.

Finally, after a hard-fought competition that had spanned more than ten years, the USAF announced on April 23, 1991 (about two months early), that the team of Lockheed/Boeing/General Dynamics had won the ATF battle and been awarded the FSD contract. It was also announced that the Pratt & Whitney YF119 ATF prototype turbofanjet engine would be produced to propel the F-22A.

The F-35A uses a 43,000-lbf (maximum thrust with afterburning) Pratt & Whitney F135-PW-100 turbofanjet engine to propel it to Mach 1.6 (1,200 mph) with a full internal weapons load. The F-35A Lightning II is also highly maneuverable with a maximum g-rating of 9. Its maximum combat weight is in the 70,000-pound class; empty weight is 29,300 pounds. The F-35A can carry up to eighteen thousand pounds of ordnance when it is configured for both internal and external weapons carriage. Its range on internal fuel is 1,380 miles; combat radius is 680 miles. *LM/Code One*

YF-23
SPECIFICATIONS

Propulsive system
Two afterburning, axial-flow, 35,000-lbf (with afterburning) Pratt & Whitney YF119-PW-100 (Model PW5000) turbofanjet engines (PAV-1); two afterburning, axial-flow, 35,000-lbf (with afterburning) General Electric YF120-GE-100 (Model GE37) turbofanjet engines (PAV-2)

Length	67 feet, 5 inches
Height	13 feet, 11 inches
Wingspan	43 feet, 7 inches
Empty weight	29,000 pounds
Gross takeoff weight	62,000 pounds
Maximum speed	2.2-plus Mn at best altitude; 1.6-plus Mn in supercruise
Maximum range	3,000 miles
Combat Ceiling	65,000 feet

SDD F-35A number one releases an inert Raytheon AIM-120C "Slammer" AAM during a weapon deployment test flight. The USAF Air Combat Command wants to procure a total of 1,763 F-35A airplanes to build its fleet of Lightning II aircraft. The F-35A is 51.4 feet long, 14.4 feet high, with a wingspan of 35.0 feet. *LM/Code One*

Now that the F-22A has emerged as the reigning world champ, the F-15 will relinquish its long-held crown. Unfortunately for the Northrop/McDonnell Douglas team, there will be no rematch in this winner-take-all competition.

Northrop (now Northrop Grumman) built two YF-23 prototype air vehicles (PAVs). The first example, nicknamed *Spider* or *Black Widow II*, was powered by two afterburning YF119 turbofanjet engines. The number two YF-23, nicknamed *Gray Ghost*, was powered by two afterburning YF120 turbofanjet engines.

Since the YF120 engines in PAV-2 were somewhat more powerful than the YF119 engines in PAV-1, PAV-2 was used to demonstrate full-out supercruise. A reported attainment of Mach 1.8, with Sandberg at the controls, has been leaked to the public, but that Mn has never been confirmed. Sandberg, who still isn't allowed to say just how fast he flew PAV-2 in supercruise, is credited with being the pilot who flew the fastest ever in an air vehicle without using afterburner. Sandberg did make the following statement: "I've never flown a better flying airplane, and I don't think I ever will. This airplane was truly a joy to fly."

PAV-1 was flown thirty-four times, totaling forty-three flight hours. PAV-2 flew only sixteen times, totaling twenty-two flight hours, with its last flight taking place on December 18, 1990.

NORTHROP GRUMMAN FB-23

In the early 2000s, the USAF began investigating an aircraft acquisition program it had dubbed the Interim Bomber program. The plan was to create an interim bomber using an aircraft that already existed. In late 2004, therefore, Northrop Grumman offered a modified version of its F-23 as the FB-23 to meet the requirement. Lockheed Martin offered its FB-22 and Boeing offered its B-1R.

Northrop Grumman planned to modify YF-23 number two (PAV-2) to serve as its proposed interim bomber. The possibility of an YF-23-based interim bomber ended with the 2006 Quadrennial Defense Review, which instead favored a long-range bomber with a much greater payload. With this, the FB-23 program ceased.

SUMMARY

As mentioned above, the Northrop/McDonnell Douglas YF-23 runner-up was, to many, the contender that should have won the ATF competition. It appears to have been stealthier than the YF-22, and much sleeker as well. In appearance, instead of employing wings attached to its fuselage, it seems that the plane's fuselage (nose section) is attached to its wings.

Bird of Prey

In the mid-1990s, the McDonnell Douglas Phantom Works tried its hand at creating a viable stealth fighter. To do this, it designed and built a unique single-engine demonstrator airplane using in-house resources. Since it resembled a Klingon "Bird of Prey" (as seen in *Star Trek* television programs and films), it was therefore dubbed the "Bird of Prey" (BOP).

McDonnell Douglas selected Fredrick Albert "Fred" Madenwald III to be primary project pilot on the BOP flight-test program. During the first high-speed taxi test, this pilot felt the airplane was unsafe, and he refused to be associated with it going forward. Rudy Haug Jr., backup project pilot, stepped up to replace him.

On September 11, 1996, Phantom Works test pilot Haug completed the first flight on the BOP at Area 51. It made another thirty-seven flights, with thirty-eighth and final flight taking place in April 1999. Haug had flown it eight times, while two other pilots flew it on the remaining runs: USAF test pilot Lt. Col. Douglas A. "Doug" Denjamin took it up for twenty-one flights, and Boeing Phantom Works test pilot Joseph W. "Joe" Felock III piloted it for nine flights. These were the only pilots to fly the BOP.

Meanwhile, on August 1, 1997, the McDonnell Douglas Corporation and its Phantom Works was absorbed by the Boeing Airplane Company and the airplane became the Boeing Phantom Works Bird of Prey, since Boeing continued to fund it (spending a reported $67 million in all). So cloaked in secrecy was the Bird of Prey, it wasn't publically announced and shown until October 18, 2002, more than three years, six months after its last flight had been flown.

The single-seat Boeing Bird of Prey was propelled by a single non-afterburning, axial-flow, 3,190-lbf Pratt & Whitney JT15D-5C turbofanjet engine. No speed demon by any stretch of the imagination, the maximum speed attained by the BOP was just 330 mph. The Bird of Prey is forty-seven feet long, with a wingspan of twenty-three feet; gross takeoff weight was approximately 7,400 pounds.

The Bird of Prey (BOP) in an anechoic chamber. *BAC*

The BOP is emulated on a test flight superimposed over a mountain range in this concept drawing. *BAC*

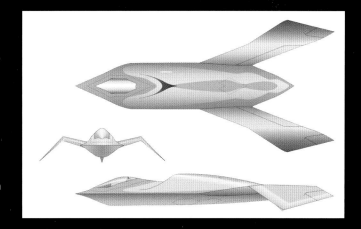

The futuristic silhouette of the Bird of Prey in three-view. *G. De Chiara © 2015*

Boeing's X-32A Joint Strike Fighter

The Conventional Takeoff and Landing (CTOL) Joint Strike Fighter (JSF) proposal for the USAF from the Boeing Airplane Company was its X-32A Concept Demonstration Aircraft (CDA). It was the first to appear and first to fly—but it failed the JSF test.

The Boeing X-32A was publically unveiled on December 14, 1999, at Boeing's facility in Palmdale, California. From there, on September 18, 2000, it made its first flight with Boeing JSF program chief engineering test pilot Fred D. "Davy" Knox Jr. at the helm. It landed at Edwards AFB some thirty miles away after its inaugural twenty-minute flight. The X-32A flew thirty-three flights in all, with its last taking place on February 3, 2001; it logged 25.2 total flight hours.

The X-32A is an unattractive airplane, having been likened to a "flying frog"; some have described it as having been "hit with an ugly stick." Yet it performed exactly as it was designed. As they say, "You can't judge a book by its cover"; this came into play during its ground evaluations and its flight tests, to no avail.

The portly X-32A featured a single nose-mounted, ventral engine air inlet with a large maw shaped like a sugar-scoop. This surely was one of its downfalls: when viewed straight-on, the face of the craft's engine is plainly visible. And if the engine face is visible to the human eye, it is doubly so for radar. The X-32A also came with delta wings and twin outward-canted, vertical tails.

The plane did meet JSF CTOL-type requirements, however, but it was judged less capable than the Lockheed Martin entry, and was passed over.

ABOVE: The CTOL Boeing X-32A CDA featured a delta wing design, whereas the X-35A used a trapezoidal wing shape. *USAF*

BELOW: The CTOL X-35A CDA (LM serial number 301) banks left during one of its test flights at Edwards AFB. It was flown twenty-seven times and accumulated 27.4 flying hours between October 24 and November 22, 2000. *USAF*

The YF120-powered YF-23 (the second of the two YF-23s to fly) reportedly hit Mach 1.6 in supercruise—on its second flight! This figure has not been confirmed, and so remains unofficial.

The YF-23's eye-pleasing and interesting design can be appreciated where the two examples are on view. PAV-1 is on display at the National Museum of the United States Air Force near Dayton, Ohio, and PAV-2 can be seen at the Western Museum of Flight in Hawthorne, California.

SUMMARY

The X-32A JSF CDA was flown by six government pilots, and, according to Knox: "The airplane is a pleasure to fly. It is already showing the precise handling qualities we expected based on hundreds of hours of simulator work."

The X-32A also served as the US Navy Carrier Variant (CV), but was never designated X-32C, as had the Lockheed Martin CV, which became the X-35C, an airplane that was built as such from the outset, while its X-35A was modified to become the X-35B.

LOCKHEED MARTIN F-35A LIGHTNING II

The Lockheed Martin Corporation F-35A Lightning II of the USAF Air Combat Command is a supersonic fifth-generation tactical stealth fighter bomber.

The F-35A is the CTOL version of the JSF that is used exclusively by the USAF ACC. The USAF plans to procure 1,763 F-35As.

The F-35A is expected to supplement the growing number of older high-time F-16 Fighting Falcons and to completely replace the fleet of A-10 Thunderbolt IIs. Working in concert with F-22As, the F-35As will help fill the void left by the now retired fleet of F-117A Nighthawks.

X-32A SPECIFICATIONS

Propulsive system
One afterburning, axial-flow, 43,000-lbf JSF119-PW-614 Pratt & Whitney turbofanjet engine

Length	45.0 feet
Height	17.3 feet
Wingspan	36.0 feet
Gross takeoff weight	50,000 pounds
Maximum speed	1.6 Mn

Range
690 to 850 miles-radius (internal fuel)

Ceiling	50,000 feet

Armament
One 20mm M61A2 cannon, two AIM-120 AMRAAMs, and two 2000 pound JDAMs (proposed)

Lockheed's X-35A Joint Strike Fighter

Lockheed Martin produced a single X-35A JSF CDA air vehicle (factory serial number 301) to compete against the Boeing X-32A JSF CDA in hope of winning the JSF competition.

With Lockheed Martin test pilot Thomas A. "Tom" Morgenfeld at the helm on October 24, 2000, the X-35A made a successful test hop, flying from Palmdale (where it was built) to Edwards AFB, where it would be flight-tested.

The X-35A flew twenty-six times before it was returned to Palmdale on November 22, 2000, to undergo modification. After this was completed, it was transformed into the X-35B Short Takeoff and Vertical Landing (STOVL) version for US Navy evaluations.

The trio of Lockheed Martin X-35 CDA aircraft—X-35A, X-35B, and X-35C—won the JSF competition on October 26, 2001. As far as the USAF's Air Combat Command was concerned, the CTOL X-35A version would be its baby to raise.

Two F35As and two F22As fly in formation to give us a unique way to distinguish the two types. *USAF photo by M/Sgt. Shane A. Cuomo*

What Happened to the F-24 to F-34 Designation Numbers?

Designation numbers F-24–F-34 for USAF fighter-type aircraft have not been used—not officially, at any rate. The eleven-number leap in designation numbers, from the YF-23 to the F-35A, was quite the surprise when the JSF competition was won by Lockheed Martin and the airplane was designated F-35 (not F-24, which was the next available "F for fighter" designation in numerical order).

There was (or is) an alleged super-secret-aircraft designated YF-24 that was flight tested in Area 51. Reportedly flown by USAF Lt. Col. (later Brig. Gen.) Joseph A. Lanni sometime between July 1995 and June 1997, when he commanded a unit simply named "Flight Test Squadron." Real or not, the aircraft is described only as a "classified prototype."

As far as the missing designations F-24 through F-34 are concerned, however, there is no documentary evidence that any such aircraft exist or have ever existed.

F-35A

On October 26, 2001, the DOD announced that Lockheed Martin had won its JSF competition, adding that Pratt & Whitney would produce its F135 turbofanjet propulsive system.

As part of the system development and demonstration (SDD) phase, Lockheed Martin ultimately produced five SDD F-35A aircraft under contract number N00019-02-C-3002. These included AA-1, AF-1, AF-2, AF-3, and AF-4; a sixth SDD F-35A (AF-5) was cancelled. No USAF serial numbers were assigned to these service test aircraft. F-35A (AA-1) completed ninety-one test flights before it was retired.

On December 15, 2006, the first SDD F-35A (AA-1) made a successful first flight at LM's Fort Worth, Texas, facility. It was piloted by Lockheed Martin chief engineering test pilot, Jon S. Beesley, who also played a significant part in the development of the F-22A.

Why was the F-35A designated as such, instead of as F-24A? This eleven-digit leap requires an explanation. If the status quo had been maintained, the F-35A would have been designated F-24A, since the Northrop/McDonnell Douglas YF-23 ATF prototype was the last USAF fighter produced before it.

F-35A SDD
PRODUCTION BREAKDOWN:

AA-1:	December 15, 2006	John S. Beesley (Lightning 01)
AF-1:	November 14, 2009	David M. "Doc" Nelson (Lightning 06)
AF-2:	April 20, 2010	Jeffrey W. "Jeff" a.k.a. "Slim" Knowles (Lightning 02)
AF-3:	July 6, 2010	William J. "Bill" Gigliotti (Lightning 11)
AF-4:	December 30, 2010	Bill Gigliotti

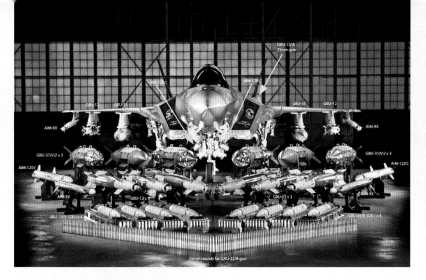

In March 2005, this writer asked Public Affairs in the JSF System Program Office to explain why this had happened. The office's succinct reply came back: "We asked and got permission from HQ USAF that we change only the X prefix to save a multitude of alterations to the reams of paperwork." This explanation goes against several other plausible explanations, which leads us to conclude the final answer has not yet been made public.

The first flight particulars on AA-1 and AF-1 through AF-4 shown below.

SUMMARY

Production of the USAF F-35A Lightning II proceeds with an ever-increasing number of planes going to its first assigned fighter wing, the 56th Fighter Wing at Luke AFB, Arizona. The first of these F-35As was delivered to Luke AFB on March 10, 2014 (USAF serial number 11-5030). The 56FW is slated to receive 144 total F-35As at the time of this writing.

On May 28, 2014, the twenty-sixth and last F-35A (11-5034) arrived at Eglin AFB, Florida, to complete the 56th Fighter Squadron, the first all-F-35A squadron, which is part of the 33rd Fighter Wing. Eglin is home to the Integrated Training Center for the F-35A, F-35B, and F-35C aircraft.

The first operational F-35A units will be the 388th Fighter Wing (72 F-35As) and the Air Force Reserve 419th Fighter Wing (number of F-35As unknown at the time of this writing) at Hill AFB, Utah, scheduled to receive its first F-35A in September 2015. The 388FW will get an initial cadre of pilots from the F-35A test and training units at Eglin AFB, Florida, Luke AFB, Arizona, and Nellis AFB, Nevada. The USAF has also selected Eielson AFB in Alaska for a future wing: three squadrons of F-35As to be the first Lightning II unit in the Pacific area of operations.

F-35A SPECIFICATIONS

Propulsive system	
One afterburning, axial-flow, 43,000-lbf Pratt & Whitney F135-PW-100 turbofanjet engine	
Length	51.4 feet
Height	14.4 feet
Wingspan	35.0 feet
Empty weight	29,300 pounds
Gross takeoff weight	70,000 pounds
Maximum speed	1.6 Mn
Ceiling	50,000 feet (estimated)

Maximum range
Unlimited with aerial refueling

Armament
One 25mm GAU-22/A cannon, two AIM-9s, and two AIM-120C AMRAAMs

Payload
Two 2000-pound GBU-31 JDAMs, small diameter bombs (SDB), and variations of internal/external ordnance

TEN

THE LONG-RANGE STRIKE BOMBER & NEXT-GENERATION AIR DOMINANCE FIGHTER

> Only air power can defeat air power. The actual elimination or even stalemating of an attacking air force can be achieved only by a superior air force.
>
> —Maj. Alexander P. de Seversky, USAAC

We have reached the year 2024. Two days ago, on July 24, the US Department of Defense issued a short press release telling of a new kind of bomber now in its advanced test phase. This is the text of the release:

> Several months ago, an unoccupied heavy bombardment air vehicle circumnavigated the globe—flying nonstop, longitudinal rather than latitudinal, via an undisclosed number of aerial refuelings. This marks the first time that any aircraft has circled the world flying north to south rather than east to west.
>
> This air vehicle, code-named Senior Hula, has been officially designated B-3A. It was developed under the Long-Range Strike Bomber (LRSB) program

This illustration depicts the next-generation long-range strike aircraft as imagined by Northrup Grumman. *Northrup Grumman*

Teammates Boeing Airplane Company and Lockheed Martin Corporation released this image showing what their version of the 2018 Bomber might have looked like. With an all-flying wing design, it appears to have a four-engine propulsive system. *LM*

Artist's conception of a pair of Boeing X-45Cs dropping their ordnance. The X-45C was developed into the *Phantom Ray*. *BAC*

first implemented in July 2014, when the USAF sent out its RFPs to the industry.

The first B-3A made its first flight on December 17, 2021, the 118th anniversary of the Wright brothers' first flight. Several others have flown since then, and they too are undergoing thorough evaluations. The B-3's initial operational capability (IOC) is expected by mid-2026.

The B-3A is being built by a consortium of airframe contractors, rather than by a single contractor or contractor team. Another consortium of powerplant contractors is providing the craft's propulsive system.

Still highly classified, the B-3A is a dedicated stealth bomber with very-low-observables, and it is expected to fully replace the venerable B-52 by 2040.

While the foregoing is fictitious, it's not totally out of the realm of possibility. The US Air Force is currently planning to procure a real LRSB to replace its fleet of B-52s. Moreover, the USAF is seeking a next-generation air dominance (NGAD) fighter type as well as a next-generation turbofanjet engine to propel it.

The previous pages offer a great deal of information about past and present bombers and fighters. But what about forthcoming bombers and fighters?

The USAF keeps future requirements fixed in its sights, in particular newer and more advanced combat aircraft, such as bombers and fighters. The USAF Air Force Materiel Command (AFMC) continues to investigate next-generation bomber and fighter aircraft in a process that has taken on many different guises, without an ample budget to get much done.

This "buck's-down" situation has been a liability for USAF planners. That is, until one current data point came to light: the development of advanced stealth-type fighter aircraft in other countries like China and Russia.

The US Department of Defense now realizes that America will need new and advanced combat aircraft to meet and defeat any future threats from potential adversaries. Beginning with its FY2015 budget, the USAF will at last receive adequate moneys to move forward on its goal to field a next-generation bomber and a next-generation air dominance fighter.

FORTHCOMING BOMBER

The Air Force Global Strike Command arm of the US Air Force currently has two strategic bomber fleets at its disposal: the Boeing B-52H Stratofortress, and the Northrop Grumman B-2A Spirit. At the same time, the Air Combat Command arm of the USAF has the multi-role Boeing B-1B Lancer.

The B-2 fleet is minuscule, with only a maximum of twenty aircraft capable of performing combat missions. The B-1 fleet is substantially larger than the B-2 fleet, with as many as sixty-five aircraft available for combat. The B-52 fleet continues to dwindle, with no more than seventy-five aircraft available for action. At this writing, the B-2 fleet is the youngest, at twenty-one-plus years, followed by the B-1 fleet (twenty-eight-plus years) and the B-52 fleet (fifty-three-plus years).

The Air Force Global Strike Command recognizes the need for a new strategic bombardment system, since all three fleets continue to age and diminish in size (because of mishaps and attrition).

THE NORTHROP GRUMMAN LONG-RANGE STRIKE BOMBER

On July 9, 2014, the USAF released its request for proposals (RFP) for the Long-Range Strike Bomber (LRSB). The details of the RFPs hadn't yet been disclosed at this writing, but in all likelihood it required the use of an advanced propulsive system, avionics, ISR, and a diverse carriage of weapons—both conventional and nuclear.

On October 27, 2015, US Secretary of Defense Ashton B. "Ash" Carter, Secretary of the Air Force Deborah Lee James, and Chief of Staff of the US Air Force Gen. Mark A. Welsh III announced the awarding of the Long-Range Strike Bomber contract in the press briefing room at the Pentagon. Following their remarks, Assistant Secretary of the Air Force for Acquisition Dr. William A. LaPlante; Gen. Robin Rand, commander, Air Force Global Strike Command; and Lt. Gen. Arnold W. Bunch, Jr., military deputy, Air Force Acquisition answered questions from the attending media.

During the briefing it was announced that Northrop Grumman Corporation had won the right to produce the LRSB, which wasn't given a designation at that time, and that it was awarded the engineering and manufacturing development (EMD) and early-production contract.

Still highly classified, no photographs were released and no specifications were given—not even regarding its propulsive system or its designation.

Northrop Grumman design for an advanced bomber. *NG*

What is known is that the LRSB is to be a long-range strategic bomber for the US Air Force Global Strike Command (AFGSC). It is to be a stealthy heavy-payload bombardment aircraft capable of carrying and precisely delivering any air-dropped weapon in the US arsenal. Its initial operational capability (IOC) is expected in the mid-2020s. The USAF plans to purchase 100 LRSBs at a cost of $564 million each in fiscal year 2016 dollars.

The LRSB program had been preceded by a number of programs including the Regional Bomber, Interim Bomber, 2037 Bomber, 2018 Bomber, and Next Generation Bomber (NGB) programs. Lockheed Martin Corporation and Boeing Airplane Company—specifically their Skunk Works and Phantom Works, respectively—had joined forces to compete for the LRSB contract, as they had on the 2018 Bomber and Next Generation Bomber (NGB) programs. Northrop Grumman went at alone and eventually prevailed

"Boeing and Lockheed Martin are bringing together the very best of the two enterprises, and the rest of industry, in support of the Long-Range Strike Bomber program, and we are honored to support our US Air Force customer and this important national priority," said Dennis Muilenburg, president and chief executive officer of Boeing Defense, Space & Security. "Stable planning, along with efficient and affordable development and production approaches, enables our team to reduce development risk by leveraging mature technologies and integrating existing systems." The partnership believed they would be able to produce unique and affordable solutions that could not be achieved on their own.

"Building on decades of manned and unmanned weapon systems experience, we're proud to bring our collection of technologies, capabilities, and resources to affordably design, develop, produce, and sustain the bomber program," said Orlando Carvalho, executive vice president of Lockheed Martin Aeronautics. "We're confident that our team will meet the well-defined system requirements and deliver a world-class next generation Long-Range Strike Bomber to the US Air Force within the budget and timeframe required."

When RFPs for the LRSB program were released to the industry in 2014, the USAF announced it would award a contract in spring 2015. "The new bomber will be a long-range, air-refuelable, highly survivable aircraft with significant nuclear and conventional stand-off and direct-attack weapons payload," a USAF statement read. "The LRSB will provide operational flexibility across a wide range of military operations."

"Northrop Grumman's design, production, and sustainment of the B-2 Spirit stealth bomber, the bomber most recently produced for the US Air Force, positions [Northrop Grumman] well for the LRSB program," spokesman Brandon R. "Randy" Belote III said. "We are very interested in working with the Air Force to provide this critical capability for the nation."

Its work with the USAF can now proceed.

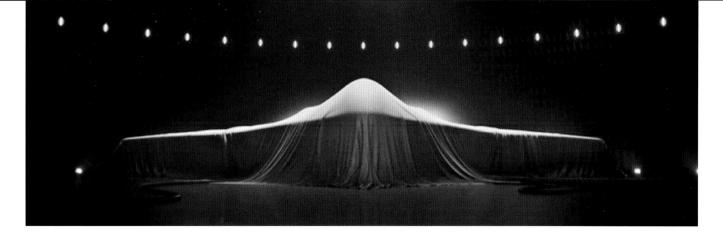

SUMMARY

After the announcement, Wesley G. "Wes" Bush, chairman, chief executive officer and president, Northrop Grumman, stated, "The Air Force has made the right decision for our nation's security."

Bush went on to add, "As the company that developed and delivered the B-2 Spirit stealth bomber, we look forward to providing the Air Force with a highly-capable and affordable next-generation Long-Range Strike Bomber.

"Our team has the resources in place to execute this important program, and we're ready to get to work."

Boeing and Lockheed Martin responded to the decision, as well: "The Boeing and Lockheed Martin team is disappointed by today's announcement. We will have further discussions with our customer before determining our next steps. We are interested in knowing how the competition was scored in terms of price and risk, as we believe that the combination of Boeing and Lockheed Martin offers unparalleled experience, capability and resources for this critically important recapitalization program."

Whether the LRSB team of Boeing/Lockheed Martin would decide to protest the decision wasn't clear as this book went to press.

FORTHCOMING FIGHTER

Fighters take a tremendous amount of abuse while pulling maneuvers at several g's almost every day while in training and/or combat. After so many hours of these day-to-day poundings, they have to undergo expensive and time-consuming structural rebuilds, or they are simply recycled. Unlike their bomber counterparts, which fly for many decades (witness the B-52), the lifespan of a fighter is much shorter. It is for this reason that new fighters are sought more often than new bombers.

THE NEXT-GENERATION AIR DOMINANCE FIGHTER

In March 2014, the USAF requested $15.72 million in its budget proposal for FY2015. This money was to be allocated for a new next-generation, air dominance research and development project that would lay the foundation for an acquisition program in FY2018.

Occupied vs. Unoccupied Bombers and Fighters

Fighter aircraft have been occupied since before World War I and their respective successes are well known. The possibility of fielding unmanned fighters in the near future grows stronger with each day.

As of 2015, there are several unoccupied combat-type air vehicles undergoing flight and weapon test activities. It is a matter of time before one or more of these will be adapted to aerial combat.

One of these unoccupied combat-type air vehicles is the privately funded Boeing Phantom Ray. On April 27, 2011, the fighter-sized Phantom Ray unmanned airborne system (UAS) made its first flight at Edwards AFB. It flew for a second time just eight days later, on May 8. Boeing unveiled the Phantom Ray at its St. Louis, Missouri, facility on May 10, 2010, after only two years of development. It is thirty-six feet long with a fifty-foot wingspan. Its propulsion comes from a single non-afterburning, axial-flow, 7,000-lbf General Electric F404-GE-102D turbofanjet engine.

This engine was earlier selected for use by the Boeing X-45C Joint Unmanned Combat Air Systems (J-UCAS), which were discontinued as they evolved into the Phantom Ray.

Whether the next generation bomber or fighter will be occupied or unoccupied (or both) remains unclear at this time. Almost certainly, though, fleets of unoccupied combat aircraft will someday take to the air, whether they are attack, bomber, or fighter aircraft—or a combination of all three.

The turbopropjet-powered General Atomics MQ-9 Reaper is a most effective unoccupied aerial vehicle (UAV) that continues to seek and destroy enemy targets. Formerly known as Predator B, the Reaper can carry up to 3,800 pounds of ordnance—800 pounds internal, 3,000 pounds external. It was developed from the MQ-1B Predator *B* and developed into the Predator C Avenger. *USAF photo by Staff Sgt. Brian Ferguson*

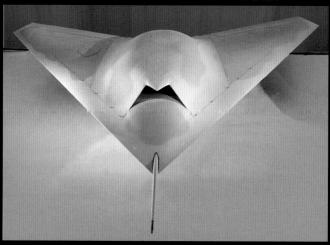

ABOVE: Boeing unveiled its Phantom Ray on May 10, 2010, and it made its first flight on April 27, 2011 at Edwards AFB. It was a development of the DARPA/USAF/USN X-45C J-UCAS, which was cancelled on March 2, 2006. The Phantom Ray—designed and built by the Boeing Phantom Works—is a flying-wing with a span of fifty feet and a length of thirty six feet. Powered by a single non-afterburning, General Electric F404-GE-102D turbofanjet engine, the Phantom Ray's top speed was 614 miles per hour (Mach 0.8). *BAC*

LEFT: The General Atomics Predator C Avenger is an ongoing, unsolicited, company-funded program for which GA-ASI hopes to gain production orders. It has six external hard points for a wide variety of ordnance, and its internal weapons bay can house up to 3,500 pounds of ordnance. The Avenger has an eighteen-hour endurance. *GA-ASI*

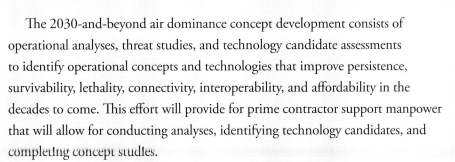

The 2030-and-beyond air dominance concept development consists of operational analyses, threat studies, and technology candidate assessments to identify operational concepts and technologies that improve persistence, survivability, lethality, connectivity, interoperability, and affordability in the decades to come. This effort will provide for prime contractor support manpower that will allow for conducting analyses, identifying technology candidates, and completing concept studies.

The fleet of USAF fifth-generation F-22 Raptors has been operational for some ten years as of this writing, and, according to Lockheed Martin, the USAF fifth-generation F-35A Lightning II won't begin to become operational (that is, capable of meeting its IOC) until the first operational squadron is equipped with between twelve and twenty-four aircraft. This will not happen until August 2016 (objective) and December 2016 (threshold).

Nevertheless, the USAF has initiated its search for an F-X aircraft, a "next-generation air dominance" type, to replace the F-22 after 2030. Boeing, Lockheed Martin, and Northrop Grumman have already shown conceptual designs of their respective F-X ideas. Nevertheless, these early concepts will change many times over before an airframe contractor is selected to produce the next-generation air dominance fighter that is capable of truly becoming operational.

SUMMARY

A future air dominance fighter in the 2030-and-beyond period may include laser-beam control systems for operations in the flight regime, for altitudes ranging from sea level to more than sixty-five-thousand feet and speeds from Mach 0.6 to 2.5. Pilots of such aircraft may be ground-based. A fighter of this design is far in the future, but the fire has been lit, and its eventual creation has begun to give off a little heat. When and if these plans unleash a raging inferno remains to be seen.

EPILOGUE

This book's chapters were written to be short, concise, and factual. The sixty-plus years of research performed by this writer on past, present, and future jet-powered bomber and fighter aircraft presented herein continues.

As time passes, new information comes to light, largely due to the declassification of information and the release of documentary papers and/or photographs; as these become available, they will be used in future editions of this work. The once tightly held secrets from the past will eventually be told and the individual stories behind these aircraft will come into sharper focus. Until then, the text and illustrations presented here will have to suffice.

The development of turbojet-, turbopropjet-, and turbofanjet-powered bombers and fighters has come a long way since an XP-59A nicknamed "Squirt" first lifted off from the dry lakebed at Muroc more than seventy years ago. Since then, jet-powered fighter speeds have increased from just over 400 mph to 2,200-plus mph, and from Mach 0.60 to 3.35. Even the cumbersome turbojet-powered XB-70A reached speeds in excess of Mach 3 (2,000 mph), which was an extraordinary accomplishment when one considers that the XB-43, America's first jet-powered bomber, struggled to attain speeds higher than Mach .75 (500 mph).

The awesome capabilities enjoyed by today's jet-powered bombers and fighters in USAF service rely on the achievements and disappointments of the thousands of dedicated people who participated in all aspects of their creation—from their preliminary design to their closing flight tests.

We should not forget the propulsion systems that propel them. From the diminutive 1,250-lbf Model I-A of 1942 to the 43,000-lbf F135 in use today, gas turbine engine development has likewise come a long way. In fact, without a suitable propulsion system, none of the bomber and fighter aircraft that have gone into production could have lifted off and performed as they have.

Finally, a great deal is owed to the test pilots, flight test engineers, and crewmen who tested these aircraft. Special thanks are owed to those who lost their lives while attempting to prove the worth of the aircraft they were evaluating. Without their tragic losses, many of the bomber and fighter aircraft described here would not have reached the heights and broken the technological and military barriers that they did. That is an appropriate epitaph for anyone.

APPENDIX

TIMELINE: SIGNIFICANT DATES IN THE DEVELOPMENT OF USAF X-BOMBERS AND X-FIGHTERS

May 15, 1941: USAAC Commanding Gen. Henry H. "Hap" Arnold witnesses first flight of Gloster G.40, Great Britain's first jet-powered airplane, made arrangements to develop its propulsive system in the United States.

June 20, 1941: US Army Air Corps (USAAC) becomes the US Army Air Forces (USAAF).

September 4, 1941: The USAAF offers General Electric a contract to build under license a British-designed and developed gas turbine (turbojet) engine in the United States.

September 5, 1941: The USAAF offers the Bell Aircraft Corporation a contract to build three airplanes to be powered by the aforementioned General Electric turbojet engines.

September 30, 1941: Bell awarded a contract to build three XP-59A airplanes, each to be powered by two General Electric turbojet engines, with first delivery promised in eight months from date of contract.

February 1942: USAAF Col. Benjamin W. Chidlaw and Lt. Col. Ralph P. Swofford of the Air Materiel Command (AMC) at Wright Field, Ohio, who were on an extended tour of the western United States to find an adequate flight test site, choose Muroc Dry Lake as the ideal location to test the then secret Bell XP-59A. (It had been called Muroc Dry Lake as a convenience for many years since Muroc Army Air Field—later Muroc Air Force Base—now Edwards AFB, bordered it. But its proper name is Rogers Dry Lake, which wasn't widely used until the 1960s. Thus, until the 1960s, Rogers Dry Lake was called Muroc Dry Lake for the most part.)

April 18, 1942: General Electric I-A (Type I) turbojet engine test run for the time; first turbojet engine to run in the United States.

October 1–2, 1942: Bell test pilot Robert M. "Bob" Stanley makes successful first flights at Muroc AAF on the first of three Bell XP-59A Airacomet airplanes.

April 18, 1943: Bell test pilot Jack V. Woolams flies a safe test hop on the second of thirteen Bell YP-59A Airacomets at Muroc AAF; it flies before number one.

November 8, 1943: Muroc Army Air Base renames Muroc Army Air Field.

January 8, 1944: Lockheed test pilot Milo G. Burcham flies a successful first flight on the Lockheed XP-80 at Muroc AAF.

June 10, 1944: Lockheed test pilot Anthony W. "Tony" LeVier makes the first flight on the prototype Lockheed XP-80A airplanes at Muroc AAF.

August 1, 1944: LeVier successfully flight-tests XP-80A prototype number two at Muroc AAF.

February 11, 1945: Convair test pilot Frank W. Davis successfully test flies the first of two XP-81 aircraft at Muroc AAF with an interim piston engine in its nose, where a turbopropjet engine would later be installed.

February 25, 1945: Bell test pilot Jack V. Woolams successfully pilots the first of two XP-83 airplanes on its first flight from Niagara Airport in Buffalo, New York; the second example first flies on October 19, 1945.

September 12, 1945: Northrop test pilot Harry H. Crosby pilots the Northrop XP-79B for its first and last flight. The airplane goes out of control during this flight and crashes to destruction; Mr. Crosby is killed during his bailout attempt.

December 21, 1945: First of two Convair XP-81 airplanes makes its first flight with the design nose-mounted GE Model TG-100 (XT31-GE-1) turbopropjet engine installed.

February 28, 1946: USAAF test pilot Major Wallace A. "Wally" Lien successfully flight-tests the first of two Republic XP-84 Thunderjet aircraft at Muroc AAF.

March 21, 1946: USAAF Strategic Air Command (SAC) established with headquarters at Bolling Field, District of Columbia.

March 21, 1946: USAAF Tactical Air Command (TAC) established with headquarters at Langley Field, Virginia.

March 27, 1946: USAAF Air Defense Command (ADC) established with headquarters at Mitchel Field, New York.

May 17, 1946: Douglas Aircraft's experimental XB-43 Jetmaster successfully flight-tested at Muroc AAF by Douglas test pilot Robert P. "Bob" Brush.

March 17, 1947: North American Aviation test pilot George W. Krebs successfully flight-tests the experimental North American XB-45 Tornado airplanes at Muroc AAF.

April 2, 1947: Convair test pilot Ellis D. Shannon first flies the Convair XB-46 from the Convair plant in San Diego, California, to Muroc AAF.

May 27, 1947: The Douglas XB-42A (a modified XB-42) makes a successful first flight from the Douglas plant in Santa Monica, California, to Muroc AAF with Douglas test pilot Russell W. "Russ" Thaw at the helm.

June 22, 1947: Martin's experimental XB-48 makes a successful first flight in Baltimore, Maryland, with Martin test pilot Orville E. "Pat" Tibbs at the helm.

September 18, 1947: US Army Air Forces (USAAF) becomes US Air Force (USAF)—a whole and separate service, no longer a part of the US Army.

October 1, 1947: Experimental North American XP-86 Sabre makes a successful first flight at Muroc AAF with NAA test pilot George S. "Wheaties" Welch at the controls.

October 21, 1947: Prototype Northrop YB-49 flies from Hawthorne to Muroc AAF successfully with Northrop test pilot Max Stanley in control.

December 17, 1947: Boeing test pilot Robert M. "Bob" Robbins Boeing makes a successful first flight on the experimental XB-47 Stratojet in Seattle, Washington.

February 12, 1948: Muroc Army Air Field unofficially renamed Muroc Air Force Base.

March 5, 1948: Curtiss XP-87 Blackhawk, an experimental night fighter, makes a successful first flight at newly named Muroc AFB with Curtiss test pilot B. "Lee" Miller under the cockpit canopy.

June 11, 1948: "P for Pursuit" prefix changed to "F for fighter"; "F for Photographic" changed to "R for Reconnaissance"; "TP for Trainer Pursuit" changed to "TF for Trainer Fighter."

August 16, 1948: Northrop test pilot Fred C. Bretcher flies the experimental Northrop XF-89 Scorpion successfully at Muroc AFB.

August 23, 1948: Experimental McDonnell XF-85 Goblin first flight at Muroc AFB with McDonnell test pilot Edwin F. "Ed" Schoch on the control stick.

September 18, 1948: Convair test pilot Ellis D. "Sam" Shannon successfully pilots the Convair XF-92A Dart for the first time at Muroc AFB.

October 20, 1948: Experimental McDonnell XF-88 Voodoo makes first flight at Muroc AFB with McDonnell test pilot Robert M. Edholm at the controls; the second XF-88 example flew on April 26, 1949.

March 26, 1949: Prototype Convair B-36D, a converted B-36B, makes its first flight at Fort Worth, Texas; it is the first version of the B-36 to feature a supplemental propulsive system comprised of four non-afterburning General Electric J47-GE-19 turbojet engines.

April 16, 1949: Prototype Lockheed YF-94 makes a successful first flight at Van Nuys airport with Lockheed test pilot Tony LeVier seated under glass.

May 9, 1949: Republic test pilot Carl A. Bellinger completes a successful first flight at Muroc AFB on the first of two Republic XF-91 Thunderceptor airplanes.

June 3, 1949: Lockheed test pilot Tony LeVier completes the first flight on the experimental Lockheed XF-90 at Muroc AFB.

October 28, 1949: Martin test pilot Pat Tibbs completes the first flight on the experimental Martin XB-51 at Baltimore, Maryland; the second example flies on April 17, 1950.

December 8, 1949: Muroc Air Force Base renamed Edwards Air Force Base in honor of USAF Capt. Glen W. Edwards who had lost his life in the crash of a prototype Northrop YB-49 on June 5, 1948.

December 22, 1949: Prototype North American YF-86D Sabre makes its first flight at newly, but unofficially named, Edwards AFB, with NAA test pilot Joseph A. "Joe" Lynch in command.

January 19, 1950: Lockheed chief test pilot Tony LeVier completes first flight at Edwards AFB on the prototype Lockheed YF-94C Starfire; formerly designated YF-97A.

January 24, 1950: Prototype North American YF-93A makes successful first flight at Edwards AFB with NAA chief test pilot George Welch at the helm.

January 27, 1950: Muroc Air Force Base ceremonially and officially renamed Edwards AFB.

May 4, 1950: Northrop test pilot Fred C. Bretcher pilots the Northrop YRB-49A prototype from Hawthorne, California, to Edwards AFB on its first flight.

June 3, 1950: Prototype Republic YF-96A Thunderstreak (later redesignated YF-84F), with interim Allison J35 engine, makes its first flight at Edwards AFB with Republic test pilot Oscar P. "Bud" Haas behind the control stick.

September 28, 1950: Prototype YF-94B flies out of Van Nuys airport for the first time; it flew after the YF-94C.

February 14, 1951: Prototype Republic YF-84F Thunderstreak (formerly designated YF-96A) makes its first flight with design Wright J65 Sapphire engine at Edwards AFB after its modification; same plane that first flown on June 3, 1950 with J35 engine.

February 3, 1952: Republic YRF-84F Thunderflash prototype makes its first flight at Edwards AFB with Republic test pilot Carl A. Bellinger in control.

April 15, 1952: Boeing chief test pilot Alvin M. "Tex" Johnston completed a successful first flight on the service test Boeing YB-52 Stratofortress at Seattle, Washington.

April 18, 1952: Prototype Convair YB-60 makes a successful first flight at Fort Worth, Texas, with Convair test pilot Beryl A. Erickson in command.

October 2, 1952: Boeing chief test pilot Tex Johnston makes the first flight on the experimental Boeing XB-52 Stratofortress at Seattle, Washington.

November 18, 1952: Convair notified that its XB-58 program would move forward and Boeing notified that its XB-59 program has been cancelled.

December 9, 1952: Republic test pilot Russell M. "Rusty" Roth achieves supersonic speed at Edwards AFB in level attitude flight while piloting the first of the two XF-91 Thunderceptor airplanes.

April 10, 1953: Prototype North American YF-86H Sabre Jet makes its first flight at Edwards AFB with NAA test pilot Joel Robert "Bob" Baker under its cockpit canopy.

May 25, 1953: NAA chief test pilot George Welch completes the first flight on the prototype North American YF-100 Super Sabre at Edwards AFB going supersonic in level attitude flight.

July 20, 1953: Martin test pilot Pat Tibbs completes a successful first flight on the first production (service test) Martin B-57A Canberra in Baltimore, Maryland.

October 24, 1953: Convair chief test pilot Richard L. "Dick" Johnson successfully flight-tests the prototype Convair YF-102 Delta Dart at Edwards AFB.

March 4, 1954: Experimental Lockheed XF-104 Starfighter makes its first flight at Edwards AFB with Lockheed chief test pilot Tony LeVier in its seat.

May 7, 1954: Republic test pilot Rusty Roth flight tests the prototype Republic YF-84J Thunderstreak II for its first time at Edwards AFB.

June 28, 1954: First preproduction Douglas RB-66A Destroyer makes its first flight at Edwards AFB with Douglas test pilot George R. Jansen at the controls.

September 29, 1954: McDonnell chief test pilot Robert C. "Bob" Little successfully pilots the first preproduction McDonnell F-101A Voodoo on its first flight at Edwards AFB.

December 19, 1954: Prototype Convair YF-102A Delta Dagger makes its first flight at Edwards AFB with Convair chief test pilot Dick Johnson in control.

October 22, 1955: Republic test pilot Rusty Roth makes a successful first flight on the Republic YF-105A Thunderchief prototype at Edwards AFB.

September 10, 1956: Prototype North American F-107A completes its first flight at Edwards AFB with NAA test pilot Bob Baker in the pilot's seat.

November 11, 1956: Convair chief test pilot Beryl A. Erikson successfully demonstrates the Convair B-58 Hustler on its first flight at Fort Worth, Texas.

December 26, 1956: First full-scale development Convair F-106A Delta Dart completes its first flight successfully at Edwards AFB with Convair chief test pilot Dick Johnson at its controls.

March 27, 1957: McDonnell chief test pilot Bob Little makes a successful first flight on the special test McDonnell NF-101B Voodoo in St. Louis, Missouri.

August 21, 1957: The Republic XF-103 program cancelled.

June 30, 1959: Prototype Northrop N-156F Freedom Fighter makes its first flight at Edwards AFB with Northrop test pilot Lewis A. "Lew" Nelson in control.

September 23, 1959: The North American F-108 program cancelled.

Spring 1961: The Bell D-188 ("YF-109") program cancelled.

November 1962: The McDonnell F-110A demonstrator, a modified U.S. Navy F4H-1, makes its first flight out of Lambert Field in St. Louis, Missouri, with McDonnell chief test pilot Bob Little at the controls.

May 27, 1963: McDonnell chief test pilot Bob Little successfully completes the first flight on the prototype McDonnell YF-4C Phantom II in St. Louis, Missouri.

July 31, 1963: Prototype Northrop YF-5A Freedom Fighter makes its first flight at Edwards AFB with Henry E. "Hank" Chouteau at the control; officially named Freedom Fighter August 9, 1962.

August 7, 1963: Lockheed Skunk Works test pilot James D. "Jim" Eastham makes the first flight on the Lockheed YF-12A at Area 51, Groom Lake, Nevada.

September 21, 1964: Experimental North American XB-70A Valkyrie makes a successful first flight with NAA chief engineering test pilot Alvin S. "Al" White in command; second example also flown by Al White on July 17, 1965.

December 21, 1964: General Dynamics' chief engineering test pilot Dick Johnson completes the first flight on the first preproduction General Dynamics F-111A at Fort Worth, Texas.

July 30, 1967: Demonstration General Dynamics FB-111A makes its first flight at Fort Worth, Texas, with General Dynamic' test pilot Val Prahl at the controls.

July 27, 1972: McDonnell Douglas test pilot Irving L. "Irv" Burrows makes a successful first flight on the first preproduction McDonnell Douglas F-15A Eagle at Edwards AFB.

February 2, 1974: Prototype General Dynamics YF-16 completes first flight at Edwards AFB with General Dynamics' chief test pilot Phillip E. "Phil" Oestricher at its controls.

June 9, 1974: Northrop chief test pilot Hank Chouteau flies the prototype Northrop YF-17 Cobra successfully at Edwards AFB.

December 23, 1974: Prototype Rockwell International B-1A completes first flight, Palmdale, California to Edwards AFB with Rockwell International chief test pilot Charles C. "Charlie" Bock Jr. at the helm.

June 30, 1977: President James Earl "Jimmy" Carter Jr. cancels the Rockwell International B-1A program.

December 1, 1977: Lockheed Skunk Works test pilot William C. "Bill" Park successfully flight-tests the prototype Lockheed Have Blue XST at Area 51; the second example flies on July 20, 1978.

March 31, 1980, USAF Aerospace Defense Command (ADC) inactivated; aircraft reassigned to USAF Tactical Air Command (TAC).

June 18, 1981: Prototype Lockheed YF-117A makes a successful first flight at Area 51 with Lockheed Skunk Works test pilot Harold C. "Hal" Farley at its controls.

October 2, 1981: President Ronald Wilson Reagan orders that the Rockwell B-1 production program be reinstated and that one hundred Rockwell B-1Bs be built.

February 5, 1982: Northrop test pilot Richard G. "Dick" Thomas completes first flight on demonstrator Northrop Tacit Blue air vehicle at Area 51.

August 30, 1982: Demonstrator Northrop F-20A Tigershark completes first flight at Edwards AFB with Northrop test pilot Russell J. "Russ" Scott at the controls.

May 18, 1983: USAF releases RFP for the Advanced Tactical Fighter (ATF).

May 26, 1983: RFP for ATF amended to include "increased emphasis on low-observables" (stealth).

October 18, 1984: Rockwell test pilot Mervin L. Evenson makes successful first flight, Palmdale to Edwards AFB on service test Rockwell (now Boeing) B-1B Lancer.

December 11, 1986: Demonstration McDonnell Douglas (now Boeing) F-15E Strike Eagle completes a successful first flight in St. Louis, Missouri, with McDonnell Douglas project test pilot Gary L. Jennings at the controls.

July 17, 1989: First full-scale development Northrop B-2A Spirit completes first flight, Palmdale to Edwards AFB with Northrop chief test pilot Bruce J. Hinds in command.

August 27, 1990: Northrop chief test pilot Alfred P. "Paul" Metz makes a successful first flight at Edwards AFB on the prototype Northrop YF-23 at Edwards AFB.

September 29, 1990: Prototype Lockheed YF-22 completes first flight, Palmdale to Edwards AFB with Lockheed chief test pilot David L. "Dave" Ferguson in command.

June 1, 1992: USAF Strategic Air Command (SAC) and Tactical Air Command (TAC) both disestablished and made part of Air Combat Command (ACC), which is established on the same day with headquarters at Langley AFB, Virginia.

June 1996: DOD releases RFP for Joint Strike Fighter (JSF).

September 11, 1996: McDonnell Douglas test pilot Joseph "Joe" Felock III makes the first flight on the demonstration McDonnell Douglas (now Boeing) "Bird of Prey" air vehicle at Area 51.

September 7, 1997: First of nine Engineering, Manufacturing and Development (EMD) Lockheed Martin F-22A Raptor aircraft (Raptor 01) makes a successful first flight in Marietta, Georgia, with Lockheed Martin chief test pilot Paul Metz under the cockpit canopy.

December 15, 2006: First System Design and Development (SDD) Lockheed Martin F-35A Lightning II makes a successful first flight at Fort Worth, Texas, with Lockheed Martin chief test pilot Jon S. Beesley at the controls.

August 7, 2009: Air Force Global Strike Command (AFGSC) activated with headquarters at Barksdale AFB, Louisiana; responsible for ICBMs, B-52Hs, and B-2As.

July 8, 2010: Boeing F-15 Silent Eagle demonstrator airplane (modified F-15E demonstrator) makes a successful first flight at St. Louis, Missouri, with Boeing F-15 chief test pilot Daniel "Dan" Draeger in control.

March 2014: USAF releases initial funds for its upcoming Next-Generation Air Dominance (NGAD) fighter program.

July 9, 2014: USAF releases its RFP for a Long-Range Strike Bomber (LRSB) aircraft program; RFP goes to Northrop Grumman and the team of Boeing and Lockheed Martin.

October 27, 2015: USAF selects Northrop Grumman to produce its Long Range Strike Bomber.

BIBLIOGRAPHY AND SOURCES

ARTICLES

Dean, Jack. "Back to the Future." *Wings*, October 1986.

Jenkins, Dennis R. "North American F-108 Rapier." *Airpower*, September 2004.

Koehnen, Richard C. "Bell's No Name Fighter." *Airpower*, January 1982.

Landis, Tony. "Mach 3 Masterpiece." *Airpower*, May 2003.

———. "Putting the OX before the Cart." *Airpower*, May 2002.

Mailes, Yancy, and Tony Landis. "Sting of the Scorpion." *Wings*, February 2005.

Pace, Steve. "Airacomet." *Wings*, October 1987.

———. "Birds of a Feather." *Airpower*, July 1988.

———. "F-22: Air Combat Master?" *Combat Aircraft Monthly*, November 2013.

———. "Fighter on the Fringe." *Wings*, April 1987.

———. "The Great White Bird." *Air Classics Quarterly Review*, Fall 1977.

———. "Northrop and the Light Heavyweight Legend!" *Wings*, August 1989.

———. "Supersonic Cavaliers." *Airpower*, November 1986.

———. "Triplesonic Twosome." *Wings*, February 1986.

Pape, Garry R., and Gerald H. Balzer. "Scorpion." *Airpower*, May 1981.

Patton, Pat. "Long Range Longshots." *Wings*, October 1988.

Sunday, Terry L. "Thunderceptor." *Airpower*, September 1986.

BOOKS

Angelucci, Enzo (author), Bowers, Peter M. (Photographer). *American Fighter from 1917 to Present*. 1987. SBN 5550235579. Hardbound. 480 pages.

Bowers, Peter M. *Boeing since 1916*. Aero Publishers Inc. 1966. Library of Congress Catalog Card number 66-11374.

Francillon, Rene J. *Lockheed Aircraft since 1913*. Naval Institute Press. 1987. ISBN 0870218972.

Knaack, Marcelle Size. *Encyclopedia of U.S. Air Force Aircraft and Missile Systems: Volume 1, Post-World War II Fighters, 1945–1973*. Washington, DC: Office of Air Force History, 1978. ISBN 0-912799-59-5.

Knaack, Marcelle Size. *Encyclopedia of U.S. Air Force Aircraft and Missile Systems: Volume II. Post-World War II Bombers, 1945-1973*. Washington, DC: Office of Air Force History, 1988. ISBN 0-912799-59-5.

Pace, Steve. *Valkyrie: North American XB-70A*. Aero Series Volume 30. 1984. ISBN 0816806101. Aero Publishers, Fallbrook, CA. Softbound. 96 pages.

Pace, Steve. *North American XB-70A Valkyrie*. Aero Series Volume 30 (Completely Revised, Second Edition). Foreword by Chief Engineer Walter A. "Walt" Spivak. 1990. ISBN 0830686207. Tab Books, Blue Ridge Summit, PA. Softbound. 96 pages.

Pace, Steve. *X-Fighters: Experimental and Prototype USAF Jet Fighters, XP-59 to YF-23*. Foreword by Laurence C. "Bill" Craigie—Lt. Gen. USAF (Retired). 1991. ISBN 0879385405. Motorbooks International, Osceola, WI. Softbound. 128 pages.

Pace, Steve. *F-117A Stealth Fighter*. Aero Series Volume 43. Foreword

by Bill Scott—Senior Engineering Editor, *Aviation Week and Space Technology Magazine*. 1992. ISBN 0830627952. Tab/Aero Books, Blue Ridge Summit, PA. Softbound. 96 pages.

Pace, Steve. *Lockheed F-104 Starfighter*. Warbird History Series. Foreword by Anthony W. "Tony" LeVier – Chief Test Pilot. 1992. ISBN 0879386088. Motorbooks International, Osceola, WI. Softbound. 128 pages.

Pace, Steve. *Lockheed Skunk Works*. Published in 1992. ISBN 0879386320. Motorbooks International, Osceola, WI. Hardbound. 128 pages.

Pace, Steve. *Edwards Air Force Base: Experimental Flight Test Center*. Foreword by Chief Historian James O. Young, Edwards AFB, Air Force Flight Test Center History Office. 1994. ISBN 0879388692. Motorbooks International, Osceola, WI. Softbound. 128 pages.

Pace, Steve. *X-Planes at Edwards*. Enthusiast Color Series. 1995. ISBN 0879389850. Motorbooks International, Osceola, WI. Softbound. 96 pages.

Pace, Steve. *Boeing North American B-1 Lancer*. Warbird Tech Series Volume 19. Foreword by B-1A Chief Engineer Walter A. "Walt" Spivak. 1998. ISBN 1580070124. Specialty Press, North Branch, MN. Softbound. 100 pages.

Pace, Steve. *F-22 Raptor: America's Next Lethal War Machine*. The Walter J. Boyne Military Aircraft Series Volume One. 1999. ISBN 071342710. The McGraw-Hill Companies, an Aviation Week Book, New York, NY. Hardbound. 129 pages.

Pace, Steve. *B-2 Spirit: The Most Capable War Machine on the Planet*. The Walter J. Boyne Military Aircraft Series Volume Four. 1999. ISBN 071344330. The McGraw-Hill Companies, Aviation Week Books, New York, NY. Hardbound. 121 pages.

Pace, Steve. *McDonnell XF-88 Voodoo*. Air Force Legends Series Number 205. 2000. ISBN 0942612965. Günter Books, Simi Valley, CA. Softbound. 56 pages.

Pace, Steve. *Bell P-59 Airacomet*. Air Force Legends Series Number 208. 2000. ISBN 0942612930. Günter Books, Simi Valley, CA. Softbound. 92 pages.

Pace, Steve. *Republic XF-91 Thunderceptor: Rocket Fighter*. Air Force Legends Series Number 210. 2001. ISBN 0942612914. Günter Books, Simi Valley, CA. Softbound. 52 pages.

Pace, Steve. *X-Planes: Pushing the Envelope of Flight*. 2003; Foreword by Archivist/Historian Raymond L. "Ray" Puffer, Air Force Flight Test Center History Office. ISBN 0760315841. Motorbooks International, St. Paul, MN. Softbound. 128 pages.

Pace, Steve. *Lockheed SR-71 Blackbird*. Crowood Aviation Series. 2004. ISBN 1861266979. The Crowood Press, United Kingdom. Hardbound. 200 pages.

Wagner, Ray. *American Combat Aircraft of the 20th Century*. 2004. ISBN 0930083172. Jack Bacon & Company. Hardbound. 758 pages.

SOURCES

Air Force Test Center/History Office

Hehs, Eric (Editor), *Code One Magazine*, Lockheed Martin Aeronautics Company, Fort Worth, TX

Martin, Tom, and Schmidt, Rachel. A Case Study of the F-20 *Tigershark*. Rand Corporation. June 1987.

Matthews, Henry Paul. PDF entitled MX-324/MX-334/XP-79B, published July 18, 2013.

MX-324 and MX-334 Memorandum Report TSEAL-2-4302-20-5, dated January 1, 1945, Appendix—I History.

Nankivil, Mark. Greater St. Louis Air and Space Museum; data and photographic support.

National Museum of the United States Air Force—USAF Aircraft & History.

Stoff, Joshua. Cradle of Aviation Museum.

Technical Evaluation – MX-1554-A, Section II (page 19), Republic Model AP-57, March 27, 1951 (Document No. 51S-43239-A).

ACKNOWLEDGMENTS

This reference could not have been produced without the contributions provided by the many accomplished and knowledgeable individuals listed below, to whom I give my deepest thanks for their unselfish assistances to this work.

My heartfelt thanks go to Gerald H. Balzer, J. F. Baugher's American Military Aircraft website and his USASC-USAAS-USAAC-USAAF-USAF Aircraft Serial Numbers—1908 to Present website, Peter M. Bowers, Walter J. Boyne, Dr. Ira E. Chart (Western Museum of Flight Research Library), Giuseppe De Chiara (color artworks), Lt. Gen. Laurence C. Craigie (USAF, Ret.), Ryan Crierie (Standard Aircraft Characteristics Archive, Alternate Wars website), Ron Downey (Aviation Archives), Harold C. Farley, Jozef Gatial (color artworks), Jeannine M. Geiger (Archives Technician, Air Force Test Center History Office), Arlen "Gus" Goss, Eric Hehs, Lockheed Martin Aeronautics Company (Editor, *Code One Magazine*), Denny Lombard (Lockheed Skunk Works, retired), Michael J. Lombardi (Corporate Historian)—The Boeing Company, Lt. Gen. David J. "Marshall" McCloud, Mark Nankivil (Greater St. Louis Air & Space Museum), Sandra Lee "Sandy" Mabra (proofreader), David W. Ostrowski Collection (proofreader), Stan Piet (Glenn L. Martin Maryland Aviation Museum), Raymond L. Puffer (former Archivist, Air Force Test Center History Office), Benjamin R. Rich (former President of the Lockheed Skunk Works), Erik Simonsen (color artworks), Terry L. Sunday (Director of Development, War Eagles Air Museum), Chris Wamsley (North American Aviation, retired), Alvin S. White (NAA Chief Test Pilot on the XB-70A program), Wikipedia, Stephane Beaumort, Evan L. Mayerle , Richard C. Koehnen, Dr. James O. Young (former Chief Historian, Air Force Test Center History Office).

ILLUSTRATION CREDITS

The hundreds of artworks and images shown throughout this reference were provided by many aircraft artists and/or illustrators, aviation history aficionados, authors and historians, aerospace companies and corporations, branches of the US Armed Forces, Department of Defense, friends, modelers, museum archivists, directors and curators, National Archives and Records Administration, photographers, powerplant companies and corporations, societies, and civilian and government websites. To avoid hundreds of distracting credit lines throughout this work, they have been replaced with the acronyms shown in alphabetical order.

PHOTOGRAPH AND ARTWORK CREDIT ACRONYMS:

Aircraft Engine Historical Society (AEHS)

Air Force Test Center/History Office (AFTC/HO)

Boeing Airplane Company (BAC)

Boeing Corporate Archives (BCA)

Peter M. Bowers Collection (PMB Collection)

Cradle of Aviation Museum (COAM)

Department of Defense (DOD)

General Atomics-Aeronautical Systems Inc. (GA-ASI)

Giuseppe De Chiara (Artwork by Giuseppe De Chiara)

Jozef Gatial (Artwork by Jozef Gatial)

Glenn L. Martin Maryland Aviation Museum (GLMMAM via Stan Piet)

Greater St. Louis Air & Space Museum (GSLASM via Mark Nankivil)

Lockheed Martin (LM)

Lockheed Martin/*Code One Magazine* (LM/*Code One*)

National Advisory Committee for Aeronautics (NACA)

National Aeronautics and Space Administration (NASA)

National Museum of the United States Air Force (NMUSAF)

Naval Aviation Museum (NAM)

Northrop Grumman (NG)

Pratt & Whitney (P&W)

San Diego Air and Space Museum Archives (SDASM Archives)

Stan Piet Collection (SP Collection)

Erik Simonsen (Photo/Illustration by Erik Simonsen)

The F-35 Lightning II Program website (JSF.mil)

The Porthole to Texas History via Lockheed Martin (PTH via LM)

United States Air Force (USAF)

United States Navy (USN)

Chris Wamsley (CW)

Western Museum of Flight (WMOF)